商务·行政办公·工业 设计

Design for business, administrative and industrial purposes.

（德）耐特 编
张晨 译

辽宁科学技术出版社

图书在版编目（CIP）数据

商务、行政办公及工业建筑设计 / (德) 耐特编；
张晨译. -- 沈阳：辽宁科学技术出版社，2014.9
ISBN 978-7-5381-8823-3

Ⅰ. ①商… Ⅱ. ①耐… ②张… Ⅲ. ①商务－服务建筑－建筑设计②行政建筑－建筑设计③办公建筑－建筑设计④工业建筑－建筑设计 Ⅳ. ① TU24 ② TU27

中国版本图书馆 CIP 数据核字 (2014) 第 196491 号

出版发行：辽宁科学技术出版社
（地址：沈阳市和平区十一纬路 29 号 邮编：110003）
印 刷 者：沈阳天择彩色广告印刷股份有限公司
经 销 者：各地新华书店
幅面尺寸：240mm×325mm
印　张：42
插　页：4
字　数：80 千字
出版时间：2014 年 9 月第 1 版
印刷时间：2014 年 9 月第 1 次印刷
责任编辑：陈慈良 杜丙旭
封面设计：于天睿
版式设计：袁　殊
责任校对：周　文

书　号：ISBN 978-7-5381-8823-3
定　价：268.00 元
联系电话：024-23284360
邮购热线：024-23284502
E-mail: lnkjc@126.com
http://www.lnkj.com.cn

H.M. NELTE

建筑美学与能效设计

——商务、工业及行政办公建筑

AESTHETICS EFFICIENCY

BUILDINGS FOR BUSINESS INDUSTRY ADMINISTRATION

目录

CONTENTS

前言

INTRODUCTION

从左至右：建筑师海恩：大众－手工玻璃制造，德雷斯顿，克里斯托夫·马克勒教授：dmc2，哈脑，栅栏＋后部：卡布里科，门兴格拉德巴赫，福里克·莱切特建筑师：布莱科特化工广场的湿盐库–4a建筑工作室。生产和管理大楼，MBE钢铁建筑，营业中心，乌斯坦豪斯，柏林

From left to right: Henn Architekten: Volkswagen's 'Gläserne Manufaktur', Dresden – Prof. Christoph Mäckler Architekten: dmc2, Hanau – Gatermann + Schossig: Capricorn, Mönchengladbach – Frick.Reichert Architekten: Damp Salt Store in the Chemical Industrial Park of Bleicherode – Architektenbüro 4a: Factory and Offices Of MBE, Eisenach – Nalbach + Nalbach: Ullstein House Business Centre, Berlin

建筑美学与能效设计

我们的城市和乡村不断地、无情地被工业区破坏，人们普遍混淆了“廉价”和“经济效率”的含义，以此作为经济目标，为建筑的无名、平庸和丑陋辩护。

建筑批评家克利斯托夫·汉克尔斯伯格曾讽刺地称这些区域为“工业荒原”。除了劳动合同规定的工作时间，无人愿意在那里停留。

建筑物及其中间（剩余）的空地毫无争议地被认为是社会无人区。

投资于经济建筑的每一欧元和花在计划和设计的每一天都被认为是过分奢侈的。

通过对商贸建筑和工业建筑衰落的基本研究发现，当今社会建筑集中表现出各种不同的不规则性。在激烈的市场经济竞争中，城市和农村政府放弃了他们的建筑特点。高度资助或仅仅赠予的广大地区免遭城市或地区质量设计者的影响。企业家自己确实由于自己的短见、不明智的建造决策而影响（破坏）了自己的未来环境，因为生产的再次变化，迫使企业搬迁到另一个地方，通常在城市或乡村永久地留下了含有有毒物质的废墟。

当企业向公众展示他们的建筑时，他们大多展示的是总部，却很少能看到工厂状况。最小的预算、死板的施工计划和线性的而非合作的计划程序，影响了最自然的建造方式，即阻碍了为建筑工程开发好的建筑方案。尽管如此，建筑还是改变了工业建筑领域中的古典建筑的风格特点，保留了自由建筑的风格。

工业建筑多使用新技术、新建筑体系和材料，而且对新建筑理念的探索绝对是一件令人激动的事情。以此为目标，共同努力的结果是使工厂生产成本大大降低，具有比例适当的内部空间、迷人的结构和宜人的工作环境。怎样做才能使人们认识到这一点，并且能够开创“工业文化”新时代？所有有关工业建筑的共同努力和增加对其他企业目标的共同理解，是取得长期成功的唯一光明道路。

美观且实用的建筑

除了对工业建筑的经济和技术方面的讨论之外，美学因素在19世纪初就已经被讨论过。例如，路易斯·凯特尔，在1802年第一次规定每一个施工建筑，建筑美学效果在满足实用目的之外，首要原则就是以最简单的形式体现出美观；1834年，哥特弗雷德·塞姆泊抱怨说：美观的观念仅仅体现在仿古上。

“我们需要新的东西”，他说，“给予我们的甚至更古老，和我们的时代需求相差甚远。因

于尔根・赖卡特教授
埃森

前言

INTRODUCTION

建筑美学与能效设计

AESTHETICS IN INDUSTRIAL BUILDING AND OFFICE CONSTRUCTION

Aesthetics in Industrial Building and Office Construction

Our cities and rural areas are increasingly marred by inhospitable industrial estates. Wide-spread confusion over the economic goals 'inexpensive' and 'economically efficient' are used to justify anonymity, banality and ugliness.

Architectural critic Christoph Hackelsberger once ironically called these areas 'industrial steppes'. Nobody would willingly stay there for longer than the contracted hours of work.

The buildings and the intermediate (residual) spaces are generally accepted without complaint as a social no-man's-land.

Every single euro invested in excess of the absolute minimum of the 'economic' building, and every additional day spent in planning or constructing are regarded as undue extravagance.

Findings from basic research into the decay of the aesthetic quality of buildings for business, trade and industry reveal a concentration of different irregularities in our society. In the rat-race for market-economical competitiveness, urban and rural authorities sacrifice their architectural identity. Vast areas, highly subsidized or simply given away for nothing, are thus removed from the influence of urban or regional planners committed to quality design. The entrepreneurs themselves literally 'block' (i.e. spoil) their own future by short-sighted, strategically unwise decisions to build, as already the next change in production will often force companies to move to another location, often leaving behind them permanently denaturated wastelands in cities or rural areas.

When industrial enterprises wish to present themselves to the public with their buildings, they do so mostly with their headquarters. The state of their factories, however, is rarely presentable.

Minimal budgets, tight construction schedules and linear, instead of co-operative planning procedures prevent the most natural way of building, i.e. of developing good architectural solutions for a construction project. It is, after all, in the field of industrial building that architecture has retained a measure of freedom removed from the stylistic fashions the classical types of building are subject to.

Industrial architecture is open to new technologies, construction systems and materials; and the quest for new building concepts is an absolutely thrilling affair. The results of joint endeavours in this direction can be extremely cost-efficient production plants, well-proportioned interior spaces, fascinating structures and pleasant working environments. What must be done to

make people aware of this and to promote a new era of 'industrial culture'? The joint efforts of all those involved in industrial building, and a growing mutual understanding of the others' aims are the only promising path to long-term success.

Beauty and Functional Building

Apart from the economic and technical aspects of industrial architecture, aesthetic factors were discussed already at the beginning of the 19th century. For Luis Catel, in 1802, the first rule for the beautiful in architecture required every building to be constructed in the simplest form corresponding to its purpose; and in 1834 Gottfried Semper complained that the concept of beauty focused too exclusively on antiquity.

'We want something new,' he said, 'and are given something even older and farther removed from the needs of our time. Thus we are expected to conceive and create order from the point of view of the beautiful and not just to see beauty where our vision is partly darkened by the mist of far-away places. Art knows only one master called need.'

Five hundred years before Semper, Thomas Aquinas, the leading religious philosopher of the Middle Ages, had formulated the guiding principle of beauty being the glory of truth, thus pointing to honesty, conclusiveness and clarity as the roots of beauty. At the beginning of the 20th century,

the Deutscher Werkbund, an association of artists, craftsmen and industrialists founded in Munich in 1907, addressed the phenomenon of industrialization as a new field for artistic creation.

The leading theoretician of the Werkbund, Hermann Muthesius, called for a 'neue Sachlichkeit', or new functionality, for the outer form being the result of the inner nature of an object and not of attaching decoration to it taken 'from another world'. The Werkbund recognized the quality of the technical architecture of the 19th century –

此，希望我们从美的角度构思并创造新形式，而不是远方的迷雾阻挡了我们的视线而看到美。艺术唯一的大师就是需求。”

森帕之前的500年，托马斯·阿奎那是中世纪主要的宗教哲学家，形成了“真实就是美”的指导性原则，因此指出，真实、明确和清晰是美的根基。在20世纪初，德意志制造联盟（一个由艺术家、建筑师和企业家于1907年在慕尼黑建立的协会）宣称：工业化现象是艺术创新的新领域。

制造联盟的主要理论家赫尔曼·穆特休斯呼吁“新即物主义”或“新功能主义”，因为物体的外在形式是内在本质的表现而不是选自另一个世界的附加装饰品。制造联盟认识到19世纪艺术史上工艺建筑的质量并未被认为是“具有艺术性的”，而被认为是无意识美学成果的典范，这一思想在20世纪20年代被鲍豪斯建筑学院坚持并得到进一步发展；沃尔特·格罗庇乌斯认为，工业建筑是住宅建筑的样板，他坚持认为：很久以来工程师就已认识到工厂在装配和制造产品时，总是寻求用最小的劳动力、时间、原材料和资金生产最大产出的最简捷方案，然而建筑行业为住宅建筑采取同样措施才刚刚开始不久。

第二次世界大战后，鲍豪斯这一代人的解决方案转向了空洞的公式和套路。对“实用主义”的社会批判，海蒂·伯恩特说：“我们批判今天的实用主义，因为它用一种有限的方式解释实用性，只接受既定的、狭隘的、最简单的形式。它首先忽略了经验中的主观表达因素和所使用的材料带来的愉悦感。”

奥托·艾舍认为，和传统建筑相比，工业建筑的优点是独立性且不可能固定在一种建筑风格上。对艾舍来说，特殊的研究和发展方法、工作实践和思想能创造出一种“逆建筑”。

他说目前保留了自由风格的工业建筑又占据主导地位，因为它把自由看做真正发展的前提条件。在这里，没有对文化标准的束缚，在这里，广场不仅仅作为广场而建造，玻璃罐也不仅仅是作为器皿而制造。

没有人预先知道将来的结果怎样，奥托·艾舍认为这种建筑是令人兴奋的。追溯到古罗马时代，在《建筑十书》中，建筑师维特鲁威用三个“必须”对“美”下了定义，即必须实用、必须稳固和必须美观。维特鲁威以后的2000多年，尤其在工业建筑，三个必须具有较高的相关性，其互相制约性高于任何其他建筑领域。

雅致的外观

建筑总是折射出蕴含在其中的某种力量。不幸的是，大多数丑陋的工业和贸易建筑也是如此。

公司的业主看起来好像把建筑仅仅作为沉重的必需品，他们宁可不要。然而建筑是最基本的，而且是对社会文化长期存在的表达。

对每一项建筑工程来说，有技巧的、有创新的设计方案，包括形式、材料和色彩，从内到外每一个细节都应体现出整体的雅致观念，比如外立面的比例或色系成为另外的设计特征。

对于总体外观，更重要的首先是构造学的发展和建筑总体分布和分类。严重有欠缺的设计常以成本效益为借口，由于高成本压力，慎重考虑之后放弃了美学效果。

本书中列出的著名工业和商业建筑对比提供了辩驳证据。

汽车大王亨利·福特，当时以精打细算而闻名，证明如何将成本效益最大化与典型的艺术感染力结合起来，给人留下深刻的印象。

福特的宣言：好的设计是合算的，说明早在20世纪20年代和30年代，他就意识到具有良好工作环境的设计是决定总利润的因素，因为它有助于减少疾病引起的旷工和人员流动，并把公司的工人长期团结在一起。“优雅”的工业建筑和办公建筑是通过采用基本结构参数，将简单、和谐和多样化统一而产生的。

同时，将偶然性和无序性相对照，城市建筑和个体建筑相重叠、交叉，建筑结构产生了一种整体和谐。

which art history did not classify as 'artistic' – and regarded its 'unconscious' aesthetic achievement as exemplary, a thought which the Bauhaus School adhered to and developed further in the 1920s. Walter Gropius saw industrial building as a model for residential architecture. He maintained that the engineer – long since aware of fabrication and factory-made products – had always sought the most concise solution producing maximum results for a minimum of human labour, time, material and money, whereas the construction industry had only started shortly before to take the same course for residential architecture.

After World War II, the solutions by the Bauhaus generation turned into empty formulas and clichés. In a sociological critique of 'functionalism', Heide Berndt said, 'We criticize today's functionalism for interpreting functionality in an all too restricted way and only accept the simplest representation of narrowly defined purposes. Above all it neglects the representation of subjective elements of experience, of the pleasure in and the use of materials.'

Otl Aicher saw independence and the impossibility of fixing on one architectural style only as the advantages of industrial building compared to the traditional architectural agenda. For Aicher, the required special methods of research and development, practical work and thinking produced a kind of 'counter-architecture'.

Industrial building today, he said, is again the domain of an architecture that has kept its freedom because it sees freedom as a prerequisite of every true development. Here, there are no fetters of a supposed cultural canon. Here, neither the square is made for the sake of the square, nor the glass jug for its own sake.

One does not know beforehand what the next result will look like; this architecture is exciting, said Otl Aicher. Back in ancient Rome, in his ten books on architecture, the architect Vitruvius already defined beauty in three categorical 'musts', i.e. utilitas (usefulness), firmitas (stability) and venustas (beauty). Over two thousand years after Vitruvius, these categories are still highly relevant, especially in industrial architecture where they condition each other more than in every other field of building.

Graceful Appearance

Architecture always reflects the forces that contributed to bringing it about. Unfortunately, this also applies to the majority of ugly industrial and trade buildings.

It seems as if their corporate clients saw in them merely onerous necessities they would much rather do without. Yet architecture is the most basic and long-lasting expression of a society's culture.

A skilful and creative design solution involving forms, materials and colours has to be developed for every building project. This overall concept should appear graceful down to every last detail and joint – inside and out – with creative details like facade proportions or colour schemes being secondary design characteristics.

More important for the overall appearance are first of all the tectonics of volume development and the distribution and subdivision of building masses. Sorely deficient design is often excused with the 'alibi' of the required cost-efficiency, and the deliberate renunciation of aesthetic appearance is assigned to high cost pressure.

The examples of outstanding buildings for business and industry presented in this book provide the rebutting evidence.

Even the automobile king Henry Ford – known in his time for his penny-pinching habits – proved impressively how to combine maximum cost-efficiency with exemplary aesthetic appeal.

Ford's dictum, 'good design pays', shows his awareness of the fact that, as early as the 1920s and 1930s, well-designed workplaces were considered determining factors for overall profitability as they helped to reduce sickness absenteeism and personnel fluctuation and to bind workers to the company in the long-term.

'Graceful' industrial and office architecture is created by applying the basic parameters of structural order, simplicity and a balance of unity and diversity.

In contrast to the haphazard and chaotic side-by-side, overlapping and intertwining structures in both cities and individual buildings, the principle of structural order creates a wholesome harmony.

This is achieved by consistently interrelating the whole and its parts, by the logical conclusiveness of every element in its relation to the whole structure reminding one of a living organism.

For the architectural forms and the functions of the different elements of load-bearing structure, skin and interior spaces to be immediately legible and understandable, it is essential that the entire 'fabric' of a construction is clearly articulated. Following its own design grammar, the structural order extends in a logical way to floor plan, section and elevation so that one is able intuitively to grasp the interaction of the different parts.

There is no need for a special artistic kind of cladding or masking in buildings for business and industry. High aesthetic quality therefore by no means requires high costs.

The principles of simplicity, of reduction to essentials must not be confused with banality, lack of imagination and primitiveness of the average industrial and administration buildings.

The required cost-efficiency goes extremely well with austerity – which means neither cladding nor ornamentation. Unfortunately, the great number of corporate 'architectural performances' hastily assembled with parts from construction markets tell a different story, confusing simple with simpleton. Yet it is precisely the intelligent reduction of forms, materials and colours that can produce powerful aesthetic impressions. The substantial quality of an architecture freed from superfluous exterior paraphernalia affects those who look at it more deeply and lastingly than the presentation of cheap 'eyewash' showmanship.

Our powers of aesthetic perception receive countless bits of information in the field of tension

潘茨卡·平克建筑公司：北莱茵－威斯特法伦的政府大楼　柏林

Petzinka Pink Architekten: Representation of the State of North-Rhine Westphalia at the Seat of Government, Berlin.

这是通过将整体和部分有机联系起来而产生的，和整个结构有关的每一元素之间具有的逻辑决定性提醒人们它是一个活生生的有机体。

具有不同承重结构元素的建筑形式和功能，表面和内部空间清晰且易于理解，使建筑结构完整地连接在一起是很重要的。按照本身的设计法则，按照一定的结构逻辑方式，可形成平面图、剖面图和立面图，这样人们就能凭直觉掌握不同部分的相互作用。

对商业和工业建筑没有必要进行专门的美学方面的装修。因此，高质量的美学效果决不需要花费高成本。

简单这一原则，本质上不能和平庸、缺乏想像力、普通工业和管理建筑的粗糙混为一谈。

成本效益要求的绝对伴随着节俭，就是说既不装修也不装饰。不幸的是，大多数公司建筑行为都是匆忙地将建筑市场的零件组装在一起。这告诉我们一个不同的故事，把简单理解为傻子。明智地减少形式、材料和色彩能给人留下强有力的美学印象。没有多余外部设备的高质量建筑对人们产生的影响，比推销廉价眼药水技巧的影响更深远。

我们对美学的感知力可以收到无数关于单一性和各种无序性之间达成和谐的信息。而且同一性和多样化是相互依赖的。他们是两个极端，每一个工程项目都必须重新调整以保持二者的平衡。

如果建筑物普通单调，就会立刻产生厌烦效果，形式过多令人感到不安，人们会认为是杂乱无章的。

过度夸张的、短期流行的风格，不久就被人们遗忘了。多年以后，仅剩下困惑。最理想的解决方案是为建筑和城市结构形成一个永久有效的、连续性的设计框架。

很长时间以来，建筑物的高度和材料标准都是固定不变的，但是为创新设计留些空间，会使原方案未来实施成为可能。

特征和情感特点

相对于一个无关紧要的地方来说，很快会被人们忘记。不可模仿性表示一个地方的形态和结构在意识上是难以忘记的，似乎使建筑物不出名、不重要是很多工业建筑和办公建筑蓄意达到的目的。

怎样才能解释大多数建筑没有任何让人难忘的设计特征？为了弥补这一不足，一些公司用一些奇怪的设计灵感来吸引公众的注意力，但这种意外惊喜的作用很难长久给一个建筑定性。

只有通过想像力，把特定的功能、当地具体情况、构成部分的精心选择与建筑物的形状结合起来，才能形成具有无与伦比设计质量的建筑。这种“增加附加值”的建筑不需要更多的资本投资，但比通过报纸、广播或电视这种昂贵的广告方式更有效。它是通过完整的计划和共同开创性的施工进行的，是建筑物“使命”的宣言。

在全球化网络世界里，每一个商业或工业企业都不会提供看起来无意义的东西。如果他们这样做，就是严重的管理失误。这一失误很难通过节省成本而使之合理化。

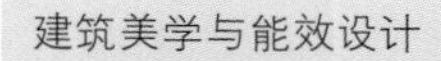

于尔根·赖卡特教授
埃森

前言

INTRODUCTION

建筑美学与能效设计

AESTHETICS IN INDUSTRIAL BUILDING AND OFFICE CONSTRUCTION

布莱纳及其合伙人公司：贝茜，柏林－同步加速器

Brenner & Partner: Bessy, Berlin – Synchrotron Accelerator.

between monotony and multifarious chaos. Yet unity and diversity are interdependent; they are the necessary extremes the balance of which has to be readjusted for every building project.

If and when the scales tip towards regularity and monotony, the immediate effect is boredom, while agitated multiformity is experienced as chaotic.

The flustered extravaganzas of short-lived fashions are soon forgotten and, after just a few years, simply embarrassing. The ideal solution is to formulate permanently valid, sustainable design frameworks for both architectural and urban structures.

Building heights and a material canon fixed over long terms, but leaving scope for creative designs will make orginal solutions possible also in the future.

代表建筑：大众汽车垃圾收集中心，卡塞尔

Opus Architekten: Refuse Collection Centre of Volkswagen Factory, Kassel.

艾斯特沃德广场的商业中心，柏林

Kleihues + Kleihues: Commercial Centre on Elsterwerdaer Platz, Berlin.

Identity and Emotive Quality

Inimitability stands for the fact that the shape and structure of a place is 'consciously' memorable – as opposed to an insignificant place that is quickly forgotten. Often it seems as if anonymity and insignificance were the deliberate aim of many industrial and office buildings.

How else could one explain that most of them do not have any memorable design features? To counteract this, some companies try to attract public attention with fanciful design gags, but it is extremely rare that formal willfulness defines true architectural character.

It is only by imaginatively combining specifically commissioned functions, individual local situation and deliberate choice of components and the shape of the building that inimitable design quality is created. Such 'added-value architecture' needs much less capital investment, but has a much bigger advertising power than expensive campaigns via the press, radio or television. It is the manifestation of the 'mission' of a building, erected by the entire planning and construction team which worked together creatively.

There is no business or industrial company that can afford to appear insignificant in a globally networking world. If they do, such serious management mistakes can hardly be legitimized by saving on cost in the wrong place.

Raymond Loewy, one of the pioneers of industrial design, once said that ugliness didn't sell. Said fifty years ago, this only referred to products, but today, the whole enterprise has to 'sell itself', both internally and externally. As with the product, however, a little bit of styling on the face of the works buildings doesn't do the trick. Graceful buildings touch an inner chord in us so that we as viewers develop a positive bond with them. This emotive quality leads to our deeper perception of space, material, colour and light – far beyond structural and functional relationships. The type and shape of spaces, their orientation and visual relationships and other details – all these, when thoughtfully designed, immediately make us understand the architectural structure and make it easy to use the buildings. The way materials are used and combined in a factory or corporate headquarters building conveys a sense of high industrial precision or handcrafted artistic spontaneity, so that architectural parallels to the quality of the owner's products are immediately apparent, or at least deducible.

一位工业设计的先驱雷蒙德·罗维曾经说过：丑陋的设计是不会成功的。据说50年前这仅仅指产品，但是今天，整个企业必须销售自己，包括内部销售和外部销售。雅致的建筑能触动我们内心深处的心弦，我们作为观赏者就会和建筑物产生一种积极的合力，这一情感特点导致我们对空间、材料、色彩和灯光更深的感知认识，这一认识远远超越结构和功能之间的关系。如果设计中富含思想、多种空间的类型和形态、它们的位置和视觉关系变化及其他内容，所有这些能使我们立刻了解建筑结构，更易于使用。在工厂或公司总部建筑中，材料的使用和结合方式表达了一种高度的工业精确性或工艺艺术的自发性。因此，很显然可以推断出：建筑设计类似于企业的产品质量。

高效的工商行政办公大楼

决定建筑形态的评判标准是什么？什么样的特点影响了它的外观和适应性？

显然，它的质量控制必须预先由总设计决定，进而制定更详细的计划。一般情况下，因为不能准确获得相关的基础参数，所以缺少建筑计划制定的过程。当按照工作计划书和工业生产过程的标准数据集体生产产品时，建筑设计多多少少带有偶然性。在技术设计中，市场革新产品来自于各种标准的考虑——市场需求、技术、生产成本、合格的质量、耐用性和生产时间。然而，在建筑行业，建筑要素很少得到有序发展，通常都用目光短浅的方法，在生产和物流结构确定之后才能确定承重和大厅结构及其内部设施。没有工作计划书的长期建筑投资就像俗语所说：没见实物而瞎买东西。相反，我们必须学会把建筑要素的总投资看做一种估价行为，这将有助于避免计划失误。对不同产品和生产过程的可能方案进行模拟测试，包括这些方案分别对不同工厂建筑结构和技术装置的影响。

此外，不断提高质量和降低成本的最大可能性是对生产过程（或业务经营）的整体分析和制定全面计划。这既适用于制定建筑计划，又适用于发展技术装配（生产机器和热风空调）。

工业建筑规划

从空间角度看，类似于工程设计，工业建筑的设计是由数量（硬件）和质量（软件）因素决定的。

01 图是视觉象征，总括了当前特有的需求。从他们的设计理念开始，建筑师必须明确掌握能够形成公司特征的因素，比如与技术、能源和生态有关的厂房建筑以及具有柔性的软件因素：内部设备和装潢、员工交流环境和外观。两方面都有助于取得具有协同效果的设计方案。

视觉效果

每一项工程的产生都有实干家、怀疑者和犹豫不定者。何谓视觉效果？如何捕捉并确定下来？

特别是在工程的初始阶段，在第一个车间形成并展示中，把所有提到但没讨论的方案集中在一起是很重要的。这样把许多不同的观点和不同成员的目标集合成一个统一的目标，每个人都以此为核心。这一核心就是一种希望——不久的将来付诸实施的、对每一项工作或建筑以期实现的功能或提供服务的最好可能方案。一项工程的持续进展不仅要明确硬件概念，而且包括软件因素——对设计方案情感上的接收。逻辑和“直觉”的统一是取得一致意见的重要前提条件。这需要很多勇气和毅力来探索新道路，开发全新观念。由此产生的魅力是能使人们长期接受的决定性因素。

技术

在建筑行业，技术意味着在某一特殊应用领域的处理方法。在工业生产中，技术是根据最小投入生产出最大产出的原则，与商品生产和解决工作问题相联系。比较而言，建筑计划和结构居次要地位。

原因之一就是建筑者和承包商固守传统。

于尔根・赖卡特教授
埃森

前言

INTRODUCTION

高效的工商行政办公大楼

EFFICIENT INDUSTRIAL, BUSINESS AND ADMINISTRATION BUILDINGS

Efficient Industrial, Business and Administration Buildings

What are the structural criteria determining the morphology of a building? What characteristics influence its performance and adaptability?

Obviously, its con-trolling qualities must already be fixed by the overall design, and later by detailing. The usual architectural planning processes are lacking because the related basic parameters have not been defined precisely enough. While an industrial product is developed in teamwork following job specifications and the benchmark figures of the industrial production process, buildings are designed more or less incidentally. In technical design, marketable innovative products are developed from ideas processed along multiple criteria – market requirements, technology, production costs, formal quality, durability and production time. In the construction industry, however, the systematic development of building elements rarely takes place. Load-bearing and hall structures as well as interior fixtures are determined only after the production and logistic structures have been fixed, often in a rather short-sighted way. A long-term building investment without job specifications resembles the proverbial buying a pig in a poke. Instead, we must learn to regard the sum total of a building's elements as an assessable performance. This could help to

沟通

COMMUNICATION

愿景

VISION

技术

TECHNOLOGY

生态环境

ECOLOGY

图 01

Fig. 01

适应性

FLEXIBILITY

特征

IDENTITY

协同作用

SYNERGY

能源

ENERGY

不像其他领域，建筑行业中新材料的使用及与此有关的、取得高性能体系的方法很难取得进展。与技术产品相似，建筑一方面显现出效率的相似性和相互依赖性、生命周期和外观，另一方面也有生产成本和维修成本。

至此，我们主要集中于单一的建筑物特点。技术和工业生产原则的应用，将重点转向整个实施过程。

能源

由于世界能源，如煤、石油和天然气是有限的，因而要求我们不断使用可再生的自然资源，如：太阳能、水能、风能和生物燃料。国民经济能源需求中大约有三分之一被商业和工业消耗，三分之一用于采暖，三分之一用于交通。

尤其是工业建筑工程内以上三个领域，提供了节约能源的可能性。这使能源消耗成为关键设计参数，这一参数要求所有厂房建筑具体到每一个细节。在选择位置时，在总体规划和确定建筑结构时，明智的设计首先要根据太阳和方位来选址。将气候和技术结合起来，建筑结构及所提供的服务为最优化使用能源提供多种机会。

巧妙地组织物流能减少交通方面的能源消耗。建筑过程中的采暖、通风、空调产生的废能源能满足生产加工所需。这就是说，内部空间只需要很少的机械/人工采暖、制冷、通风和采光。

生态环境

“生态平衡”也可以简单表述为“不破坏自然平衡”。很显然，工业建筑行业中的“生态建筑”是自相矛盾的。

建筑总是意味着对自然的破坏，但并不意味着毁灭性的破坏。建筑的生态意识表明高效生产和经营。按照生态平衡的观点，就是减少因建筑施工造成的环境污染。工业建筑有可能采用不同生态方法以节省资源和避免浪费。如果有大的开发项目，例如工业区，设计者应该将绿色区域和树木繁茂地区相互连接起来形成有很多物种的生物群落，免受环境和噪音污染。所有不同单个元素，比如空隙、下层土、水层、大厅、承重结构、墙面、屋顶和技术系统，一开始就应作为重点考虑。在计划制定的开始阶段，设计者和生产计划者都应决定优先考虑哪一要素。这就要求建筑项目的所有要素必须从一开始就参与计划制订过程。

适应性

今天不可能预测将来的生产过程和所需要的面积，或者一个工业企业将进行的结构变化。一个厂房建筑最大可能地适应未来变化是公司运转适应能力的试金石。必须明确每个建筑部门的适应性和改装能力，明确每个供给材料和处理系统的适当效率。主要关于净宽、净高、楼板的载重、建筑水平、垂直扩建的结构准备。材料供给和处理系统必须考虑内部不同的空气调节方法。这要求在建筑过程中必须考虑外露的易安装设备。它们发生变化时，不影响正在进行的生产。

沟通

在传统厂房设计中，计划、决策和执行是独立的过程。现在理想模型是“手和脑”相结合。这一新的团队工作方法需要一种易于沟通的建筑，使管理者早发现生产过程的错误。除了易于纠正和利于连续的质量管理，建筑还要为工人、主管和经理提供保持联系的机会，使他们充满责任感地积极参与工作，形成他们自己的工作环境。员工在不断提高产品质量的同时，也希望他们的工作环境中有休闲性质，也就是说包括休息空间。

研究表明，企业采用的革新思想80%来自内部非正式的讨论和计划外个人交流。为此，工厂应再细分成私人、半私人和公共区域以及分散讨论区。车间氛围促使了设备的自发使用和快速更新。走廊、陈列室和楼梯也应提供新的交流机会。从内部设计来说，建筑升值了。它们促进了特别的、非正式的思想交流，成为新工厂重要的思想库。

avoid planning mistakes and to test simulated potential scenarios for different products and production processes, including their logistical effects on the respective factory's architectural structure and technical installations.

What is more, the greatest potential for increasing quality and reducing costs lies in the holistic analysis and integral planning of manufacturing processes (or business operations). This applies both to the architectural planning and the development of technical installations (production machinery and HVAC).

Specifications for Industrial Buildings

From a spatial perspective, similarly to process planning, the designing of industrial buildings is determined by both quantitative (hard) and qualitative (soft) factors.

Fig. 01 is a 'symbolical visualization' giving an overview of current specific demands. Starting from their design idea, architects must define the factory building with regard to technology, energy and ecology, and also to soft factors like flexibility of interior fixtures and furnishings, staff communication and exterior appearance as a contribution to corporate identity. Both perspectives add up to a synergetic design approach.

Design Vision

With every project there are the doers, the doubters and the ditherers when it comes to: what does a 'design vision' look like? How to catch and fix it?

Especially in the initial stages of a project, in the first workshops and presentations, it is essential to unite all those involved once the alternatives worked out that far have been discussed. Ideally, the many different ideas and aims of the different team members are then brought together in a common vision everybody rallies around. This vision is the wish – projected into the near future – for the best possible solution to every single job or function the building is expected to accomplish or serve. A sustainable vision will not only define hard facts, but also include soft factors of human emotional acceptance into the design solution. The union of logic and 'gut feeling' is an important prerequisite for any consensus. It may require a lot of courage and persistence to tread new paths and develop new overall concepts, but the fascination they cause is the decisive factor for their sustained acceptance.

Technology

In the construction industry, technology means process methodology in a particular field of application. In industrial production, technology is seen in connection with manufacturing goods and solving job problems according to the principle of maximal output/performance at minimal effort. In comparison, the planning and construction of buildings therefore often appears backward.

One reason for this could be the traditional steadfastness of builders and contractors. Unlike in other fields, new materials and the method of combining materials to achieve higher-performance systems only hesitantly gain ground in the con-struction industry, despite the fact that in analogy to technical products, buildings also show a similar interdependence of efficiency, lifespan and appearance on the one hand, and production and maintenance costs on the other.

While so far the focus has been on the single-object character of a building, applying the principles of technical, industrial production shifts the focus to its 'holistic' performance over its entire life cycle.

Energy

As the world's energy resources like coal, oil and natural gas are finite, natural renewable resources like solar, hydrological, wind energy and biofuels should increasingly be used. About a third of the energy requirements of our national economy is consumed by business and industry, another third for heating purposes and another third for traffic/transportation.

Industrial building projects, in particular, offer enormous potential for saving energy in all three areas. This makes energy con-sumption a key design parameter for every factory building right down to every last detail. In choosing a site, in masterplanning and defining architectural structures, intelligent design means, above all, orienting the building(s) depending on its (their) relation to sun and wind directions. Architectural structures and services therefore offer multiple opportunities for optimizing the use of energy through integrated climatic and technical systems.

Energy consumption for transportation is reduced by skilfully organized logistics, while the energy needed for the production processes can make use of waste energy from a building's HVAC systems. In practice, this means that interior spaces need only a minimum of mechanical / artificial heating, cooling, ventilation and lighting.

Ecology

The term 'ecological balance' may also be formulated as 'undisturbed natural balance'. Used in the context of the construction industry, it soon becomes clear that the term 'ecological building' is self-contradictory.

Building always means a disturbance of nature, but does not necessarily mean its destruction. Ecologically conscious building therefore denotes the highly efficient production, operation and reduction of architected environments in the context of natural balance. Industrial buildings have the potential for different ecological approaches like saving resources and avoiding waste. In the case of large new developments such as industrial estates, designers should interconnect green and wooded areas to create biotopes with a wealth of species, and should

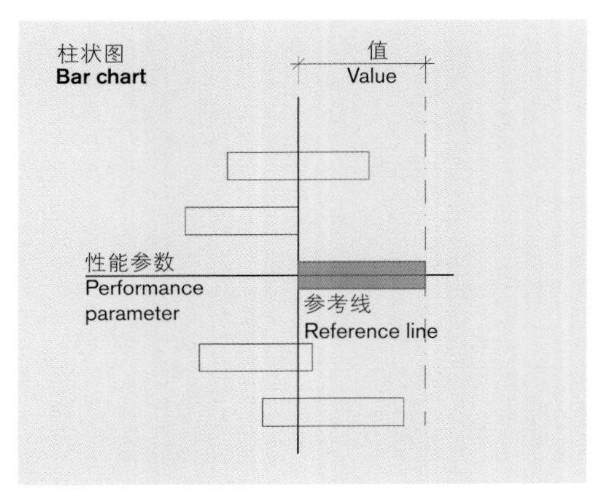

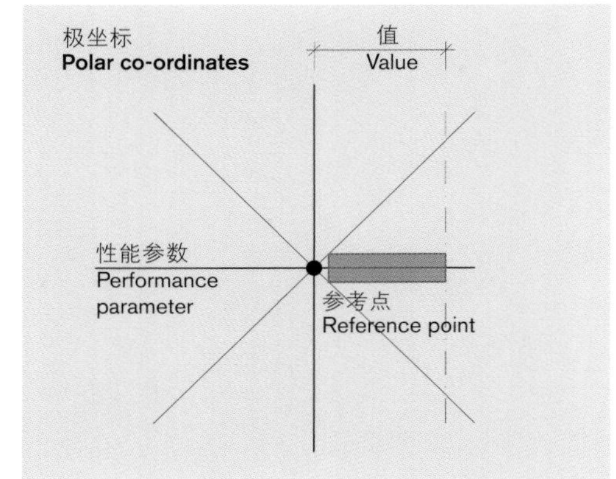

图 03：有柱状图和极坐标的极性结构剖面图。

Fig. 03: Structural polarity profile with bar chart and polar co-ordinates.

特征

许多工厂是无特征的建筑物，这是一个不容忽略的事实。因为较高质量的工作车间是不经济的，所以这类建筑被认为是合理的。

当时以吝啬著称的亨利·福特就不赞成这一错误判断，认为它会带来严重后果。按照他的说法："好的设计是合算的"，由此形成了自己企业的建筑特征。20世纪20年代，福特工厂是高质量的建筑并拥有高利润生产设备。

目前在全球化市场上，产品种类变化多端，让公司的员工和顾客很容易识别公司的特征越来越重要。

一个企业从他的竞争者中脱颖而出，公司的整体形象（公司特征）不仅是由产品决定的，而且由工厂的外观和内部形象决定。现在许多公司开始认识到公司历史所具有的潜力，试图据此形成标识并从中产生自信。

这也有可能形成公司建筑的设计理念，公司建筑尤其会给有竞争力的企业留下良好的第一印象。这样，公司建筑就成为顾客、劳动力和材料供给的统一标记和提示物。

在各种媒体中，它代表优秀的广告，成为吸引顾客，让顾客了解有关社会、文化和公司承担社会责任的线索。

协同优势

协同优势是指不同力量、因素或机制共同作用并产生协作效应。在工业建筑中，协同优势指在整体最优化过程中把表面矛盾的物体联合起来的能力。

协同优势直接决定了项目的效率，以此为目标制定的协同优势和设计方法，称之为"协同优势工厂计划"，这是一个关键因素。

协同设计能够利用积极因素，减少不利因素，找出提高合作效率的方法。然而一般计划制订过程很难取得这一结果。要求从一开始，协同优势就必须作为合作计划共同制订和执行。

形式服从功能

在这个方法中，形式和功能之间的比率是多少？在建筑形式形成过程中，形成了两个对立的建筑理论。

在 19 世纪末，美国建筑师和理论家路易斯·沙里文提出了"形式服从功能"的口号。认为作为正规的建筑方案，功能是必要的关键因素。鲍豪斯建筑学派鼎盛时期，实用主义建筑师使建筑摆脱了折衷主义风格的束缚。20世纪后半期，很多建筑师反对普通的"盒子式"建筑，坚持"形式服从功能"的口号，他们提倡设计寓于变化并且更加正式。他们把工程项目设计成事先确定的几何结构。

关于工厂建筑的适应性，形式和功能这两种方法都无能为力，因为它们仅仅说明了自然与人和建筑的功能、形式之间复杂影响的一个方面。建筑工程中经常出现的问题是当前的功能和形式是否长期可行。临时方案或昙花一现的美学时代精神不适合强健的固定形式。这就是为什么需要整体方案，即在功能和形式上都是整体的。采用什么方法才能将大量重要元素和复杂问题中的部分方案、补充方案有机结合起来。"形式服从功能"意味着从"形式服从功能"这一推断中发现方案的过程。

图01 和图02是"实用"工业建筑的例子。他们正式的具体设计显然不是事先固定的，但空间方案设计却适合所需的功能。以下图所示的建筑样例为基础，开始使用新的建筑技术。

能源消耗最优化，并且要符合生态参数。对每一个建筑区域和建筑级别，通过确定适应性标准来获取所必需的弹性程度。内部设计和光线应适合人与人之间的交流、沟通。总体来说，工业建筑应有助于培育企业文化和特征。图表中复杂参数，证实了在计划制定者之间进行交流是非常有用的。行为动作的可视性使讨论其他的选择方案变得很容易。类似于生物有机体中的DNA，图03是具有承重结构（骨架）、墙面（皮肤）和主介质（动脉）的建筑计划的例子，或称之为"基因代码"。这种情况下，把大厅划分为具有不同宽度 a、b 的格子。特

于尔根·赖卡特教授
埃森

前言

INTRODUCTION

高效的工商行政办公大楼

EFFICIENT INDUSTRIAL, BUSINESS AND ADMINISTRATION BUILDINGS

provide protection from air pollution and noise. All the different individual design elements like air space, subsoil, bodies of water, halls, load-bearing structures, facades, roofs and technical systems should initially be considered as equally important. At the beginning of the planning stage, both architects and production planners should decide which of these elements should be given priority. This presupposes that all those involved in a building project participate in the planning process from the earliest possible moment.

Flexibility

Today it is almost impossible to predict future production processes and the areas they will require, or the structural changes an industrial enterprise will un-dergo. The greatest possible adaptability of a factory building to future changes is a touchstone of a company's operational flexibility. This entails defining the adaptability / convertibility of every architectural section and the adequate efficiency of every supply and disposal system. This mainly concerns span widths, clear heights, loading of floors, and structural provisions for horizontal and vertical extensions to the

促进沟通

PROMOTING COMMUNICATION

形成视觉效果

CONCEIVING A VISION

利用技术

UTILIZING TECHNOLOGY

与生态相结合

INTERLINKING WITH ECOLOGY

图 02

Fig. 02

建立适应性

ESTABLISHING FLEXIBILITY

赋予特征

GIVING IDENTITY

寻求协同优势

SEARCHING FOR SYNERGY

最佳使用能源

OPTIMIZING ENERGY

定的工作行为标准、墙面开口、工作区较适宜的温度或内部结构的透明度都生动地表露出来，而且很容易理解。能看到生产大厅格状工作间或内部走廊。

极性剖面图

大楼和施工程序的整体分析和完整计划能最大可能地提高质量和降低成本。这既适用于建筑设计又适用于生产装置、供暖和空调系统计划。工业建筑和设备管理的每一个设计师和工程师都必须认识到这种可能性。为此，不同环节的作用不再分别进行判断和计划。为了编制建筑的“基因代码”以提高效率，必须对基本理念、建筑设计、技术装置和将来使用进行整体考虑。

考虑到影响工程的所有不同因素，传统方法不再有效。设计者不仅要考虑全部工程，还必须更加注意未来使用者的工作习惯和行为。由于建筑物是一次性产品，必须满足大量不同需求。早早地起草合同条款，即明确规定所有人都易于理解的计划矩阵是很重要的。通常认为这有助于讨论和消除在实际施工时出现的问题，这些问题在工程施工时解决为时已晚。分析计划制定参数应标明下列区域（从内到外）：过程（生产经营）、结构、介质（生产供给/热风空调）、内部工作和外观。

结构的重要功能参数包括柱网（大厅/标准楼板）、固定的活负荷和悬浮负荷、特殊载重区、净高（大厅/标准楼板）、防火措施/防火安全、建造期、扩建选择、成本效益。外围的重要功能参数包括外立面和建筑容积的比例、保温、光线区、大门、房门、窗的适应性、通风孔、噪音、防火措施/防火安全、生态环境、热能利用、建筑期、扩建选择、成本效益。

介质系统的重要参数包括储备规定、储备的分配、电源/连接设备的适应性、系统的模块化、提高能源效率的可能性、控制选择（中央控制）、建造期、扩建选择、成本效率、内部系统的重要参数包括面向生产大厅透明的标准楼板墙、内部重建的规定（大厅和标准楼板）、模块化、防火措施/安全、建造期、扩建选择、成本效益。

协同优势工厂计划

如何处理以上所提到的工业建筑设计参数，如何建立并确定工程合同结构，这些问题引出了空间设计方法学。

对当前工业建筑的批判显示了汽车行业工作流程的技术水平，它和工业区所使用的计划制订及建筑方法有巨大不同。20世纪30年代，建筑思想家的先驱里查德·巴克明斯特·福勒用“文化滞差”来描述传统建筑行业的落后。传统建筑行业反对更先进的其他行业，例如汽车或航空业，据说这一行业技术领先20年。

对工业建筑惯常做法的进一步研究表明，它的做法与先进行业“数字”工作方法完全不同。通常它是按照线性连续的方法分别界定项目部门，然而，汽车行业采用“同时施工”节省时间的方法。结果形成了用于加工、选址、建造和服务的一系列互相独立的方案，图04显示了不同工程部门未能有效合作，同时由于建筑的大部分直接表现出大量的“交互界面”问题，产生大家熟知的建筑风险。

空间项目早期一体化

从过程和空间角度看，合作性质的新厂房计划的质量是由空间项目早期一体化确定的（过程、位置、大楼和提供的服务）。

图05显示的是三维结构，从草图（假设）到最后设计（定型），是通过不断地共同讨论选择方案，不断评价所作出的决定形成的。每一项目目标，例如适用性或功能效率在每一个合同中都进行明确规定，再转换成三维模型并采用整体设计进行检验。

统一的三维模型
数字模型

目前，可利用计算机辅助设计/计算机辅助制造技术来发展并不断提高一体化的三维数字模型，使工程计划在整个过程和循环中控制整个空间质量并达到最优化。

building. The supply and disposal systems must allow for different interior air-conditioning methods. This makes it necessary to conceive exposed, easily accessible installations, independent of other systems in the building so that they can be altered without disturbing ongoing production.

Communication

In the traditional factory, thinking (planning), decision-making and imple-mentation were isolated processes. Today the ideal is the 'hand and mind' model. This new teamwork approach requires a communicative architecture which allows supervisors to detect faults in production early on.

Apart from facilitating rectification and constant quality controls, the architecture provides opportunities for workers, foremen and managers to keep in touch, work in a responsible, participatory way and contribute to shaping their own working environment. Employees with ever higher qualifications increasingly also expect their workplaces to have recreational qualities, i.e. to include spaces for leisure. Studies have shown that 80% of the innovative ideas implemented in enterprises came up in informal internal discussions and unscheduled personal exchanges. To further this process, the factory should be subdivided into private, semiprivate and public areas as well as decentralized forums. A workshop atmosphere will facilitate spontaneous uses and quick changes of furnishings. Corridors, galleries and stairways also open new communicative potential. Revaluated in terms of their interior design, they promote ad-hoc informal exchanges and will thus become important 'think-tanks' of the new factory.

Identity

The nondescript architecture of many industrial buildings is a reality that cannot be overlooked. It is often 'legitimized' with the argument that a higher-quality workplace (the result of a more thought-through design) is uneconomical.

Henry Ford, known in his day for being a niggard, disproved of this misjudgement of serious consequences. With his motto, 'good design pays', he also shaped the architectural identity of his enterprise. Ford factories of the 1920s were highly profitable production facilities of high architectural design quality.

In the current globalized markets with their indeterminable variety of products, the identity of a company easily recognizable by both staff and customers becomes increasingly important.

The qualities which make an enterprise stand out from among its competitors and the entire image of a company – its corporate identity – are not only determined by its products, but also and especially by the interior and exterior appearance of its factory building(s). Many firms are currently becoming aware of the potential of their corporate history, and are trying to foster identification with and self-confidence drawn from it.

This may also lead to design ideas for the company's buildings which support the first positive impression particularly competitive enterprises have made. In this way, the corporate architecture becomes a physical reminder and a sign of unity of customers, workforce and suppliers.

In a multi-media environment, it represents an excellent advertisement and gives the attentive viewer clues as to the social, cultural and public responsibilities the company has taken.

Synergy

Synergy means different forces, factors or organisms working together and producing a coordinated achievement. In industrial architecture, synergy means the capacity to unite apparently incompatible targets in a holistic process of optimization.

The degree of synergy achieved is a direct expression of the efficiency of the project, and this makes synergy and the design method aiming for it, called 'Synergetic Factory Planning™', a key factor.

Synergetic designs detect potential improvements to networking effect, combine positive and reduce negative factors. However, such projects will rarely be the outcome of the usual sequential planning processes. From the beginning, they must be approached and developed as a co-operative planning effort.

Form Follows Performance

What is the ratio between form and function in this method? Architectural theory developed two seemingly diametrical positions on finding architectural form.

At the end of the 19th century, the slogan 'form follows function', coined by the American architect and theoretician Louis Sullivan, saw functional necessity as the key factor for developing formal architectural solutions. In the heyday of the Bauhaus, functionalist architects adhered to it to free architecture from the fetters of eclectic styles. In the second half of the 20th century many architects, reacting to the plain 'box', hoped for more variety and formal eminence of their designs by adhering to the motto 'function follows form'. This made them design programmes and processes 'into' predetermined geometries.

When it comes to adaptable factory buildings, these two strategies do little to achieve them, because they both address only one aspect of the complex interaction between nature, man and the function and form of architecture. A frequent question arising with a building project is whether the present function and form will be viable in the long-term. Temporary programmes or short-lived aesthetic zeitgeist fashions are ill suited to foster robust, solid forms. This is why holistic solutions, holistic in terms of process (function) and space (form), are in demand. What matters is to carefully, deliberately combine a number of significant elements with partial, ideally complementary solutions to complex problems. The term 'perform-

图 4：工业建筑——结构基因代码

Fig. 04: Contract item 'industrial building' – structural genetic code.

在适当的软件程序和标准数字公式帮助下，有可能取得大量的具体预测结果。

此外，通过识别并剔除不同签约公司之间潜在的“冲突”，并通过不断的质量检查，三维协作优势工厂模型可达到最优化。这就确保在早期就能发现过程、结构和服务设计之间的冲突。这些影响成本、建造期和质量的负面冲突因素，就不用在以后施工时再待解决。

数字化三维项目循环质量控制

厂房设计工程所习惯使用的静态的、以目标为导向的方法逐渐转变为动态的、以过程为导向的方法。不断获得资料的真正价值是显而易见的，尤其在闭环系统地点进行改建、扩建、新建中显得尤为重要。协作优势工厂计划模型和传统方法相比具有特殊优势。在成本估价、确定计划的相互依赖性、控制质量、生产运转方面具有很高的准确性。存储在设备管理模型中的信息很可能快速设计出新车间，为工厂的扩建提供建议或者在其他地方建立新厂房。尤其在全球化的工程体系下，以数字形式储存的项目资料的透明度和协作计划程序的建构方法为顾客提供了附加值。

ance' is meant to denote the process of finding such solutions, from which the conviction of 'form following performance' is deduced.

As an extension of fig. 01, fig. 02 illustrates the notion of industrial buildings as 'performances' with examples of implemented projects. Their specific formal design was not fixed beforehand, but the result of the spatial solution suited to the required functional performance. Based on the visions the examples started from, new

结构类型以制造工厂为例。
性能参数：净宽

Type of structure, for example of a seat manufacturing plant, performance parameter: span width.

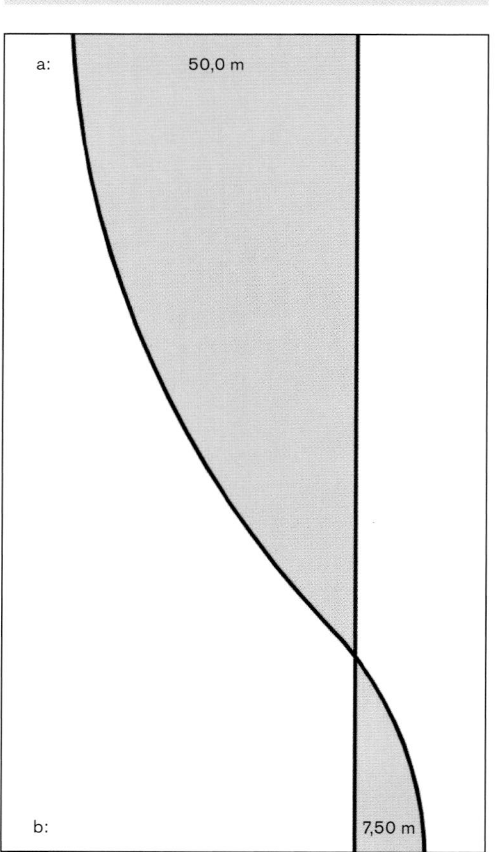

墙面类型以制药厂为例。
性能参数：开口

Type of skin, for example of a pharmaceutical plant, performance parameter: openings.

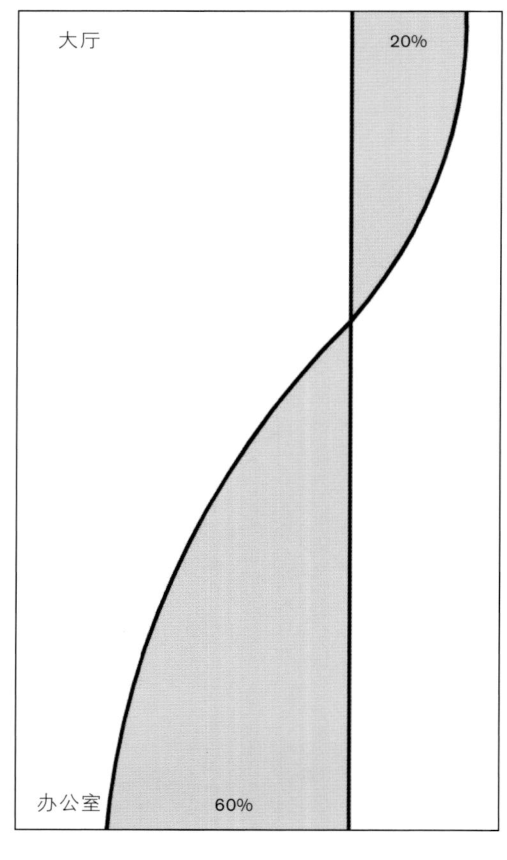

construction technologies should be used; energy consumption should be optimized and ecological parameters be meshed. The degree of flexibility deemed necessary is to be secured by defining the adaptability criteria for every area and level

介质类型以大的面包厂为例。
性能参数：降低温度

Type of media, for example of a large bakery, performance parameter: reduction of temperature.

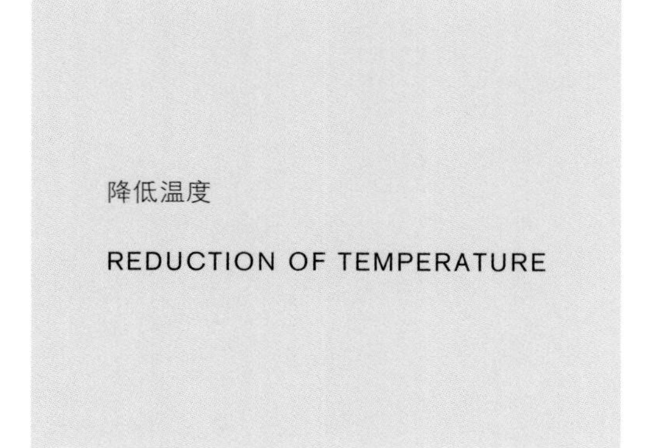

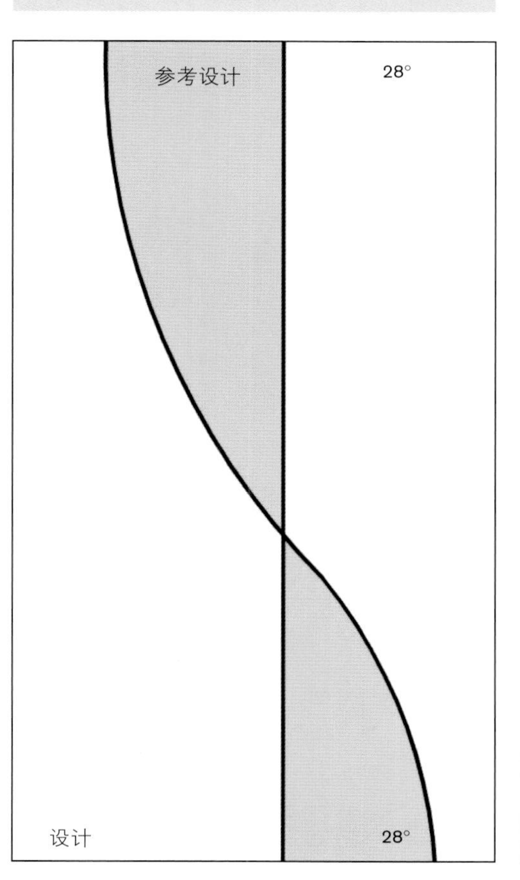

of the building. Interior design and furbishment should facilitate communication from person to person. On the whole, industrial architecture should help to foster corporate culture and identity. Graphic representations of complex parameters

内部工厂以摩托车制造工厂为例。
性能参数：透明度

Type of interior work, for example in a motor manufacturing plant, performance parameter: transparency.

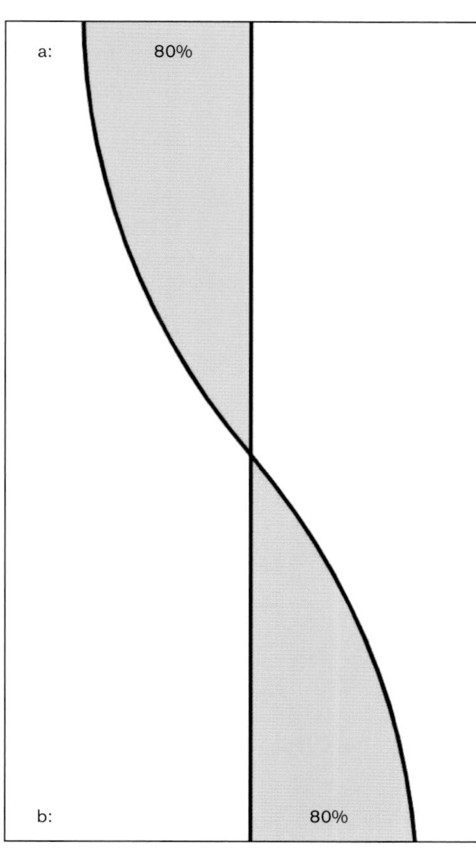

高效的工商行政办公大楼

EFFICIENT INDUSTRIAL, BUSINESS AND ADMINISTRATION BUILDINGS

n diagrams proved extremely useful for communication among planning teams. The direct visualization of such performance profiles made it easier to discuss alternative solutions. In an analogy to the DNA of biological organisms, fig. 03 shows examples of architectural plans, or 'genetic codes', for load-bearing structure (skeleton), facade (skin) and media mains (arteries). In this way, the specified work-day performance criteria for the grid dimensions a, b of hall span widths; facade openings; more pleasant temperatures at the workplace, or degree of transparency of interior structures can be presented in a vivid and immediately understandable form. Specifications for structural grids of production halls or interior galleries can thus also be grasped at one glance.

Polarity Profiles

Holistic analysis and integral planning of buildings and processes hold the greatest potential for improving quality and reducing costs. This applies both to architectural design and the planning of production installations, heating and airconditioning systems.

Every architect and engineer involved in industrial building and facility management must make it his / her responsibility to recognize such potentials. To achieve this, the different trade contributions must no longer be judged and planned separately. For the 'genetic code' of a building to be pro-grammed for efficient functioning right from its 'conception', architectural design, technical installations and future uses must be viewed holistically.

Conventional methods are no longer sufficient to take into account all the different factors that influence a project. Not only must planners view projects holistically, they must also pay more attention to the working habits and behaviour of future users. As buildings are 'one-off' products that have to meet a great number of different requirements, it is essential early on to draw up contract specifications, i.e. clearly set out a plan-

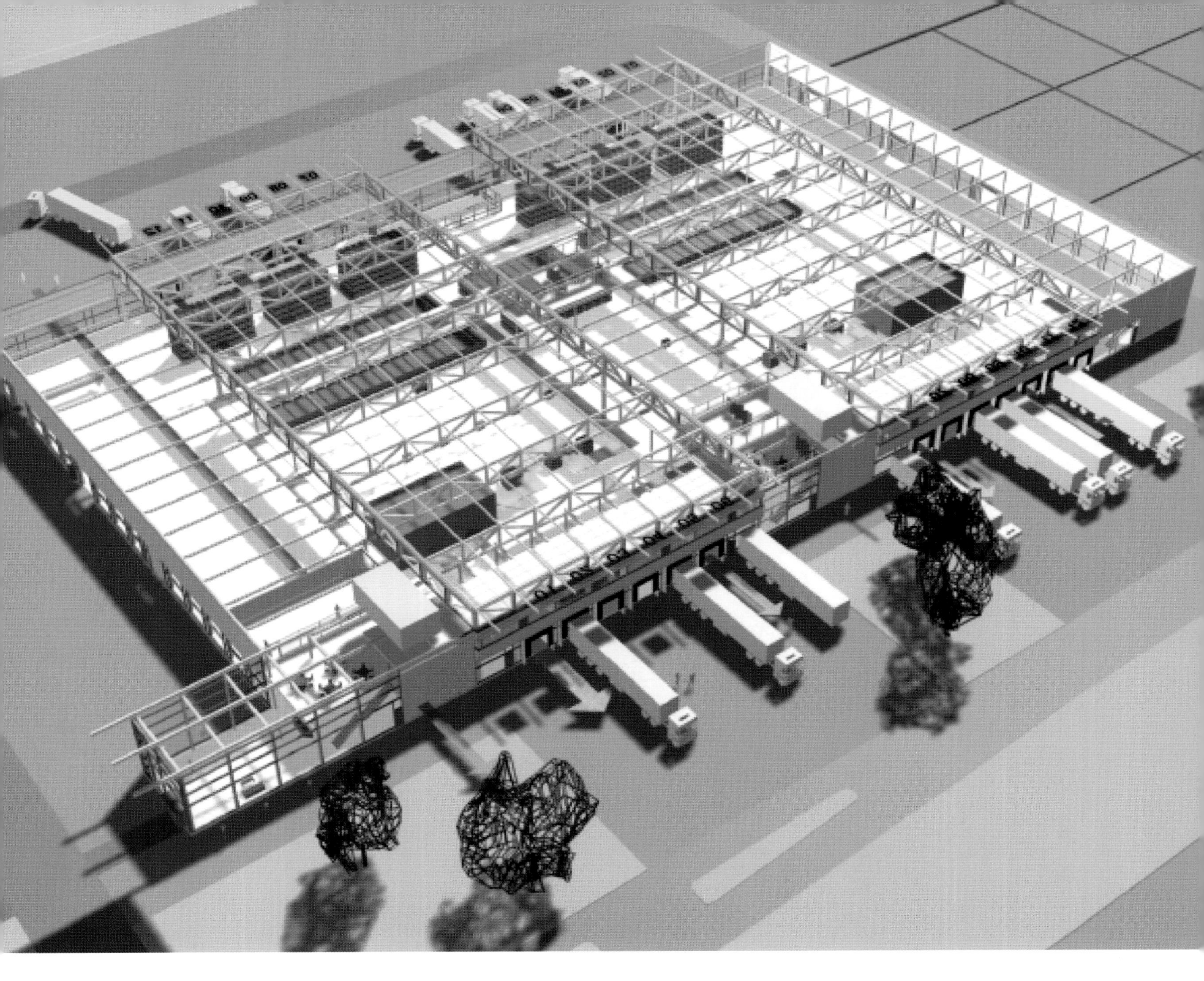

图 05：汽车生产工厂散热器生产区图。

Fig. 05: A vision for a factory producing car radiator blocks.

ning matrix easily understandable to all involved. This will help to discuss and eliminate problems which are often recognized too late in the reality of the construction site. The analysis of planning parameters identifies the following areas (from inside out): process (production operations), structure, media (process supplies / HVAC), interior work, and facade.

The important functional parameters of the structure include column grids (halls / standard floors); fixed, live and suspended loads; special load areas; clear height (halls / standard floors); fire precautions / fire safety; construction period; options for extension; cost-efficiency. Important functional parameters of enclo-sure include the ratio facade surface / building volume; thermal insulation; areas for daylighting; flexibility of gates, doors and windows; exhaust ventilation outlets; noise protection; fire precautions / safety; ecology; heating energy gains; construction period; options for extension; cost-efficiency.

Important parameters of media systems include provision of reserves; distribution of reserves; flexibility of electrical sockets / connecting units; modularity of systems; potential for improving energy-efficiency; controlling options (central controls); construction period; options for extension; cost-efficiency. Important parameters of interior systems include transparent walls towards production halls and standard floors; provision for interior restructuring (halls and standard floors); modularity; fire precautions / safety; construction period; options for extension; cost-efficiency.

Synergetic Factory Planning™

The questions of how to deal with the design parameters for industrial buildings outlined above, and how to structure the project-determining contract specifications leads on to the subject of space-planning methodology.

A critical review of current practices in industrial building reveals serious differences be-tween the state-of-the-art workflow of the automobile industry and the planning and construction method applied to a factory complex. Pioneer architectural thinkers like Richard Buckminster Fuller of the 1930s used the term 'cultural lag' to describe the traditional backwardness of the construction industry as against other more progressive branches of industry such as the automobile or aircraft industry which were said to be twenty years ahead.

A closer look at the usual practice of industrial architecture reveals that it is substantially different from the 'digital' working methods of progressive industries. As a rule, the separate definition of project sections is carried out in a linear, sequential way, while the automobile industry has adopted the time-saving method of 'simultaneous engineering'. The logical result is a sequence of 'insular solutions', independent of each other, for process, site, building and services. Fig. 04 shows that insufficient coordination of the different project sections is responsible for the well-known risks of building, for the most part a direct expression of the problems in the vast 'sea of interfaces'.

Early Integration of Spatially Defined Project Sections

The new quality of co-operative factory planning from both process and spatial perspectives results from an early integration of spatially defined project sections (process, site, building and services).

As illustrated by fig. 05, the three-dimensional structure is continuously refined from rough draft (assumptions) to final design (fixations), with decisions being constantly evaluated in joint discussions of alternative solutions. The targets for every project section, such as adaptability or defined functional efficiency / per-formance) can be clearly laid down in the respective contract specifications, translated into three-dimensional models and tested with the overall design.

Consistent Three-dimensional Digital Modelling

The capacities of currently available CAD / CAM database technology are used to develop and constantly update consistent integrated three-dimensional digital models in order to optimize the project in its entirety and cyclically to control overall spatial quality.

With the help of suitable software programmes and based on standard digital formats, it is possible to obtain a great number of specific assessments.

In addition, the 3D synergetic factory model can be optimized by identifying and eliminating potential 'collisions' between different contracting firms and by constant quality checks. This ensures that conflicts between process, structural and services design – which would influence costs, construction time and quality negatively – are recognized early on and do not have to be eliminated on the construction site afterwards.

Quality Control by Digital Three-dimensional Project Cycles

The customary static, object-oriented approach to a factory planning project is currently increasingly shifting to a dynamic, process-oriented one. The real value of constant availability of data becomes apparent especially in the case of conversions, extensions or new buildings on the same or another site that are de-fined and treated as closed-loop systems. The special advantage of the synergetic factory planning model as compared to conventional methods is its higher preci-sion in estimating costs, defining scheduling interdependencies, controlling quality and running the production line. The information stored in the facility management model makes it possible to quickly design a new workshop, prepare a building proposal for a factory extension, or to buildi a new factory elsewhere. The transparency of project data stored digitally and the clearly structured methodology of the Synergetic Planning™ process offers the client an added value, in particular in the case of project cycles seen in a global context.

Curriculum vitae

Professor Jürgen Reichardt, architect (certificated engineer)

1956 born in Idar-Oberstein · 1976–81 studied architecture at Karlsruhe and Braunschweig Technical Universities · 1986–88 master-class student at Hochschule für Bildende Künste Braunschweig (art college) · 1988–95 worked with agiplan, Mülheim/Ruhr on the design and construction of complex industrial plants · 1991 established Reichardt Architekten, competition wins, projects and buildings, distinctions · 1996 professor of structural design theory, industrial building and facility management at MSA Muenster School of Architecture · 2002 published book, 'Wandlungsfähige Fabrikbauten', with IFA, Hanover Technical University · 2002 research project 'Synergetic Factory Planning'™ · 2004 ran postgraduate correspondence course on building design and facility management for Giessen-Friedberg University of Applied Sciences.

工程项目

PROJECTS

建筑的北部较长，有一个像“背包”的三层建筑，是行政管理区“气候花园”，即未加热的缓冲区把它们分成不同的两层更小的办公室作为员工区，无需再进一步分割。大面积的面向生产车间的内部玻璃幕墙是透明的。

The longer northern side of the building carries a three-tier 'backpack' which houses the administration areas on levels + 3.00 and + 5.80. 'Climatic gardens', i.e. unheated buffer zones, divide them into different smaller two-level offices which function as staff units without needing further partitioning. Large interior glass walls make them transparent towards the production hall.

参观者通过一个斜坡进入大楼，这个斜坡是主要的通道。在埃克罗德特维格这一边有一个能看到倒影的池子，它是用来截流屋顶的雨水以改善入口条件。

Visitors access the building via a ramp which extends the main interior circulation axis outside. A reflecting pool runs along the building on the side of Eichrodter Weg. It serves the retention of rainwater from the roof and enhances the entrance situation.

4a 建筑工作室
斯图加特

MBE 的工厂和办公室
埃森纳赫

FACTORY AND OFFICES OF MBE METALLBAU, EISENACH

透明度和开放度

TRANSPARENCY AND OPENNESS

建筑成本	2 000 万欧元
施工期	1995–1997
总面积	2 830m²
总体积	142 000m³
使用面积	20 670m²
占地面积	10 265m²

The client, the internationally renowned façade element producer MBE, demanded high-quality architecture for its new production and administration building, i.e. a modern structure marked by lightness, transparency and openness. The production hall – an 86 m-long steel structure spanning 22 m – forms the centre of the structure. In order to provide the different working areas with ample daylight, the architects designed transparent or translucent façades. Additional skylights supplement

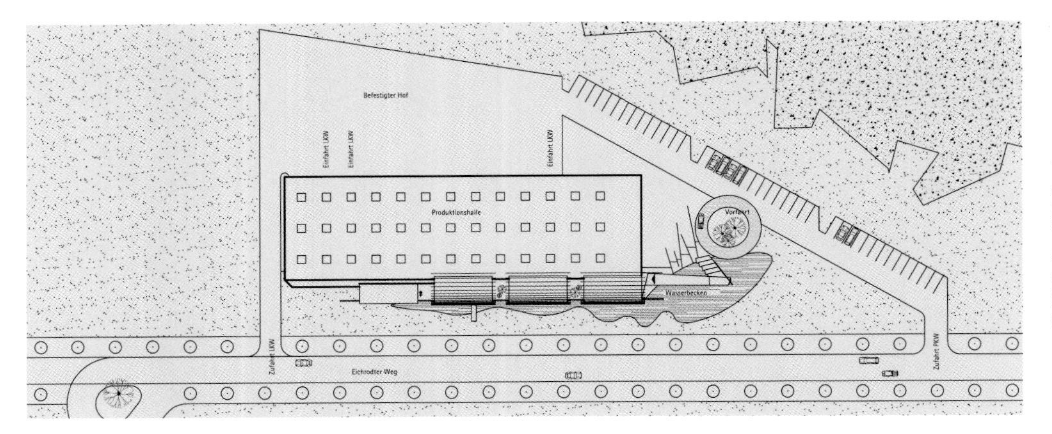

埃森纳赫东部的工业区，北部与莫尔河相邻，南部有森林。MBE的设计受这种独一无二的位置影响。这就是为什么设计这样的外立面、入口斜坡和池子，同时把大楼隐藏在周围的环境中，使建筑面向森林，面向自然。

The 'Grosse Güldene Aue' industrial estate on the eastern periphery of Eisenach is bordered by the Hörsel river to the north and a forest to the south. The MBE design was strongly influenced by this unique location. This is why the façades, the entrance ramp and the pool were designed in such a way as to lock the building into its country environment. It opens to the forest, to nature.

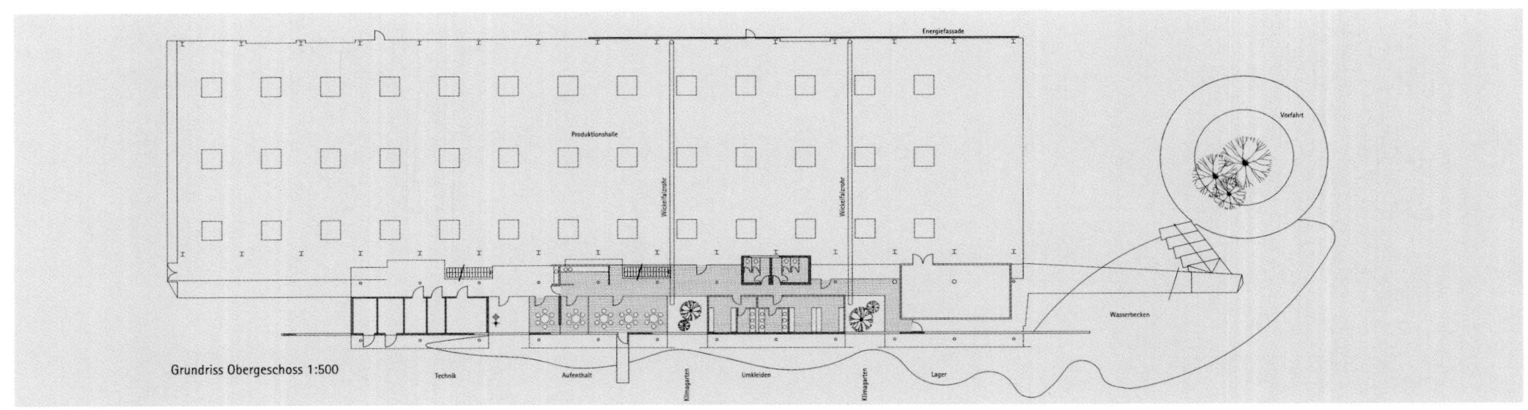

日光墙原理：

太阳把南墙黑色凿孔的护墙板加热。
温暖的表面空气通过护墙板上的小孔被吸进墙面。
热量损失由保温墙面获得补偿。
暖空气上升进入收集管。
用两层螺旋缝管，温度传感器控制，将暖空气输送到“气候花园”中。

'Solar Wall' Concept:

· The sun heats up the black perforated sheeting on
· the south façade.
· The warm surface air is then sucked inside through
· the fine sheeting holes.
· Heat losses through the insulated façades are thus
· compensated.
· The warm air rises up into the collector duct.
· The warm air is transported into the 'climatic gardens'
· through two folded spiral-seam tubes with integrated
· ventilators, simply controlled by temperature sensors.

只有在节省成本的情况下，公司大楼能量平衡最优化才有意义。因此，客户会选择相对能降低成本而不需额外投资的设计方案。建筑师在和太阳能技术有限公司的工程师紧密合作的过程中，开发了一种把能量效率和建筑设计质量结合起来的特制的适应气候变化的墙面。

Optimizing the energy balance of a corporate building only makes sense if it saves costs. Corporate clients therefore prefer solutions that, while not requiring much extra investment, considerably reduce operating costs. In close co-operation with the engineers of Transsolar Energietechnik Ltd., the architects developed a 'tailor-made' climatic façade which integrates energy efficiency with architectural design quality.

MBE 的工厂和办公室
埃森纳赫

FACTORY AND OFFICES OF MBE METALLBAU, EISENACH

能源效率

ENERGY EFFICIENCY

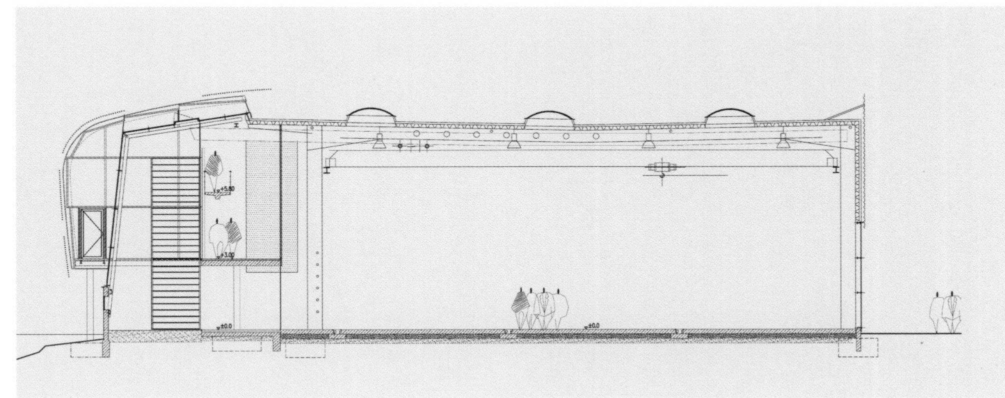

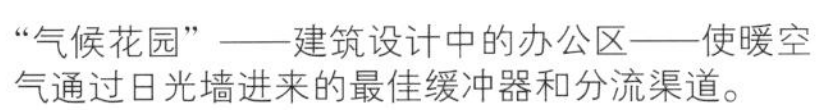
"气候花园"——建筑设计中的办公区——使暖空气通过日光墙进来的最佳缓冲器和分流渠道。

The 'climatic gardens' – the office areas in the architectural design – serve as optimal climatic buffers and distribution channels for the warm air coming in through the 'solar wall'.

建筑单位 Architektenbüro 4a, Stuttgart
业主 MBE Metallbau Eisenach GmbH
摄影 Uwe Ditz, Stuttgart

技术工程师
HLSE Schreiber Ingenieure Gebäudetechnik GmbH, Ulm
支撑结构设计 Ingenieurbüro Dipl.-Ing. Lichti und Dipl.-Ing. Laig GmbH, Möckmühl
能源技术 Transsolar Energietechnik GmbH, Stuttgart

阿尔丁杰建筑公司
斯图加特

北符腾堡州医学博士委员会管理和会议中心
斯图加特

ADMINISTRATION AND CONFERENCE CENTRE OF THE NORTH-WUERTTEMBERG DISTRICT CHAMBER OF MEDICAL DOCTORS, STUTTGART

合作建造

'CO-OPERATIVE BUILDING'

建筑成本	850 万欧元
施工期	2002–2004 年
总面积	3 800m²
总体积	13 500m³
竣工验收	2004 年
建筑标准	40m × 36m × 13m
占地面积	2 551m²

如雕刻般的东部外立面

Sculptural East Façade.

如雕刻般的建筑群把各种不同环境区分开来，将整块墙面（关闭以防周围空气散发）和完全能打开墙面（能够看到斯图加特市的全景）区别开来。一个标志性城市形象，利用可持续的能量观念和质量既满足客户的愿望又能增加自身理解。

The sculptural building mass defines the edges of its heterogeneous environment and distinguishes between massive façades (closed against ambient air emmissions) and a totally open side which affords a panoramic view over the city of Stuttgart. A marked urban presence, sustainable energy concept and quality detailing all satisfy both the wishes and the self-understanding of the client.

能够看到全景的屋顶阳台

Roof terrace with panoramic view.

能从底层看到顶层的中庭

Atrium with floor-to-ceiling glazings.

会议室

Conference room.

办公室

Offices.

会议大厅和休息室

Conference hall and foyer.

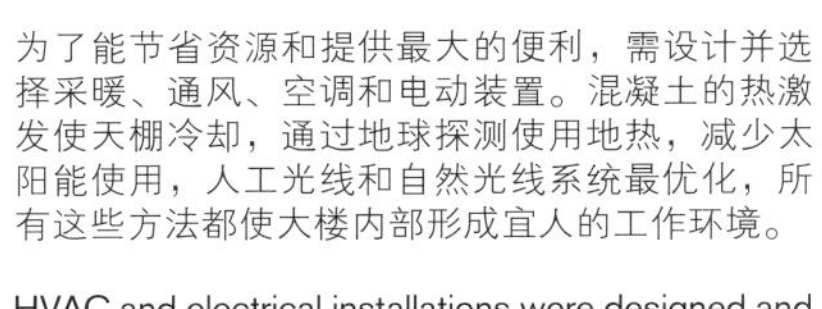

为了能节省资源和提供最大的便利，需设计并选择采暖、通风、空调和电动装置。混凝土的热激发使天棚冷却，通过地球探测使用地热，减少太阳能使用，人工光线和自然光线系统最优化，所有这些方法都使大楼内部形成宜人的工作环境。

HVAC and electrical installations were designed and chosen with a view to saving resources and creating maximum convenience. Thermal activation of concrete mass, cooling ceilings, naturally ventilated two-leaf cavity façades, use of geothermal energy via earth probes, reduction in solar heat gains and opti-mizing artificial and natural lighting systems – all these measures together have created pleasant working environments inside the building.

根据跨度很宽的屋顶天棚结构可以形成柔性平面。所有水平面钢筋混凝土的热激发会使内部温度适宜。无圆柱支撑的大厅有无黏结预应力的平顶天棚结构，目前在欧洲尚属首例。

The structure was based on wide-span flat ceilings to make flexible floor plans possible. Thermal activation of all horizontal RC elements contributes to climatizing the interior. The desired column-free hall has an unbonded prestressed flat-ceiling structure, at present one of the very first in Europe.

阿尔丁杰建筑公司
斯图加特

北符腾堡州医学博士委员会管理和会议中心
斯图加特

ADMINISTRATION AND CONFERENCE CENTRE OF THE NORTH-WUERTTEMBERG DISTRICT CHAMBER OF MEDICAL DOCTORS, STUTTGART

合作建造

'CO-OPERATIVE BUILDING'

包括测量和控制系统的技术装配都是符合工艺水平要求的，需把它调整为用户易于掌握的、易于操作和个人手工操作的形式。大楼符合国际“绿色建筑”标准。

The technical installations including gauging and control systems are state-of-the-art, while being geared to user-friendliness, easy handling and individual (additional manual) operation. The building complies with international 'green building' standards.

建筑单位 Aldinger Architekten, Stuttgart
Prof. Jörg Aldinger
Dirk Herker
业主 Bezirksärztekammer Nordwürttemberg, Stuttgart
项目 Ariane Prevedel
建筑工程管理 Wilfrid Wörner
景区规划 Christoph Luz, Stuttgart
摄影 Roland Halbe, Stuttgart

支撑结构 Mayer-Vorfelder & Dinkelacker, Sindelfingen
建筑技术 Laux, Kaiser + Partner, Stuttgart
电子技术 Werner Schwarz, Stuttgart
建筑物理 DS-Plan, Stuttgart

APD 建筑合伙公司达姆施塔特

德国电器工程师协会新办公大楼
美茵河畔法兰克福

NEW VDE ADMINISTRATION BUILDING
FRANKFURT/MAIN

建筑和城市设计

ARCHITECTURE AND URBAN DESIGN

建筑成本	约 500 万欧元
施工期	2001–2002 年
总体积	6 342m³
占地表面积	1 480m²
使用面积	640m²
竣工验收	2002 年

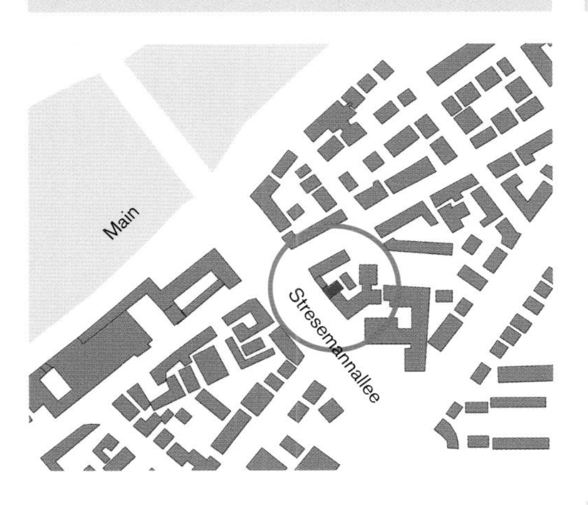

位置图

Location plan.

标准层平面图

Standard floor plan.

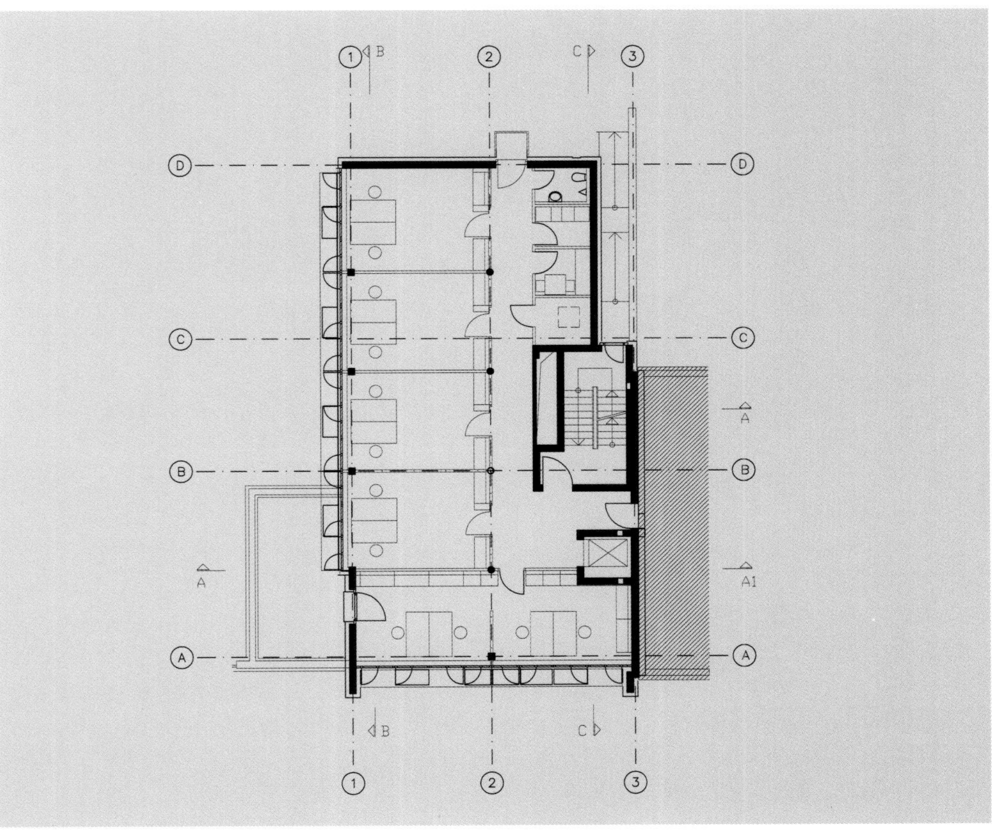

对已有的办公楼进行扩建主要在内城区，即早先的居住区。从环形区域结构这一城市设计原则出发，扩建主要定位在两个方面：一方面解决斯特莱色曼大街交通拥挤问题，另一方面使附近的建筑不受周围空气污染的影响。创新工业区的核心设计是双层外立面。

A modern extension of an existing office building was erected on the inner-city site of a former residential block. Starting from the urban design principle of completing perimeter block structures, the extension is mainly oriented in two directions: to Stresemannallee with its heavy traffic on one side and to the neighbouring buildings less subject to ambient emissions on the other side. The core design element of the innovative complex was the two-leaf façade.

APD 建筑合伙公司 达姆施塔特

德国电器工程师协会新办公大楼
美茵河畔法兰克福

NEW VDE ADMINISTRATION BUILDING FRANKFURT/MAIN

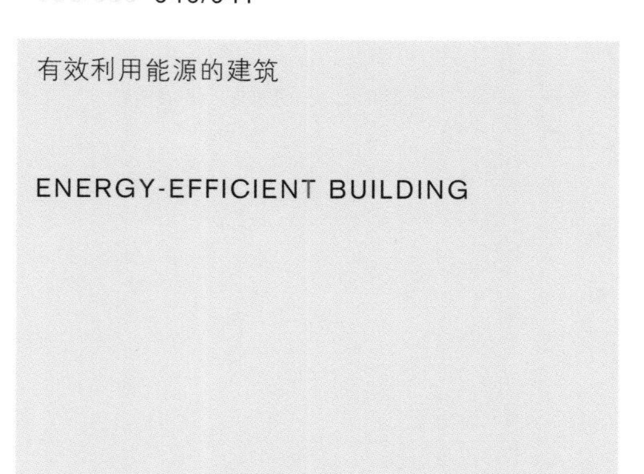
有效利用能源的建筑

ENERGY-EFFICIENT BUILDING

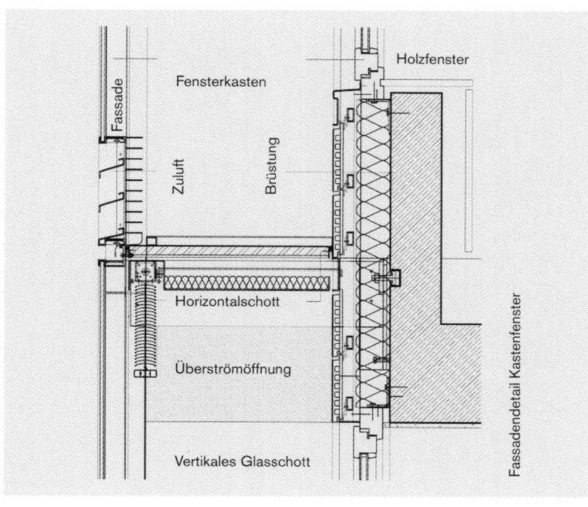

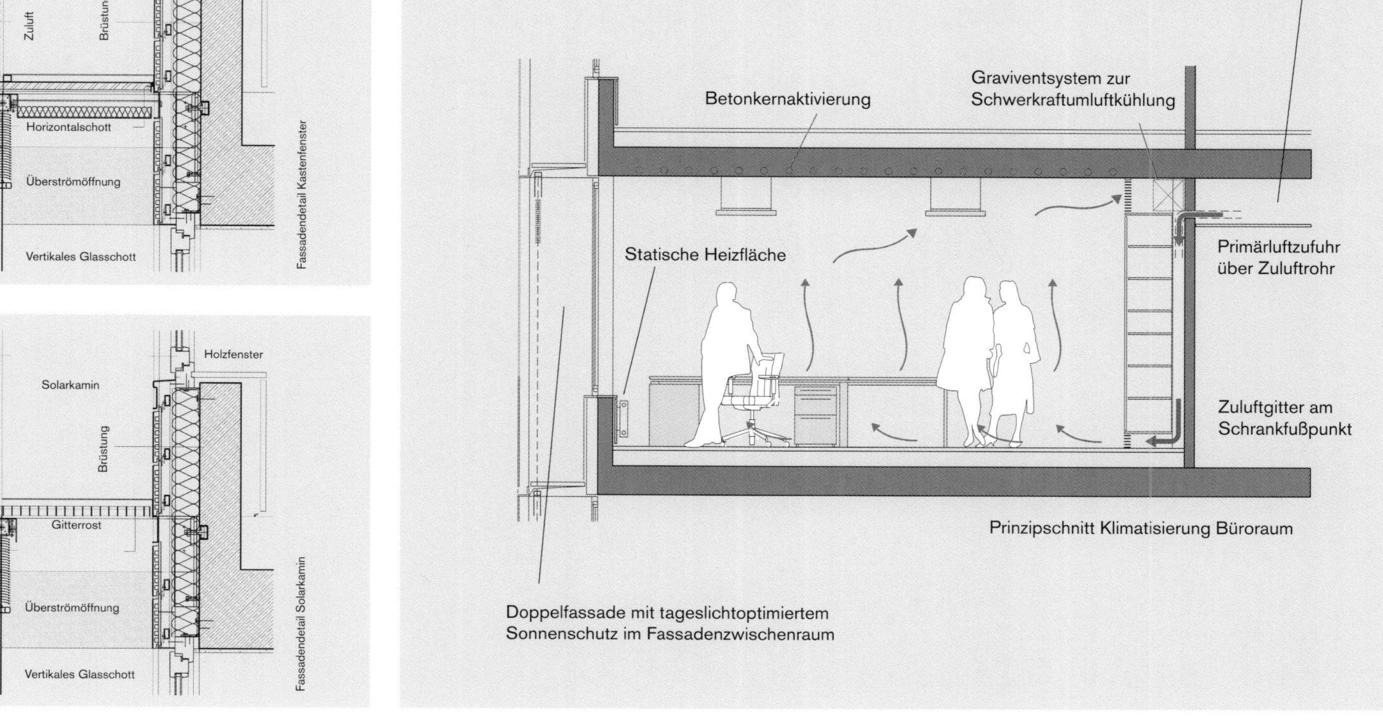

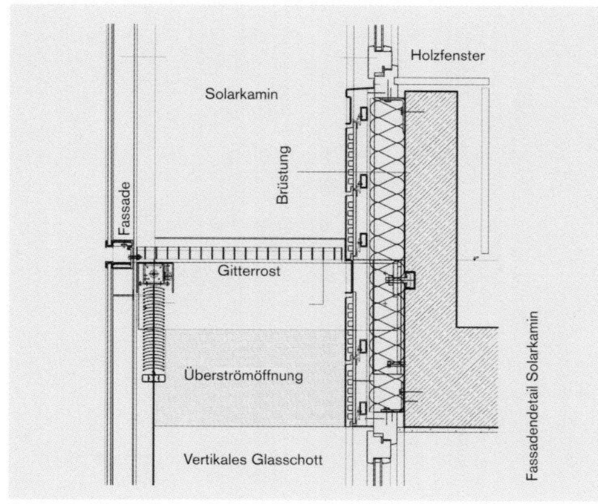

整体气候理念：
通过地面管道预处理流入的空气
夜间通过钢筋混凝土的热激发调节
通过分散冷却
即使敞开窗户也能防控噪音
其他用户对内部气候的控制

Overall Climatic Concept:
- preconditioning of ingoing air through a ground
- duct
- night-time conditioning via thermal activation of
- RC floors
- decentral cooling through gravivents
- noise protection even with open windows
- additional user control of interior climate

完整的能源理念是在和太阳能技术有限公司的工程师合作过程中产生的。它把斯特莱色曼大街上的噪音考虑进去，使双层幕墙与整体气候观念相结合，由此取得最大的能源效率和最适宜的内部温度。墙面与内部设计相配：外层的玻璃表面与砖结构幕墙相对照，幕墙的颜色质地给人留下深刻印象。

An integrated energy concept was developed in co-operation with the engineers of Transsolar Energietechnik Ltd.. It takes into account the traffic noise levels from Stresemannallee and integrates the double façade with the overall climatic concept, thereby achieving the maximum energy efficiency and optimal interior climates. The façade matches the interior design: the technoid outer glass skin is contrasted by brick curtain wall sections which impress with their colours and texture.

建筑单位	APD Architekten, Darmstadt
业主	VDE Frankfurt am Main
摄影	Sigurd Steinprinz, Wuppertal
专业工程师	
立面	Fa. Transsolar, Stuttgart
静力学	Kleinhofen und Schulenberg, Darmstadt

目前二手车交易发生了变化。一直以来，汽车制造商和销售商试图把它们的营销活动变成需好好筹备的“公共事件”。宝马二手车销售中心就是为此目的建造的。清晰、简捷的建筑体积为南法兰克福毫无特征的各种工业区树立了引人注目的里程碑。该中心的目标是每年销售7 500辆轿车，其中还包括大的修理车间。

The trade in second-hand cars is currently changing. Increasingly, car manufacturers and sales outlets try to turn their marketing activities into 'public events' that have to be properly 'staged'. The BMW second-hand sales centre was built for this purpose in the form of clear, simple building volumes that set up a striking landmark in a rather nondescript and heterogeneous industrial estate south of Frankfurt. The centre aims to sell 7,500 cars per annum and also includes a large repair shop.

主体面有150多米长。宝马和迷你车型陈列在双层的玻璃箱中，用一个细支柱支撑着顶部发光的金属面，这个支柱支撑整个建筑。

The main façade is over 150 m long. The BMWs and Minis are displayed in double-storey glass boxes, 'capped' by a shimmering metallic volume on slender stilts that 'brackets' the different parts of the building.

阿尔伯特·施拜尔及其合作有限责任公司
美茵河畔法兰克福

宝马二手车销售中心
德莱艾赫

BMW SECOND-HAND CAR SALES CENTRE, DREIEICH

建筑设计

ARCHITECTURAL DESIGN

竣工验收	2005 年 2 月
计划 / 施工	12 个月 /15 个月
建筑用途	二手车服务站
地面楼层	3
总体积	110 000m³
占地面积	24 000m²
楼层高度	3.85m
纯建筑成本	1 550万欧元，包括其他费用

通过一个斜坡和一个很宽的楼梯可以进入最顶层的销售区和陈列区，还有一个快餐店。巧妙设计的开口能够看到车间和货场。

The first upper floor with further sales and display areas as well as a snackbar is accessed via a ramp and a wide stairway. Skilfully arranged openings afford views into the workshops and delivery yards.

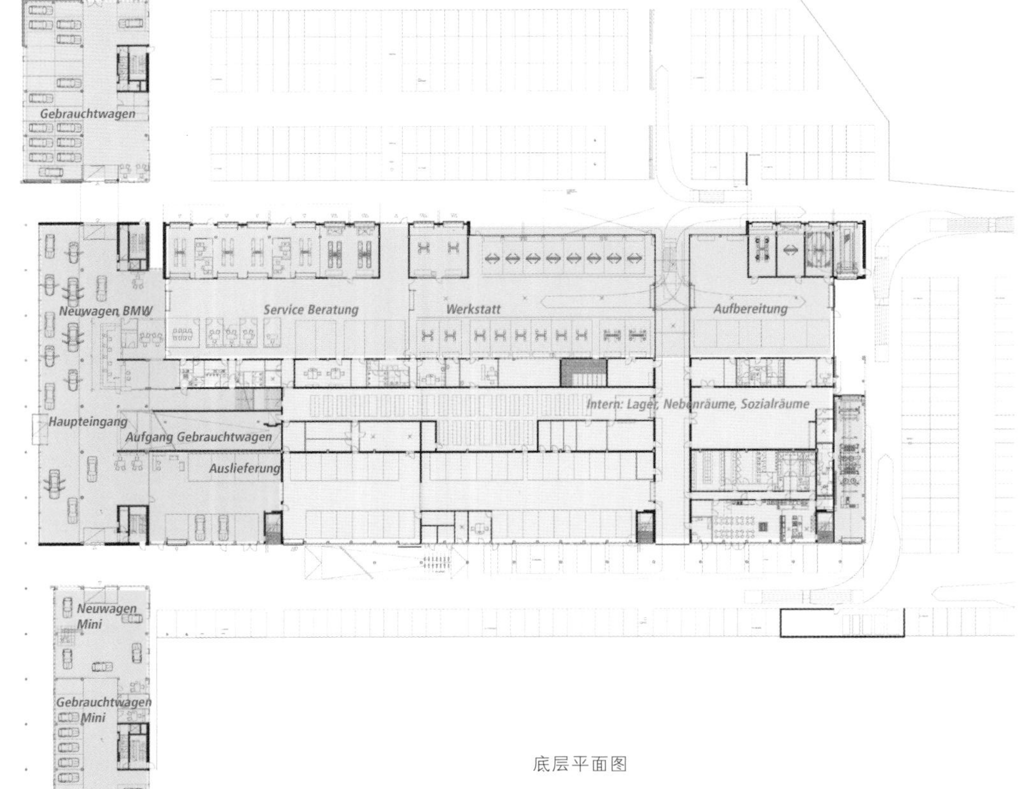

底层平面图

Ground floor plan

顺序排列的陈列室及其他部门从中间入口区一直延伸到里面，直到车间和汽车销售的货场。顾客登上一个壮观的“人行天桥楼梯”可以到达第二层及屋顶平台，这个屋顶平台也是作为汽车“陈列室”。

A sequence of showrooms and other building parts lead from the central entrance area into the depth of the site, to workshops and delivery areas for sold cars. Customers ascend a spectacular 'skywalk stairway' to reach the second upper level and the roof deck which is also used as an automobile 'showroom'.

阿尔伯特·施拜尔及其合作有限责任公司
美茵河畔法兰克福

宝马二手车销售中心
德莱艾赫

BMW SECOND-HAND CAR SALES CENTRE, DREIEICH

生态环境和效率

ECOLOGY AND EFFICIENCY

生态理念包括现场渗入雨水的排泄和水的再循环处理。结构（有水泥墙板的承重框架结构）符合绝缘规定。铝板条与主墙面成为一体用来防潮，同时巧妙、有效地突出了建筑的技术特点。

The ecological concept includes the complete drainage of rain water on site through influent seepage and the recycling of process water. The structure (load-bearing RC skeleton frame filled in with expanded concrete wall slabs) complies with insulation regulations. The aluminium slats integrated into the main façade serve as weather protection and also – subtly but effectively – emphasize the technical character of the building.

项目参与者	
委托方	BMW AG, München
承建方	AS&P – Albert Speer & Partner GmbH, Frankfurt am Main
建筑工程管理	FAAG, Frankfurter Aufbau AG, Frankfurt am Main
立面设计	AS&P – Albert Speer & Partner GmbH, Frankfurt am Main
支撑结构设计	B+G Ingenieure, Bollinger und Grohmann GmbH, Frankfurt a.M.
建筑技术	Seidl & Partner, Frankfurt am Main / Regensburg Brendel Ingenieure, Frankfurt am Main
防火 / 建筑物理	Büro Endress, Frankfurt am Main
室内建筑设计	AS&P – Albert Speer & Partner GmbH, Frankfurt am Main in Zusammenarbeit mit BMW AG, München
景区规划	WGF Werkgemeinschaft Freiraum, Nürnberg
摄影	Christoph Lison, Frankfurt am Main

紧凑的椭圆形大楼意味着最小的墙面面积，这是优化能源消耗所必需的。从第七层往上，引人入胜的住宅区能够看到法兰克福城市全景。

The compact shape of the 'Oval' building means a minimum facade surface area – the basic requisite for optimizing a low-energy concept. From the seventh upper floor upwards, attractive residential units afford panoramic views over Frankfurt's cityscape.

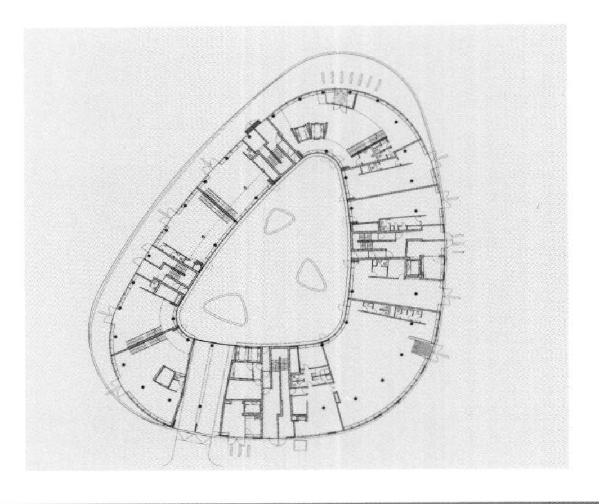

平面图

Floor plan

阿尔伯特·施拜尔及其合作有限责任公司
美茵河畔法兰克福

贝斯勒广场椭圆形建筑
美茵河畔法兰克福

THE OVAL ON BASELER PLATZ
FRANKFURT/MAIN

建筑设计

ARCHITECTURAL DESIGN

竣工验收	2004 年 5 月
计划和施工	2000–2004 年
建筑物	餐饮区 – 工艺区 – 办公区 – 住宅区
地面楼层	10
总体积	75 900m^3（层）/97 400m^3（包括地基）/115 600m^3（全部包括厅院）
总面积	21 400m^2（层）/24 200m^2（合计）
占地面积	2 200m^2
楼层高度	3.50m（办公区）/2.98m（住宅区）
建筑成本	约 3 900 万欧元（净）

位置图

Location plan

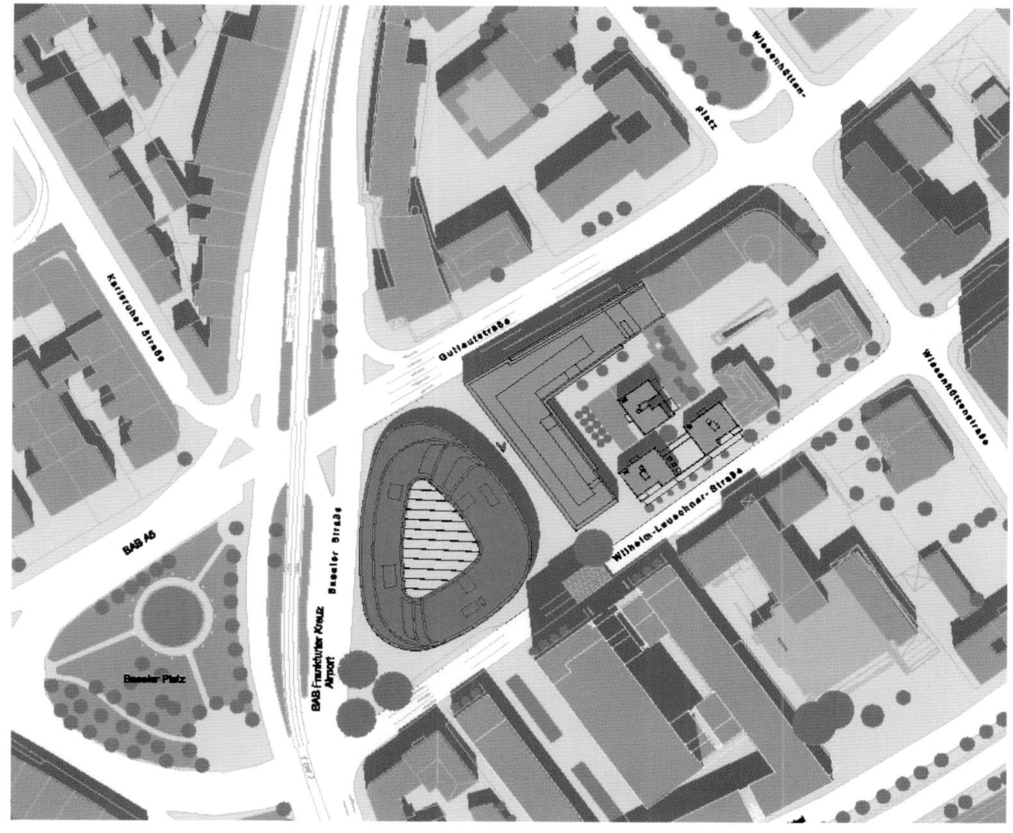

贝斯勒广场是法兰克福一个非常重要的公共广场，它是进入内部城区的南大门。广场呈椭圆形，是一个具有各种不同功能的标志性城市建筑。它的外观非常醒目，像一个圆塔，在广场建筑红线之外，同时又指向远方，好像把广场向东延伸。一层的中庭有饭店、零售商店和展览空间。一层以上有 6 层办公楼，顶上 3 层是住宅楼。

The Baseler Platz is a public square of special importance to Frankfurt, as it forms the southern gateway to the inner city area. This is the location of the 'Oval', a landmark building containing an urban mix of different functions. It is the form that makes it particularly conspicuous, like a roundtower which respects the building lines of the Baseler Platz, but also points beyond it and seems to extend the square to the east. The ground floor contains restaurants, retail and exhibition spaces arranged around an atrium. Above them the building has six office floors and three top residential floors.

阿尔伯特·施拜尔及其合作有限责任公司
美茵河畔法兰克福

贝斯勒广场椭圆形建筑
美茵河畔法兰克福

THE OVAL ON BASELER PLATZ
FRANKFURT/MAIN

结构

STRUCTURE

第7层作为气候缓冲区，因而顶上用一个25米长、3.3米宽，具有光透明结构的透明充气垫作为房顶。

At the level of the seventh upper floor the atrium was to function as a climatic buffer zone and was therefore roofed over with a light transparent structure using transparent air-filled 'cushions' up to 25 m long and 3.3 m wide.

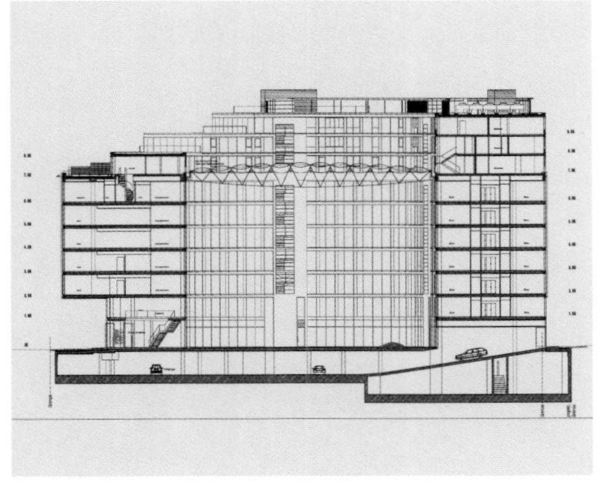

7层的钢筋混凝土框架结构的悬臂没有支柱，从1层往上有7米多高，采用未粘黏预应力楼板。这种大跨度能够很容易分割内部空间——蜂窝式、组合式、统间式的结构。

The seven-storey RC skeleton frame cantilevers column-free over up to 7 m from the ground floor structure, using unbonded prestressed floor slabs. Their large spans make it possible to partition the interior spaces flexibly – as cellular, combination or open-plan offices.

“垫子”用圆拱形的细钢管支撑，细钢管又用平行构架经V形圆拱支柱支撑。创新技术防止夏天中庭过热。每个充气垫用水平膜分成上下两个空间，膜和垫子的上半部分都是丝绢网印花的。增加下半部分的空气压力，膜就升高附着在垫子的顶部膜上，两面的圆点密集在中庭房顶的表面，作为遮阳伞。

The 'cushions' are supported by round arches of thin steel tube, in turn supported by offset parallel trusses via V-shaped arch bracings. Innovative technology prevents summer-time overheating of the atrium. A horizontal membrane divides every air cushion into an upper and a lower chamber. Both this membrane and the upper side of the cushions are silk-screen-printed. With mounting air pressure in the lower chamber, the membrane lifts and clings to the top membrane of the cushion so that the dots on both surfaces 'densify' for the atrium roof to act as a sunshade.

项目参与者 委托方	BGA Allgemeine Immobilienverwaltungs- und Entwicklungsgesellschaft mbH, Hamburg AS&P – Albert Speer & Partner GmbH, Frankfurt am Main
总设计师 / 建筑师 建筑工程管理	AS&P – Albert Speer & Partner GmbH in Zusammenarbeit mit FAAG Frankfurter Aufbau AG, Frankfurt am Main AS&P / IFFT Institut für Fassadentechnik Karlotto Schott, Frankfurt
外立面设计	B+G Ingenieure, Bollinger und Grohmann GmbH, Frankfurt a.M.
支撑结构设计	NEK Beratende Ingenieure, Frankfurt am Main
技术性建筑物装备	Sommerlad Haase Kuhli Landschaftsarchitekten, Gießen
空闲区规划	oppp consult GmbH, Frankfurt am Main
工程咨询	WISAG Service Holding GmbH & Co. KG, Frankfurt am Main
质量管理 总公司	Arbeitsgemeinschaft Oval am Baseler Platz: Wayss & Freytag Schlüsselfertigbau AG / BauBeCon Hochbau GmbH Barbara Staubach, Wiesbaden, S. 049
摄影	Christoph Kraneburg, Köln/Darmstadt, S. 046–048

大的玻璃墙面能够给内部提供最多的日光和太阳能，方格形的窗户能够打开，空气可以流通，成为机械通风的补充。

Large façade areas of glazing provide the interior with maximum amounts of daylight and solar energy, while the box-type windows can be opened to supplement mechanical ventilation.

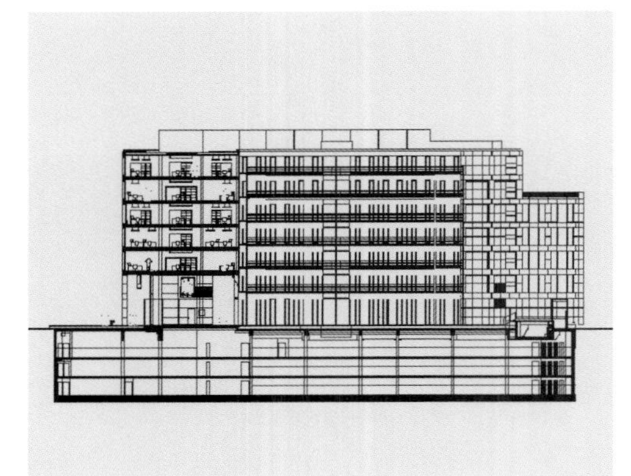

纵剖面图

Cross section

阿尔伯特·施拜尔及其合作有限责任公司
美茵河畔法兰克福

贝斯勒阿卡丹大厦
美茵河畔法兰克福

BASELER ARKADEN
FRANKFURT AM MAIN

建筑设计

ARCHITECTURAL DESIGN

竣工验收	2004 年 2 月
计划 / 施工期	2001–2004 年
建筑用途	办公大厦及地下公共停车场
地面楼层	7
总体积	42 169m^3
总面积	12 380m^2（层）/25 885m^2（合计）
出租面积	9 743m^2
GIF/BGF	0.84
楼层高度	3.50m
停车位	389
建筑成本	2 700 万欧元（净）

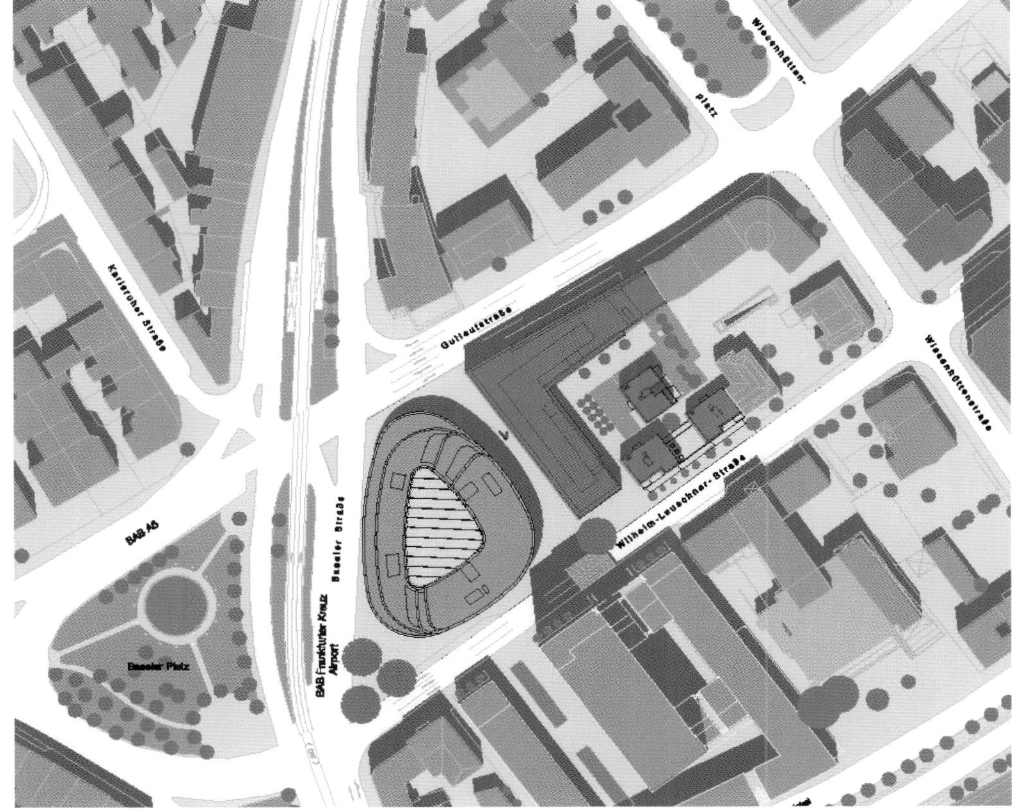

新建筑以古典建筑风格为特征：清晰、有序、统一。严格规整的墙面上的细柱和尖板石条的栏杆，有凸凹不平的变化。一排排两层高的柱子形成的“拱廊”，使大楼名符其实。他们的后面是商店和饭店，好像要使广场重归都市生活。

Clarity, order, consistency: the new building is characterized by these classical architectural qualities. Their strictly regular façades are marked by a modular system of slender columns and parapets clad with pale stone varied by recesses and shifts in plane. Two-storey high rows of columns form the 'arcades' which give the building its name. Behind them are shops and restaurants that promise to bring Baseler Platz back to urban life.

阿尔伯特・施拜尔及其合作有限责任公司
美茵河畔法兰克福

贝斯勒阿卡丹大厦
美茵河畔法兰克福

BASELER ARKADEN
FRANKFURT AM MAIN

节能办公室

LOW-ENERGY-OFFICE

新大楼大约有10 000平方米的办公空间。因为房间深度19米，能够适合各种不同用途。

The new building offers approx. 10,000 m² of flexible office space. Due to room depths of up to 19 m, it can accommodate many different uses.

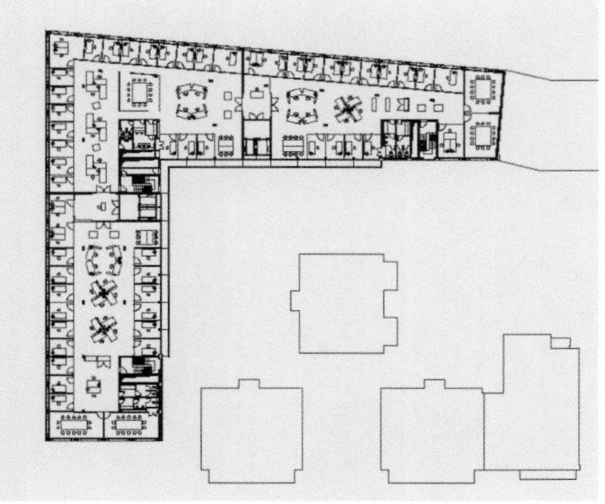

楼梯的设置又能够使每一层分成3个套房，可以分别出租。

The placement of the staircases made it possible to subdivide every floor into up to three suites that can be let separately.

大楼耗能低，主要的能源采用可再生资源。从地下80米深处，用21℃的温水注入闭合循环供热系统。冬天使用地热（通过压热泵），夏天冷却系统的废气热量被注入到地下。

The building is a low-energy house. The main element of the energy concept is the use of renewable resources. From a ground depth of 80 m, 21° C warm water is fed into the closed-cycle heating system. In winter, geothermal energy is utilized (via a heat pump), in summer, the waste heat from the cooling system is fed back into the ground.

工程参与者

委托方	FAAG, Frankfurter Aufbau AG, Frankfurt am Main
建筑单位	AS&P – Albert Speer & Partner GmbH, Frankfurt am Main
工程控制 / 工程管理	FAAG, Frankfurter Aufbau AG, Frankfurt am Main
工程发展	UPG urbane Projekt GmbH, Frankfurt am Main
外立面设计	AS&P / IFFT Karlotto Schott, Frankfurt am Main
支撑结构	Engelbach + Partner Ingenieurbüro für Baustatik, Frankfurt am Main
房屋技术方案 / 建筑物理	Lemon Consult, Zürich
房屋技术规划	Reuter und Rührgartner, Rosbach v.d.H.
周边保护 / 建筑物理	Dipl.-Ing. Dr. Sesselmann und Kollegen, Mainz
景区建筑	FreiraumX, Frankfurt am Main
基本结构	Bilfinger Berger AG Niederlassung Rohbau, Frankfurt am Main
外立面	Firma Lehr, Mainz-Hechtsheim
自然石	Hofmann Naturstein-Fassaden, Gamburg
电气	Winkler + Kolter Elektro und Fernmeldetechnik GmbH, Frankfurt am Main
摄影	Christoph Lison, Frankfurt am Main

阿尔伯特·施拜尔及其合作有限责任公司
美茵河畔法兰克福

维多利亚塔　曼海姆

VICTORIA TOWER MANNHEIM

城市位置

URBAN SITUATION

施工期	1999–2001 年
建筑用途	办公 – 管理
高度	97m
楼层	28
总体积	95 517m^3
占地面积	4 600m^2
总面积	28 664m^2
出租面积	20 800m^2
每楼层出租面积	746m^2
办公间	700（总计），250（出租）
停车位	137

标准层平面图；布局平面

Standard floor plan; Layout plan

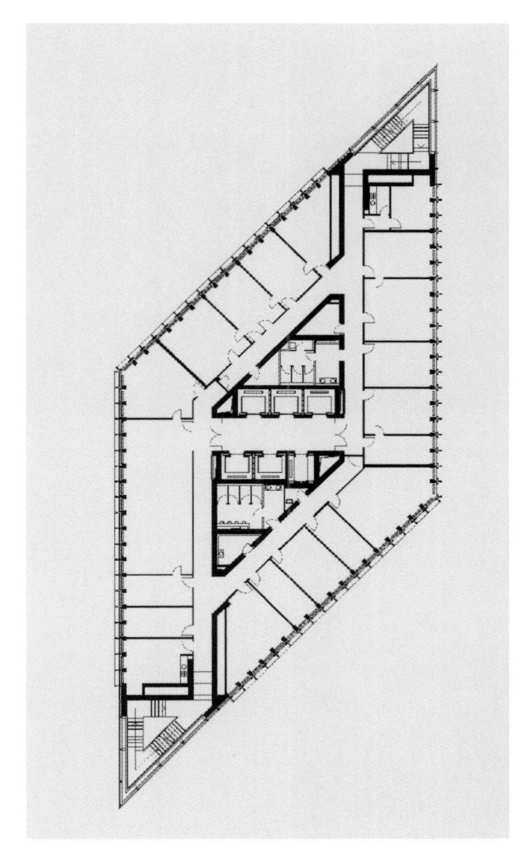

维多利亚塔（呈菱形的平面图）的一面墙与林登霍夫的主轴相连，其中一角指向城市中心。

The Victoria Tower (rhomboid in plan) was placed in such a way that one facade relates to the main axis of the Lindenhof area, while one of its corners points in the direction of the city centre.

阿尔伯特·施拜尔及其合作人事务所提出的曼海姆 21 城市设计可行性研究提出了集居住、办公、商务、文化空间和结构一体的新城区。维多利亚保险公司很幸运地在曼海姆总站附近为其西南区总部找到这块地方。它作为通向林登霍夫地区的重要大门，是建造超高层建筑的极好地方，也是发展新城区的“发令枪”。

The urban design feasibility study 'Mannheim 21' developed by AS&P proposed a new urban quarter with residential units, offices, commercial and cultural spaces and structures. It was a piece of good luck when the Victoria Insurance found a site for its new south-west regional headquarters near Mannheim central station. For several reasons, this location seemed the perfect place for a high-rise as a significant 'gateway' to the Lindenhof area and the 'starting shot' for developing the new urban quarter.

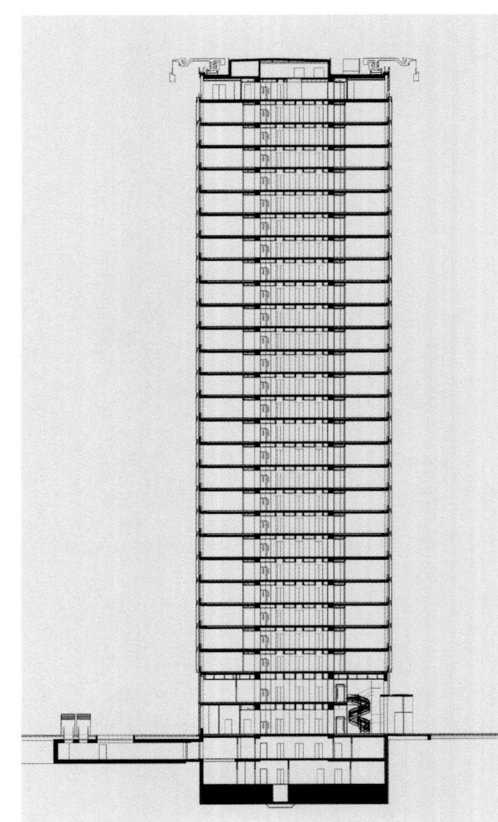

纵剖面图

Cross section

装有单层玻璃的透明楼梯和顶层办公区的框架用第二层有通风板条玻璃层来保护。这种框架式结构看起来轻巧，夜晚有灯火阑珊的灯光效果。

The single-glazed transparent staircases and top floor frame the 'block' of offices protected by a second layer of glazing with integrated ventilation slats. The frame makes the structure appear less heavy and at night it lights up to dramatic effect.

业主希望自然通风，最初由于墙面普遍的风力载荷问题和交通噪音问题，认为通过这种方式不可能取得所需的内部舒适的温度。通过动态结构模拟和在不同风力载荷和噪音下进行1：1墙面模型测试，建筑师开发出一种不同寻常的技术结构方案，通过双层夹心墙，使舒适的空气得以流通，同时消除噪音。

The client desired natural ventilation, but initially thought it impossible to achieve the required interior thermal comfort in this way due to the problems of prevailing wind loads on the facades and high traffic noise levels. By means of dynamic structural simulations and by testing 1:1 facade models under different wind loads and noise levels, the architects developed an unusual technical, structural solution based on a two-leaf cavity wall which allows a comfortable air exchange while providing adequate acoustic protection.

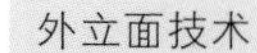

维多利亚塔　曼海姆

VICTORIA TOWER MANNHEIM

阿尔伯特・施拜尔及其合作有限责任公司
美茵河畔法兰克福

外立面技术

FAÇADE TECHNOLOGY

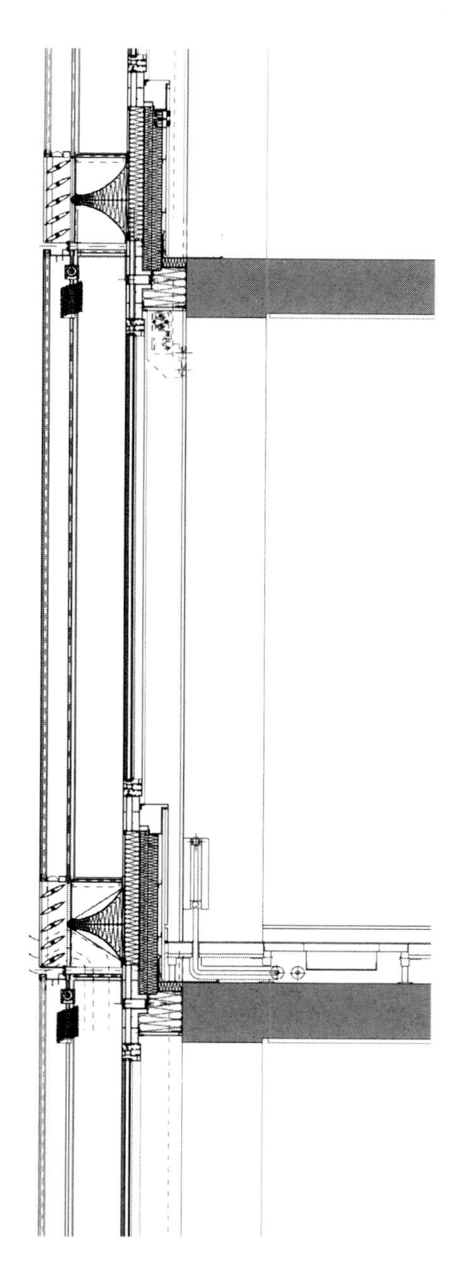

玻璃剖面构造：夏天，钢筋混凝土地板的热量足以使内部凉爽。

Vertical facade section: the thermal mass of the RC floors is sufficient to cool the interiors in the summer months.

大楼的核心，设计成坚固的塔以减少横向载荷，用最少的面积安装电梯和辅助设施。两个尖角楼梯，一层和顶层有单层玻璃，1.35米网格状墙面结构的办公区用第二层有通风板条的玻璃层来保护。

The building core, designed to stiffen the tower against lateral loads, encases the lifts and ancillary spaces using up a minimum of floor space. The two sharp-angled corner staircases, the bottom floors and the top floor have skins of single-glazing, while the 1.35 m grid-structured facade of the offices is protected by a second glass layer with integrated ventilation slats.

工程参与者

委托方	Internationales Immobilien-Institut GmbH, handelnd für iii-Immobilien-Spezialfonds der Victoria-Lebensversicherung AG; vertreten durch die MEAG, Real Estate Management GmbH, München
建筑单位	AS&P – Albert Speer & Partner GmbH, Frankfurt am Main
工程管理和接管方	V.I.M., VICTORIA Immobilien Management Düsseldorf
工程进度	Unit, Gesellschaft für Projektentwicklung mbH, Frankfurt/Main
工程控制	Bache & Partner, Hamburg
建筑管理	Knüppel & Faber Architekten, Hamburg
立面设计	IFFT Institut für Fassadentechnik Karlotto Schott, Frankfurt/Main
支撑结构设计	König und Heunisch, Beratende Ingenieure, Frankfurt am Main
HLS 技术	Horst Frank, Haustechn. Planung und Betreuung, Ludwigshafen
电气技术	Ingenieurbüro Buchwald, Sprockhövel
建筑物理	Genest und Partner, Ludwigshafen
风暴模拟	Fachgebiet Bauphysik und Technischer Ausbau, TH Karlsruhe
热模拟	Stahl, Büro für SonnenEnergie, Freiburg i.Br.
周边保护	Brandschutzberatung Haas, Edingen-Neckarhausen
特殊照明	Uli Jetzt Beleuchtungen GmbH, Backnang
内部建筑	Raum + Büro, Simone Bücksteeg, Wiesbaden
外部设备	Arcadis, Trischler & Partner, Darmstadt
摄影	Prof. Dieter Leistner, Mainz

艾特・里尔建筑设计事务所及
城市发展规划所，马丁・奇莫
达姆施塔特

赛德克化工有限公司总部
兹温根博格

SURTEC CORPORATE HEADQUARTERS
ZWINGENBERG

低耗能供热的化工工厂建筑

A CHEMICAL FACTORY BUILDINGAS A
HOUSE WITH ZERO-ENERGY HEATING

建筑成本	3 900 万欧元
计划和施工	1997–2000 年
总体积	28 380m³
占地面积	4 423m²
使用面积	3 536m²
竣工验收	2000 年

供热设备与透明中庭相结合，使得接触空气的墙面面积大大减少，这样可以扩大中庭的面积，所以整个厂区可以不安装传统供暖系统。把 3 个总功率为 35 千瓦的热太阳能系统组件分别安装在通风口。对于体积大约 25 000 立方米的内部空间来说，足以补充被动吸收太阳能热量。

By combining the facilities with glazed atria, the facades touched by environmental air were considerably reduced, while the atria were generously planted. This made it possible to do without a conventional heating system throughout the entire complex. Three heating units with a total capacity of 35 kW, installed in the air inlets, are sufficient to supplement the passive solar heating of the roughly 25,000 m^3 of interior air space.

这是新建成的赛德克化工有限公司的总部。它在欧洲的建筑中首先安装了被动式热太阳能。大楼的建成是为了满足不同的需求。作为工业区第一个低能耗供热建筑，它是节能工程的里程碑。总部设有化学实验室和厂房，并建在紧邻居民区的地方。一直以来因为环境质量的良好状态，充分证明了居民对工厂安全的信任是经得起考验的。总部的建筑成本低于平均水平，建成的大楼将多功能性、优良的空气质量、透明外墙及环保建筑内在价值集于一身。

The new SurTec headquarters is the first European building run exclusively on passive thermal solar energy. It is a synthesis of diametrically different requirements. As the first ‘house with zero-energy heating’ in industrial building, it represents a milestone in energy-efficient construction. It is a factory with chemical laboratories and production halls that was erected in the immediate proximity to residential areas and justifies the neighbours’ trust in its safety by its seminal environmental quality. Functionality, spatial quality, transparency and environmentally friendly construction are its ‘inner values’ achieved with below-average construction costs.

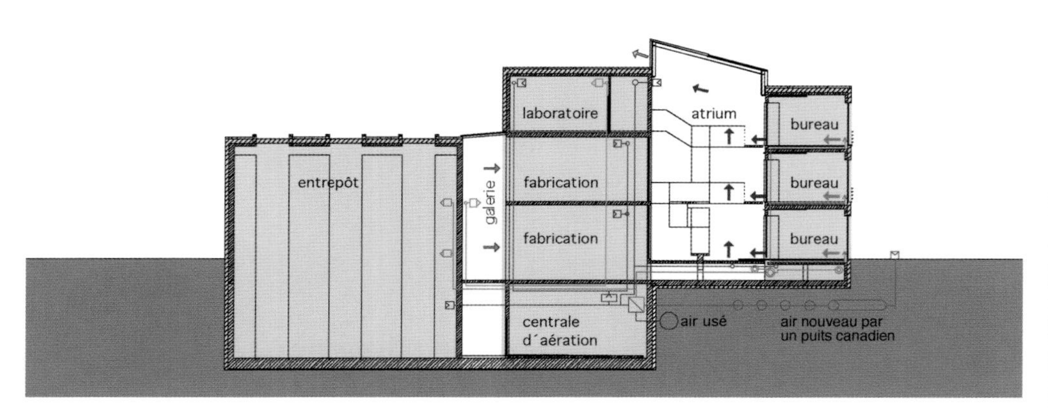

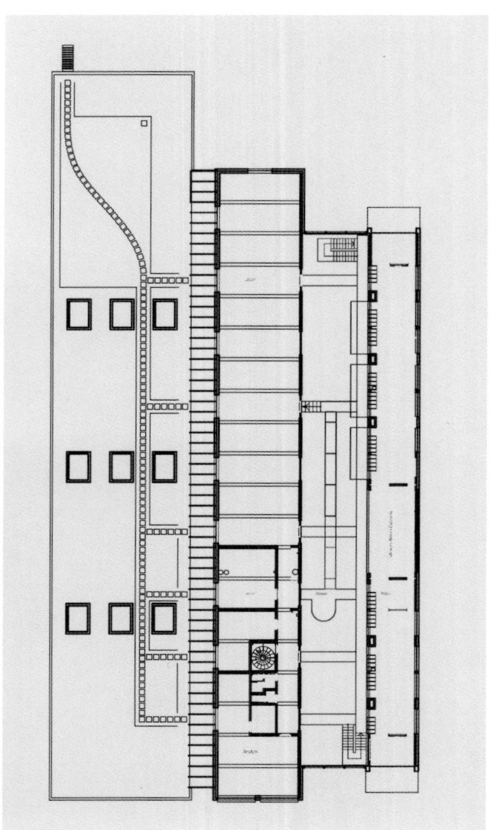

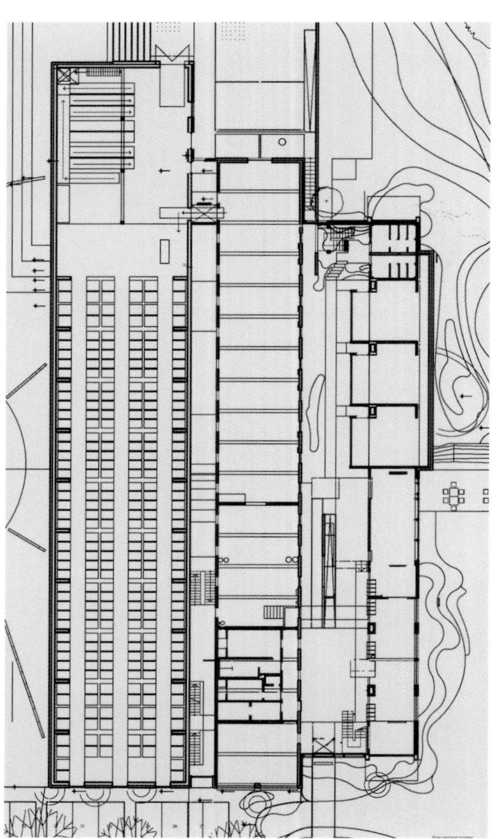

因为钢筋混凝土的外部覆盖合成绝热系统，所以不存在结构方面的热损失问题，可防止所有导热现象。中庭的窗框和梁柱型结构都含有抗热聚乙氨酯的绝热芯和背圈。首先要考虑的是建筑的节能问题，即如何采用环保的溶解建筑材料。

As the RC structure is clad with a composite thermal insulation system, there are no structural heat leaks. Every penetrating element was thermally insulated. The window frames and the post-and-beam structure of the atrium were fitted with a heat-leak-proof core insulation of polyurethane with insulated backbands. Energy-saving construction was a priority, as was the use of environmentally compatible construction materials.

因为工厂建在水源保护区，根据德国水供应条例的要求，化工产品的储藏室必须建在水密仓中。水密仓的建立使储藏室低于地面 5 米，这样与传统独立式建筑相比，外露的墙体面积减少。

As the plant is located in a water protection area, the chemicals store had to be erected in a watertight tank, in compliance with German water supply regulations. The tank made it possible to lower the store 5 m into the ground, which reduced the visible facade surfaces as compared to a conventional free-standing building.

赛德克化工有限公司总部
兹温根博格

SURTEC CORPORATE HEADQUARTERS ZWINGENBERG

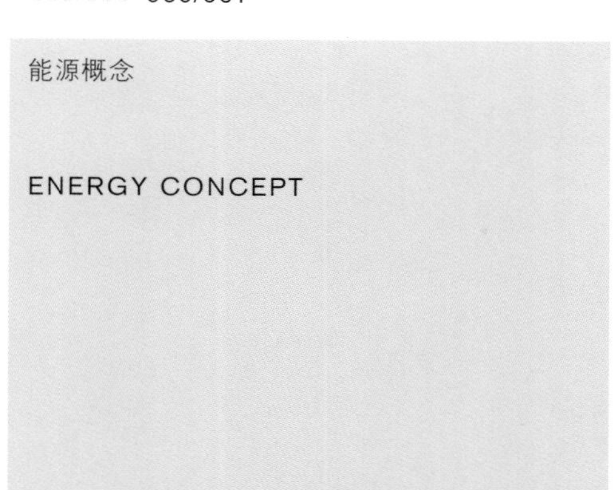

01 建筑侧面的 U 220
U 220 on the side of the building
02 挡泥板
mudguard sheeting
03 牙条 M6
threaded bar M6
04 拉铆螺母 M6
rivet nut M6
05 橡胶螺栓垫圈
rubber bolt washer
06 16 × 1.8 管道
tube 16 x 1.8
07 7 号铝杆
item 7 aluminium beam
08 1.1 号铝柱
item 1.1 aluminium post
09 4.1 号铝背带
item 4.1 aluminium backband
10 4.2 号聚氨酯 70 × 10
item 4.2 polyurethane 70 x 10
11 螺纹接头 M6
screwed connection M6
12 27 × 28 号 30 × 15 横梁
item 27/28 cross piece of U 30 x 15
13 1.2 号聚氨酯芯
item 1.2 polyurethane core
14 分层绝缘玻璃窗
laminated insulating glazing:
8 mm ESG
8 mm SZR
8 mm ESG
8 mm SZR
8 mm VSG

窗户面积：
AV 比例：0.25m−1
U中间值：0.27W/m 2K
U值窗口：0.85W/m 2K

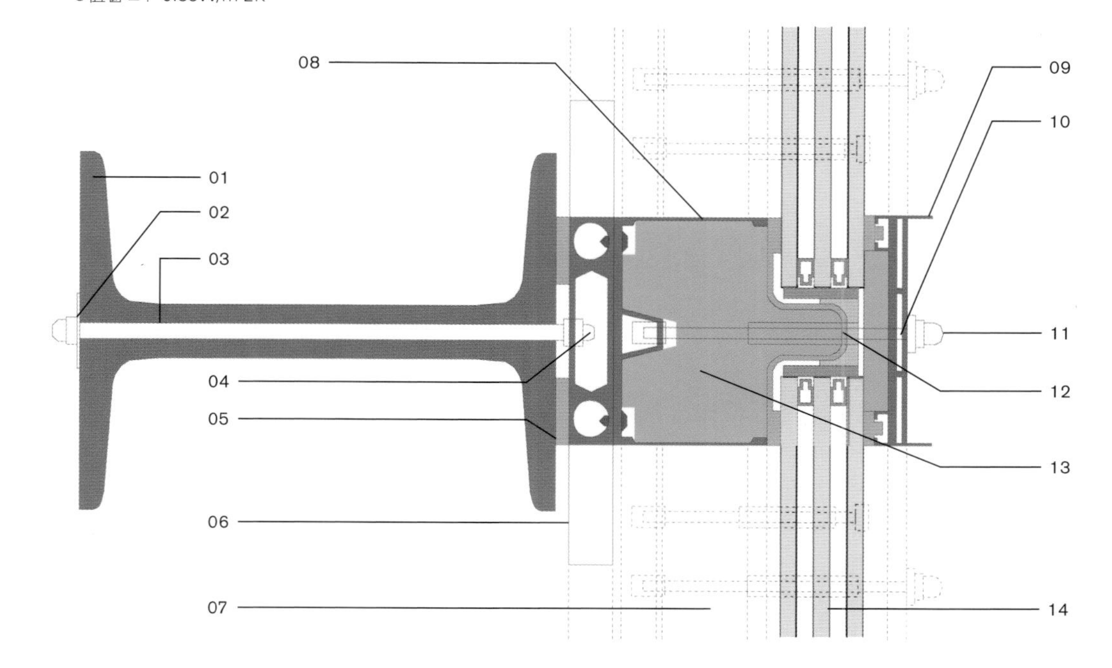

建造热能隔热系统和窗户所节省下来的成本，用来购买最先进的高能绝热玻璃墙体。通过热交换器和地热热水系统所产生的可控式通风系统及利用内部太阳能所产生的热能，使工程达到低耗能标准。太阳能的利用也意味着更多的自然光。

The costs of thermal facade insulation and windows thus saved were instead spent on a newly developed high-capacity thermal insulation of the glass facade. Zero-energy standard was achieved by using internal and solar sources of heating energy, by a controlled ventilation system via a heat exchanger and a geothermal water-heating system. Solar energy gains also mean more natural light.

建筑设计	Atelier für Architektur und Städtebau, Martin Zimmer, Darmstadt
业主	SurTec Produkte und Systeme für die Oberflächenbehandlung GmbH, Zwingenberg
摄影	Außenaufnahmen: Ulrich Reuß, Darmstadt Innenaufnahmen: Martin Zimmer, Darmstadt
专业工程师	
总装	SolarOpt-Programm des BMWi, TK3 SolarBau Monitoring koordiniert durch Architekturbüro sol°id°ar, Berlin, Dr. Günter Löhnert Passivhaus-Institut, Darmstadt TU Darmstadt
房屋技术	Atelier für Architektur und Städtebau, Martin Zimmer, Darmstadt
通风设计	InPlan, Pfungstadt
空调安装	Einzelvergabe über Architekturbüro

从“记忆园”看去，建筑的外墙设计采用玄武岩提取的自然石与透明玻璃形成鲜明对比。凸凹的条状窗户为建造交错排列式建筑的设计主旨锦上添花。

View from the 'Garden of Memory'. The façade design uses natural stone set in basalt lava contrasting with transparent glazing. Recessed and projecting strip windows enhance the design motif of the staggered buildings.

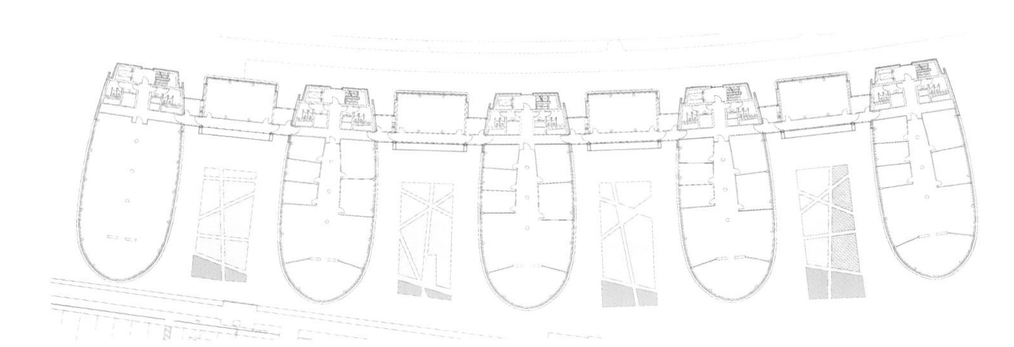

标准层平面图

Standard floor plan

位于杜伊斯堡内陆港的五子大楼，使城市拥有一个现代化多功能办公区。可以应付贸易往来的多样变化。大楼的使用者是一个健康保险公司。工程必须遵循法规和修正案。公司的扩建是建楼计划的主要考虑方面。内部办公结构应适应于多方要求。

With Five Boats on Duisburg's inland harbour, the city has received a modern flexible office block for business operations that will undergo constant changes. The user – a health insurance company – is subject to legal regulations and amendments. The additive growth of the company (especially due to mergers) was the main concern of the project. The interior office structures had to be adaptable at any time.

与土地所有者共同研究的结果是建一座可灵活使用的35层高的同一标准的大楼。根据不同使用者和部门划分为个人、工作组和开敞布置的办公室。

The result of the design co-operation with the tenant was a building that can be used very flexibly and has thirty-five identical modular storeys. They can be partitioned for different users and constellations into individual, team or open-plan offices.

LED（发光二极管）照明系统：墙面使5层楼产生背投效果。通过显示屏可随意对LED系统进行编程，使大楼呈现不同色调。660平方米的外墙表面细分为80种颜色块，所以产生不同的灯光效果和颜色组合。

LED lighting system: The façades are back-lit over five storeys. Freely pro-grammable LEDs via monitors make it possible to give the building varying colour accents. The 660 m^2 façade surface was sub-divided into eighty colour areas so that different lighting moods and colour schemes can be projected.

巴尔合作建筑设计事务所
哈根
格雷姆肖合作建筑设计事务所
伦敦

五子大楼　杜伊斯堡

FIVE BOATS, DUISBURG

设计理念：灵活性

DESIGN IDEA: FLEXIBILITY

施工期	2003–2004 年
占地面积	25 800m²
使用面积	23 500m²
竣工验收	2004 年
楼层	7（没有技术装备楼层）
用途	停车场、办公室、饭店、会议中心

滨水区外墙近景

Waterfront façade detail

因为建筑区靠海，独特的地理环境要求建筑师的设计要适应海港条件。潜在开发在于充分利用空间，这就是采用扇形透明蜂窝式构架的原因，这符合都市设计和建筑要求。外张形设计使建筑的中部留有空地。空地一边向外延伸，所以不会阻挡欣赏内港与杜伊斯堡市中心的视线。

Due to the unique waterside location, the architects developed a design which suits a harbour situation. The potential of the development site is best utilized by a fan-shaped see-through comb structure that meets both urban layout as well as architectural demands. The splayed arrangement of the buildings creates intermediate open areas which widen at one end allowing unhindered views of the inland harbour and the inner city of Duisburg.

大楼的主体设计主要考虑两点：一是港口边的地理位置，二是要建成同等质量的办公地点。办公楼建在海边，而辅助技术厂房建在对面，所有办公人员都可俯视海港，这种设计也使各部门之间可随意畅通地进行沟通，是一种极灵活的设计。

The buildings' waterside location and the aim to create workplaces of equal quality provided the main design ideas. By placing the offices on the side of the water and ancillary as well as technical spaces on the opposite side, all the office workers overlook the harbour. This arrangement also made it possible to inter-connect the departments horizontally and vertically for maximum flexibility.

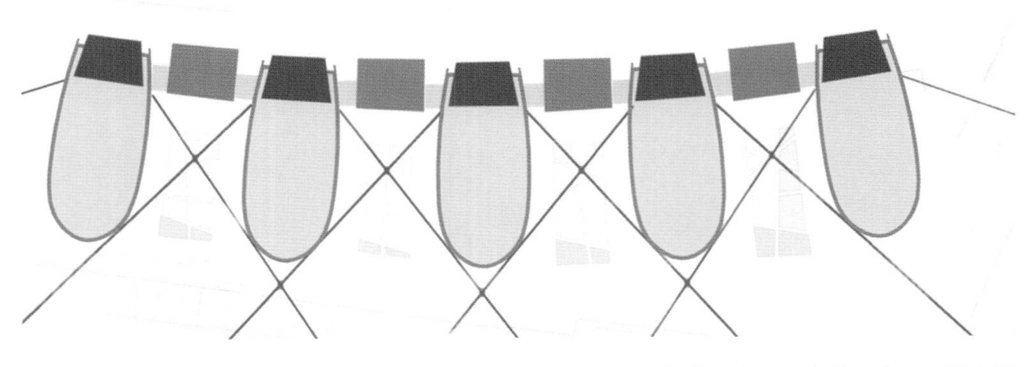

建筑内视线光轴

Sight axes in the offices

环形设计（走廊延伸至海港，楼梯可通到轮船大街）的内部办公室采用自然采光，一层有零售商店、餐饮中心和其他商业服务，二层以上是同横向侧楼相连的5个子单元楼，种有植物的庭院对一层起到衬托作用，使其成为轻松愉快的富有海洋特色的购物与娱乐场所。

The internal circulation areas (hallways oriented to the harbour and stairways to Schifferstrasse) are naturally lit. On the ground floor there are retail and catering outlets and other commercial services. The five office blocks (the boats) rise from this base, connected by transverse wings (the monitors). Planted patios turn the ground floor facilities into a pleasant shopping mall and recreational area with maritime flair.

此方案按照投资方的要求设计，能够为建筑使用者提供最大的适应性。右页图：室内墙面；左下图：可俯瞰海港的办公室；右下图：接待区。

The project was 'tailored' to the tenant's requirements and, due to its flexi-bility, offers maximum scope for further tenants. Right-hand page: interior façade; below left: office overlooking the harbour; below right: reception

巴尔合作建筑设计事务所
哈根
格雷姆肖合作建筑设计事务所
伦敦

五子大楼　杜伊斯堡

FIVE BOATS, DUISBURG

结构与能源设计

STRUCTURAL AND ENERGY DESIGN

五子大楼没有安装空调设备，可开窗自然通风。除此之外，通宵利用冷水激活混凝土地面，所以它的凉爽墙面有助于营造舒适的内部环境。

The Five Boats complex does without air-conditioning systems as the buildings are ventilated naturally by opening windows. In addition, the concrete floor masses are thermally activated overnight (using cold water) so that their cool surfaces contribute to creating pleasant interior climates.

设计组　Nicholas Grimshaw & Partners, London UK
Bahl + Partner Architekten BDA, Hagen
业主　Five Boats GmbH, Grünwald
摄影　Westwerk, Essen / Bahl + Partner
景区建筑　Müller Peuker Kampfer, Leverkusen
房屋技术　LWS Ingenieurgesellschaft für Tragwerksplanung mbH, Duisburg
静力学　Planungsgemeinschaft Haustechnik, Düsseldorf
防火安全　Dipl.-Ing. E. Püschel, Mülheim

在一片独立小楼的附近建造出有雕刻建筑效果的楼房，工程设计首先考虑特殊地区计划。

The project of a sculpturally architected ensemble in a neighbourhood of average detached houses first required the development of a special site-zoning plan.

在这个工程中，S/M/L是指代理人苏珊威特。她的丈夫马克格里格是一名设计师，马克格里格的弟弟拉斯是一名保险代理人，两兄弟在巴登——符腾州附近的博格雷登郊外购买了6 000平方米的农场。他们计划建两个块状结构的房子用于居住，一个办公楼作为马克－格里格的艾马艾斯工作室，进行工艺和通讯设计及收藏意大利和德国20世纪50年代至70年代的跑车。

In this case, S/M/L means clients Suzanne Weidt, her husband Marc-Gregor, a designer, and his brother Lars, an insurance agent. The two brothers bought about 6,000 m^2 of agricultural land on the outskirts of Burgrieden near Laupheim (Baden-Wuerttemberg), where they wanted to live (in two house cubes), work (in the office building of Marc-Gregor's 'einmaleins studio for industrial and communications design') and collect Italian and German sports cars of the 1950s to 1970s.

用有限的资金建出具有建筑超凡美感和不同建筑风格大楼，用简短而清晰的语言处处体现言简意赅的建筑特色，与周边的环境形成强烈的对比。

An aesthetically compelling ensemble of different building types was built on a tight budget in a clear reduced architectural language, which forms a deliberate contrast to the surrounding countryside and average neighbourhood on the outskirts of a village.

提图斯・伯恩哈德建筑设计事务所
奥格斯堡

集居住、办公、汽车收藏于一体的综合楼
博格雷登

S/M/L –
COMPLEX FOR LIVING, WORKING
AND AUTOMOBILE COLLECTING,
BURGRIEDEN

简约正式语言

REDUCED FORMAL LANGUAGE

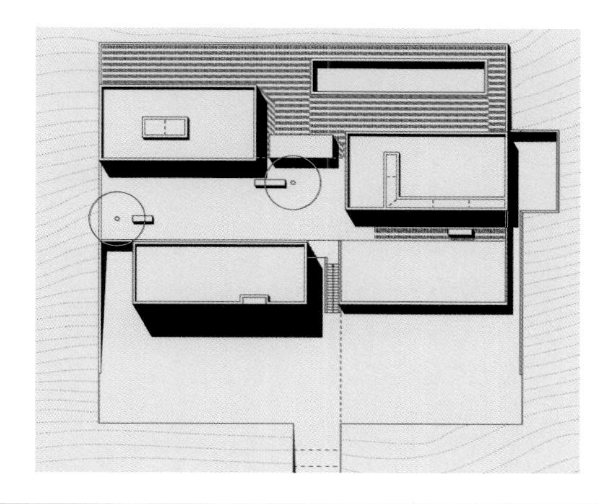

建筑成本	176 万欧元
计划和施工期	1999–2002 年
总使用面积	1 320m²
总体积	4 500m³
竣工验收	2002 年
总占地面积	6 000m²

外墙面积40m × 27m，采用混凝土建筑，一层用于拉斯威特收藏他的老爷车，外边有人工砌成的小山坡，还包括地下室和机械设备区。地基的顶部正好形成一个平地，成为一道美丽的风景，两边各有两栋楼，好像平地中间被劈开而形成的建筑雕塑品虽有裂痕却仍结实。

A 40 x 27 m hall of exposed concrete – a flat box in which Lars Weidt keeps his historic automobiles – digs into a hillside and also contains basement stores and mechanical services spaces. The roof of this base forms a clearly defined 'plateau' in a beautiful landscape and the building site of the two houses which seem to push through cuts in the plateau like perforated, yet solid architectural sculptures.

办公楼整体设计是玻璃架结构，位于高地的北面，作为艾马艾斯工作室的陈列橱，建筑的玻璃窗格两边采用双色网版印刷，达到高品质设计效果。

The office building – a fully glazed skeleton construction – 'docks onto' the north side of the plateau. The panes of this glass cube – a showcase of the 'einmaleins' studio's striving for high quality design – are screen-printed on both sides in two colours.

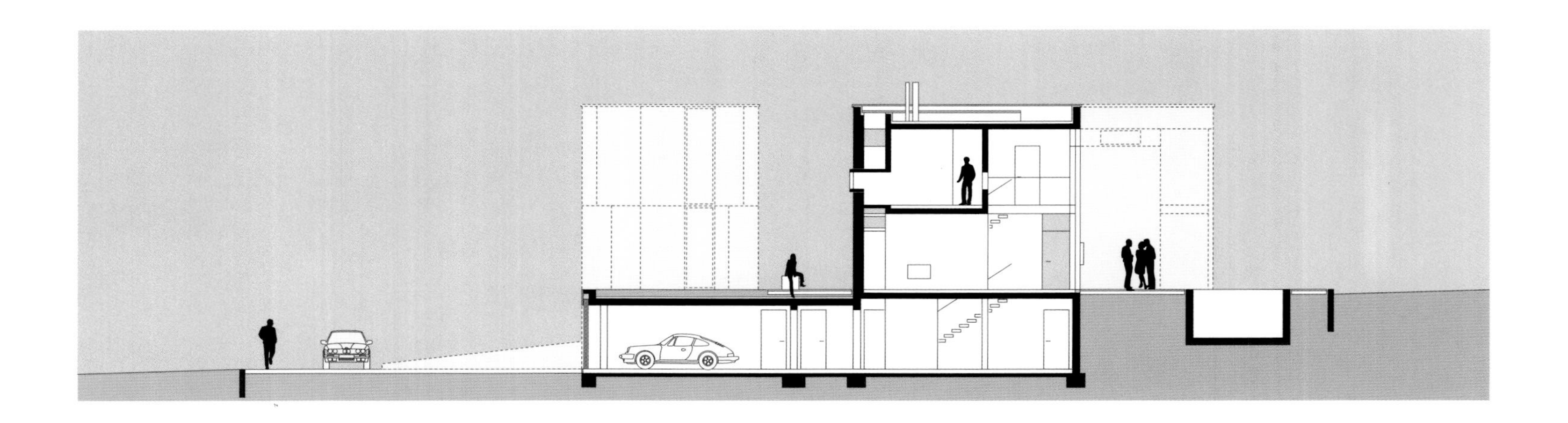

提图斯·伯恩哈德建筑设计事务所
奥格斯堡

集居住、办公、汽车收藏于一体的综合楼
博格雷登

S/M/L –
COMPLEX FOR LIVING, WORKING
AND AUTOMOBILE COLLECTING,
BURGRIEDEN

通过标准规格的变化节省成本

COST-EFFICIENCY T
VARIATION OF STANDARD SIZES

整个方案只花费小部分预算，因为很多建筑材料的不同规格被巧妙地结合在一起，有条理地形成多样性，例如：玻璃窗格需要6种不同规格玻璃，房屋需预制的钢筋混凝土板。

The entire project was implemented on an astonishingly small budget as many parts of just a few sizes were skilfully combined to create formal diversity, e.g. only six different window formats for the glass cube, and RC prefab elements for the houses.

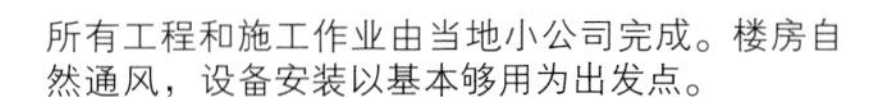

所有工程和施工作业由当地小公司完成。楼房自然通风，设备安装以基本够用为出发点。

All construction and finishing work was done exclusively by small local firms. The building is naturally ventilated throughout; service installations were restricted to essentials.

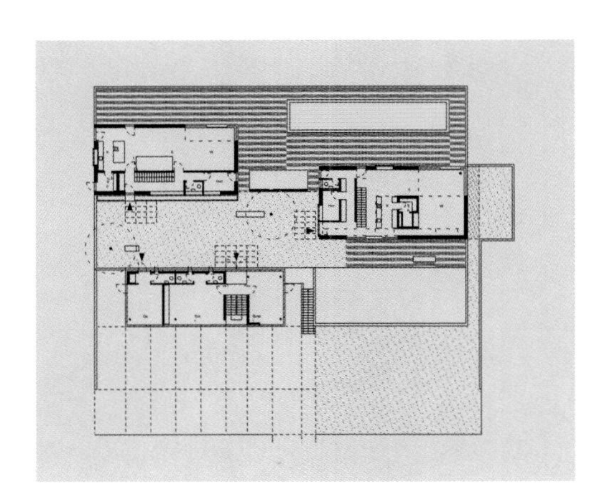

建筑整体的内部相互渗透，相互联系，灯光的相互辉映，建筑材料的选择，景色和建筑的共存使整个设计生动、活泼。

The complex interrelations and interpenetrations of interior spaces, the interplay of light, the choice of just a few materials and the symbiosis of landscape and architectural artefact enliven the whole ensemble.

建筑设计	Dipl.-Ing. Titus Bernhard Architekten BDA, Augsburg
业主	Suzanne, Marc-Gregor und Lars Weidt
制图设计	Titus Bernhard, Stefan Krippl, Helmut Schmid, Szabolcs Sóti
投资方	Allianz Immo, Stuttgart
工程总监	Helmut Schmid, Szabolcs Sóti
摄影	Klemens Ortmeyer, Braunschweig

建筑外墙安装了水平铝板条式的遮阳系统和内置防强光设备。

The façade is equipped with an exterior sun-shading system of horizontal aluminium slats, and interior anti-glare devices.

凸出的办公楼内部具有很强的灵活性，目的在于瞄准最具潜能的市场。德国金属公司老工业区成为法兰克福地平线美丽的风景。这种城市设计理念标志着从小的住宅区向北部高层建筑转换。建筑区的模型设计主要考虑大楼的高度、采光及周边环境的协调。

A salient office building with a maximum of interior flexibility – for best possible marketing – and a view of Frankfurt's skyline was erected on the former industrial site of the Metallgesellschaft AG. It is an urban design element that marks the transition from the small-scale residential quarter to the large building to the north. Building mass working models were developed to harmonize building heights, daylighting conditions and the 'dialogue' with the disparate surroundings.

与透明玻璃相比，外墙的设计可以达到泡沫所起到的原始效果，交叉式玻璃嵌条成为分段式建筑区的设计主旨。

The façade design has created an original play of foamed lava elements contrasting with clear glass. Staggered window bands take up the design motif of the stepped building mass.

BLFP– 布瑞莫·劳伦斯
弗里林豪斯

奥洛夫帕尔米大街
美茵河畔法兰克福

OLOF-PALME-STRASSE
FRANKFURT/MAIN

设计理念

DESIGN CONCEPT

建筑成本	1 770 万欧元
施工期	2000–2002 年
占地面积	21 200m²
总体积	75 000m³
竣工验收	2002 年
总使用面积	14 740m²

线性侧楼的山墙端壁面向街道，宽敞的外楼梯直通楼口，楼梯的不同构造有利于分辨方位。

The gable end walls of the linear wings face the street. Wide exterior stairs lead to the entrances and, with their different configurations, help to identify the individual addresses.

Complying with user requests, the design offers a number of different floor plans. Up to three 'suites' are grouped around the lifts and staircase core on every floor, thus providing optimal flexibility for letting.

应大楼使用者的要求，工程体现出许多不同的楼层设计，直到3楼。房间围绕电梯与楼梯之间，这样为出租提供最佳理想的灵活性。

BLFP – PROF. BREMMER · LORENZ · FRIELINGHAUS, FRIEDBERG

奥洛夫帕尔米大街
美茵河畔法兰克福

OLOF-PALME-STRASSE
FRANKFURT/MAIN

BLFP– 布瑞莫・劳伦斯
弗里林豪斯

结构

STRUCTURE

设计师设计的楼房以一系列可相互贯通的线性建筑为主要特色，主楼与 3 个侧楼由北向南穿插在一起，不管在几层楼，都可以横穿到同层的其他楼去。

The designers opted for a building characterized by a system of interpenetrating linear buildings: one being penetrated by three others, which step down from North to South, one storey at a time, to make the transition to the neighbouring houses.

建筑设计	BLFP – Prof. Bremmer · Lorenz · Frielinghaus
业主	Metros Gesellschaft zur Entwicklung moderner Bürogebäude
园艺设计	Dipl.-Ing. F.-J. Hendrix, Heuchelheim
建筑物理	Dipl.-Ing. Ralf Heinrichs, Büttelborn
供热、排气	Ing.-Büro H. + A. Klöffel, Bruchköbel
电路设计	Ing,-Büro Freudl + Ruth
塔吊安装施工	Obermeyer GmbH, NL Rhein-Main, Wiesbaden
摄影	Hans Engels, München

这座新建筑集多功能性与美学于一身，它的内部构造激发雇员的自我动力，并提供灵活的空间设计，目的在于体现出Engelhorn KGaA公司的形象及工作目标。

The new building is both functional and aesthetic. Its interior configuration encourages employees' self-motivation and accommodates flexible spatial arrangements designed to do justice to the representative, self-portraying, task of Engelhorn KGaA.

布洛切及其合伙人
斯图加特

行政后勤中心
曼海姆

ADMINISTRATION AND LOGISTICS CENTRE, MANNHEIM

建筑设计

ARCHITECTURAL DESIGN

建筑成本	900 万欧元
施工期	2001–2002 年
楼层总占地面积	12 700m^2
总体积	66 900m^3
竣工验收	2002 年
占地面积	18 000m^2
总使用面积	10 400m^2

BBP（布洛切）为零售商 Engelhorn KgaA 设计了灵活性强且技术成熟的行政后勤大楼。这幢由两部分组成的大楼采用简化的建材和均匀的颜料，轮廓十分清晰。它象征环境与效率的统一对偶性。建筑结构最醒目的部分在于 3 层出入口的接合厅，宽 3 米、长 55 米，设有玻璃通道和直跑楼梯。从网版印刷天窗穿过的日光给整幢建筑带来舒适的照明条件。

Blocher Blocher and Partners designed a flexible, technically mature administration and logistics building for the retail company Engelhorn KGaA. The impressive two-part building figure, marked by a reduced material and formal palette, represents a unified duality of atmosphere and efficiency. The most impressive part of the structure is the three-storey access 'joint' – 3 m wide and 55 m long – with glass catwalks and single-flight stairways. Pleasant lighting conditions are created by daylight coming in through screen-printed skylights.

建筑的内部实际是为了迎合外部工业美学的需要，通过对比外墙所采用的材料提高了内部环境效果，这包括外墙混凝土、落叶松木材和大面积的自然色和色调很强的红色。建筑的一面界墙减少，与办公侧楼的内部墙面形成强烈对比，采用窗户和玻璃墙分隔会议室与休息室的设计方法达到可视交流的效果。

The interior design matches the industrial aesthetics of the exterior. Contrasting surface materials 'ennoble' the interior environment. They include exposed concrete, larch, natural colours and a rich red colour accent here and there. The 'pared-down' party wall on one side forms a strong contrast to the interior façade of the office wing. Windows and glass walls separate the passage between both from the conference rooms and lounges, permitting visual communication.

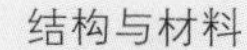

行政后勤中心
曼海姆

ADMINISTRATION AND LOGISTICS CENTRE, MANNHEIM

布洛切及其合伙人
斯图加特

结构与材料

STRUCTURE AND MATERIALS

平层与顶层空间的设计具有多功能性，用于设置会议室、展厅或办公室。在办公室，直接或间接控制的地灯突出了凉爽气氛，并提供良好的办公照明。建筑并没有采用空调装置，内部气温由通过冷却棚顶的自然通风来调节。

The multi-functional ground and top floor spaces are designed to be used as conference rooms, exhibition galleries or offices. In the offices, directly and indirectly controllable floor lamps spotlight the cooling elements which then diffuse a good working light. Instead of air-conditioning units, interior temperatures are regulated by natural ventilation through cooling ceilings.

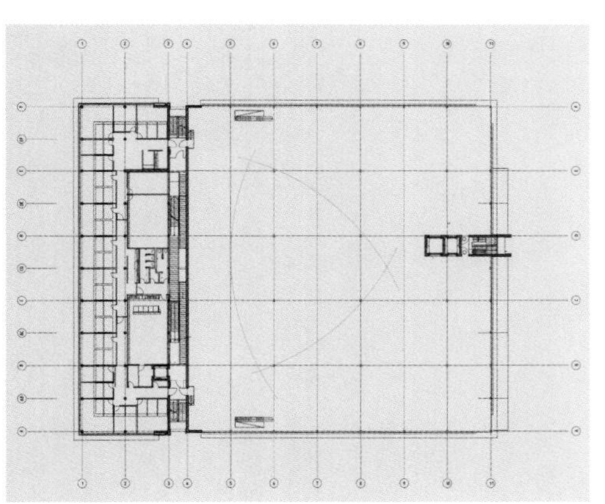

大厅内部，产品的运输采用复杂先进的悬浮轨系统，优点在于无噪音，效率高且维修率低，运用吊轨电车，而非马达驱动机械装置或电缆牵引系统。

Inside the hall, the goods are moved by means of an intricate, sophisticated suspension rail system which is quiet, efficient and low-maintenance and handles the rail-hung trolleys without motor-driven mechanisms or cable traction systems.

新建成的大楼由无烟煤外墙构成的经济型外表和亚光木材是由混凝土、钢、玻璃和木材塑造成的，并包含美学设计，而前楼的大面窗户也是如此。后勤大厅经历仅仅7个月的工程期，已开始投入使用。惊人之处在于此楼低成本钢架结构和平顶嵌板的钢架外立面，外立面包着波形护墙板。

The 'economical' appearance of the new building with its façades of anthracite, matt polished wooden material (Trespa) is shaped by the materials concrete, steel, glass and wood, but also by the aesthetic design, which includes the generously dimensioned windows of the front building. The logistics hall – which started operating after a construction period of only seven months – is impressive with its cost-efficient steel skeleton structure and coffered steel façade clad with corrugated sheeting.

建筑设计	Blocher Blocher und Partner, Freie Architekten und Innenarchitekten BDA, Stuttgart
业主	Stadtgarten Immobilien GmbH & Co. KG, Mannheim
建筑所有者	Engelhorn KGaA, Mannheim
摄影	Nikolaus Koliusis, Stuttgart
专业工程师	
固定工	Herzog + Partner Beratende Ingenieure VBI Ingenieurgesellschaft mbH, Mannheim
大厦技术计划工	T.P.I. Trippe + Partner Ingenieurges. mbH, Karlsruhe, Stuttgart u. Leipzig
金属安装工	MBM Konstruktionen GmbH, Möckmühl
木工	Ganter Ladenbau GmbH, Waldkirch
排气 / 供热	M+W Zander Facility Engineering GmbH, Mannheim

NORD/LB

罗尔斯与威尔森
明斯特

教堂附近办公楼
马格德堡

OFFICE BUILDING NORD LB –
CATHEDRAL NEIGHBOURHOOD
MAGDEBURG

城市位置

URBAN SITUATION

建筑成本	9 100 万欧元
计划和施工期	1997–2002 年
总建筑面积	31 900m²
总体积	122 000m³
使用面积	18 650m²
占地面积	120m × 62m
楼体高度	20m，最高 28m
竣工验收	2002 年
占地面积	9 100m²
办公室房间数	600
双层地下停车场	370 个停车位

银行大楼的中庭

The atrium in the bank block.

夜晚，有 24 小时自动取款机的地方安装了照明装置，精心装饰的大楼内景使周边的都市环境更有生气。

At night, the illuminated 24-hour cashpoint space, a sculpted interior animates the surrounding urban environment.

1997 年，罗尔斯与威尔森在开发马格德堡大教堂广场西区的竞争中一举获胜，这个地区具有历史的重要性和当地的复杂性，要严肃对待德国最古老的哥特式教堂，同时也是乐观面对马格德堡后共产主义未来的一个标志，所以城建计划要求有严格的城区构造和组织结构，而对角通路的设置最终赋予整个建筑以生命力。

In 1997, BOLLES+WILSON won the competition for developing the west side of Domplatz (cathedral square) in Magdeburg. The historic importance and complexity of the location required restraint respect for Germany's oldest Gothic cathedral, but at the same time an optimistic signal for Magdeburg's post-Communist future. The city plan demanded a strict block structure, the rigid organization subsequently enlivened by diagonal passages.

两座新楼建在大教堂广场西区，所在单位有北德意志银行和总商会。

The new buildings, which frame the western edge of the Domplatz, both house the principle user, the bank (the Nord LB Mitteldeutsche Landesbank) and the Saxony-Anhalt Landesförderinstitut (Chamber of Commerce).

20 米高的大楼装有双层落地窗，反射出对面巴洛克式会议大楼的壮丽景观。

The 20-m high building with its double floor window composition reflects the noble proportions of the baroque parliament buildings opposite.

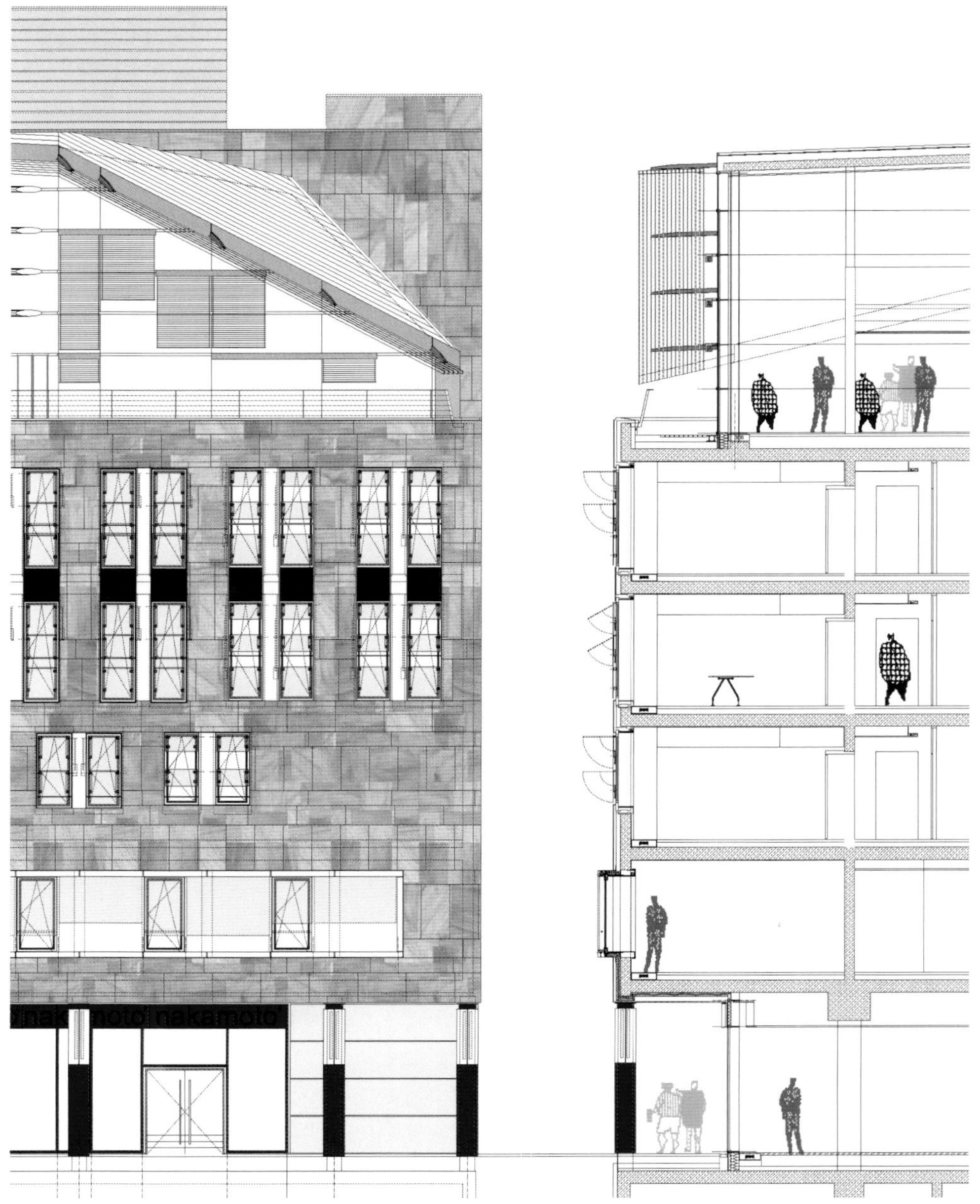

罗尔斯与威尔森
明斯特

教堂附近办公楼
马格德堡

OFFICE BUILDING NORD LB – CATHEDRAL NEIGHBOURHOOD MAGDEBURG

结构与外立面

STRUCTURE AND FAÇADE

资金紧张和规定的执行是建筑设计主要考虑的方向，内部的设计主要体现在从 18 000 平方米的办公区中设计出圆锥形循环空间。

The design theme of stringency and transgressive incident is manifested inside as conical circulation spaces, hollowed out from the mass (18,000 m^2) of offices.

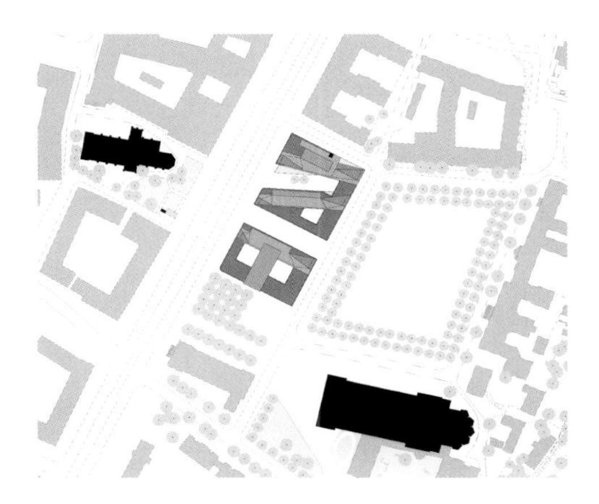

更大的内部空间也是以英式庭院方式从建筑中扩展出来的，经过精心策划提高空间质量、延展及构图透视。

Larger interior spaces are also 'carved out' of the building mass and, in the manner of an English landscape garden, offer a choreographed sequence of spatial qualities, delays and composed perspectives.

采用格林拉贾斯坦邦包装材料显示了地方的重要性，大楼内部经白色涂料粉刷后，内部楼梯及大厅显得更加高贵。这里是通向屋顶餐厅的必经之路，可以俯视整个城市。坐落在布雷特威格的银行大楼采用梯状天棚，而上面的中庭采用相应的梯状地面，成为银行内部的焦点所在。

A Green Rajasthan wrapping locates and ennobles significant locations, stairs and lobbies within the abstract white rendered interior. These are stations on the way to the roof restaurant and views over the city. The banking hall on Breiter Weg with its stepped ceiling and the atrium above it with the corresponding stepped floor are a focus within the interior of the bank.

建筑设计	BOLLES+WILSON GmbH & Co. KG, Münster
业主	NORD/LB Mitteldeutsche Landesbank
摄影	Christian Richters, Münster
网址	www.bolles-wilson.de/nordlb
专业工程师	
工程控制	Arge Saleg, Magdeburg und NILEG, Hannover
塔吊安装	ASSMANN Beraten + Planen GmbH, Magdeburg
房间技工	NEK Beraten + Planen GmbH, Groß Glienicke
照明电路设计	ag Licht Ges. beratender Ing. für Lichtplanung, Bonn
防火安全	HHP Beratende Ingenieure GmbH, Braunschweig

这座突出的大厦由 3 个单元组成，形成双 T 形结构，采用框架形多孔墙面，外面包着汝拉区所产的浅色石灰石。大厦四边呈圆形，显出楼体特点。

The striking building mass of three wings forming a double-T has a restrained perforated façade clad in pale travertine from the Jura. The four corners of the building are rounded and thus emphasize its 'body' character.

这个方案是为了法兰克福海塞事故保险公司的行政办公总部而设计的，地点位于里布斯托克区，紧邻商品交易会场和西区，总体设计由里布斯托克区划分 B 计划所决定的。建筑呈倾斜双 T 形结构。

The project is an office building as the administration headquarters of the Hesse Accident Insurance in Frankfurt/Main, in the Rebstock area in the direct neighbourhood of the trade fair grounds and the City West. The floor plan was determined by Peter Eisenman's urban grid for the Rebstock Area Plan B; the building is a double-T type with oblique angles.

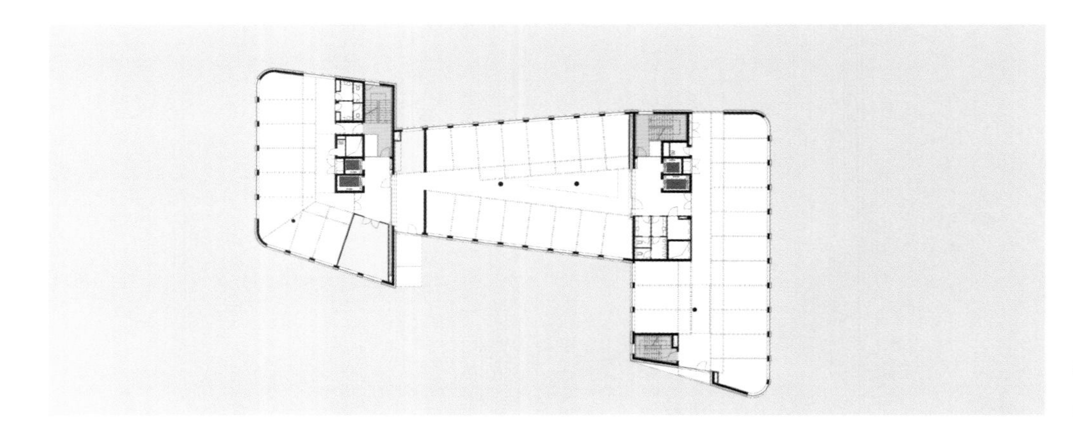

标准层平面图

Standard floor plan.

博朗与沃格特
美茵河畔法兰克福

海塞事故保险公司
美茵河畔法兰克福

UNFALLKASSE HESSEN /
HESSE ACCIDENT INSURANCE
FRANKFURT/MAIN

建筑与城市设计

ARCHITECTURAL AND URBAN DESIGN

建筑成本	1 670 万欧元
计划和施工期	2003–2005 年
总体积	53 500m³
占地面积	14 050m²
使用面积	6 730m²
竣工验收	2005 年 12 月

城市设计图
图片已得到里布斯托克公司的许可
一楼平面图（右）

Urban design pictogram
Photo published by kind permission of Rebstock Projektgesellschaft.
Ground floor plan (right)

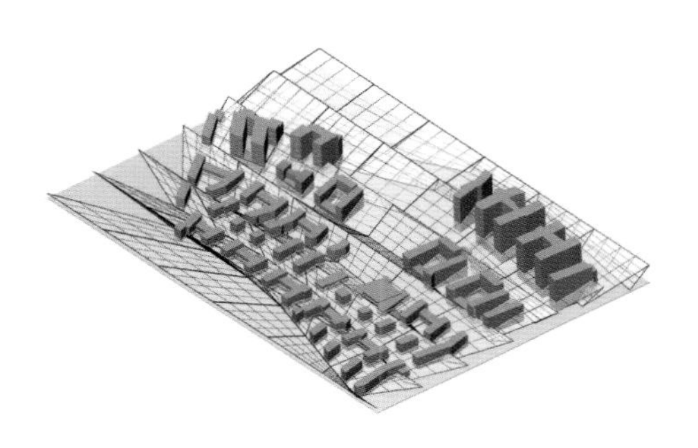

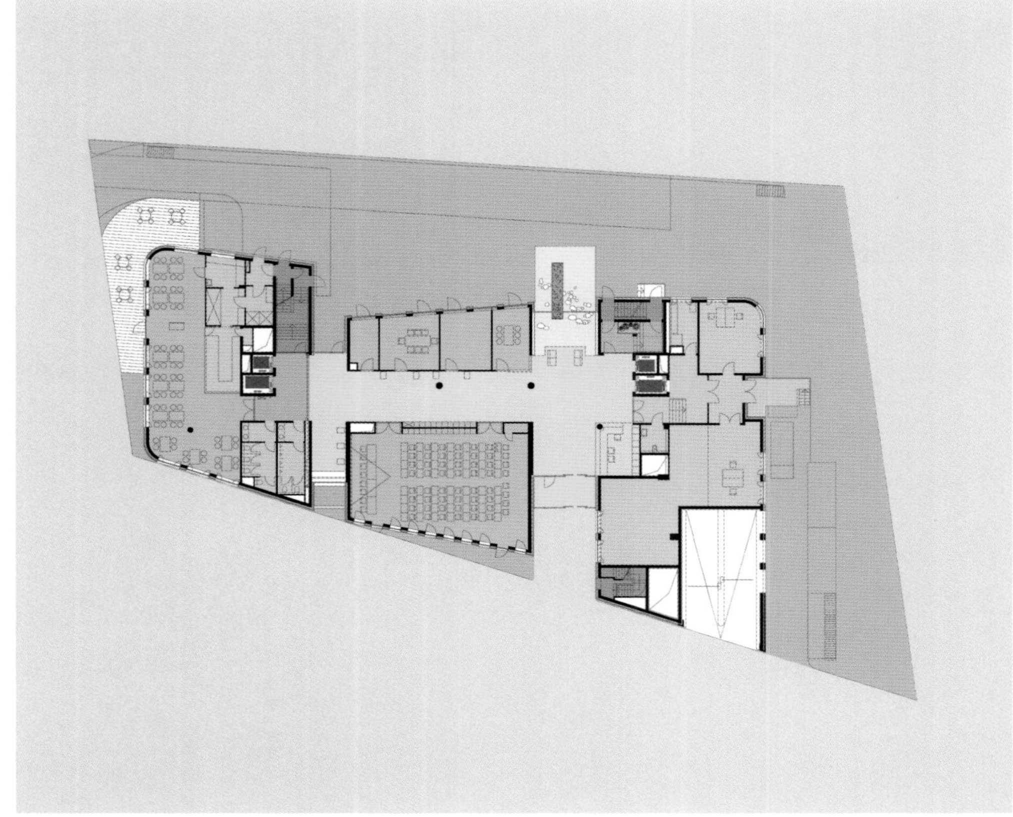

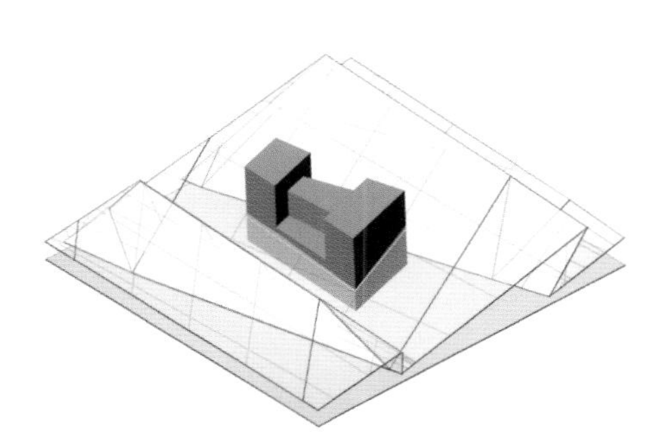

办公楼由两边的 9 层侧楼和中间梯形 7 层楼且邻街附带一层楼高的前厅构成，两个主要循环心板可通往各办公室。平面图是基于坐标方格长 2.7 米、宽 14.1 米的空间设计的，可划分小单元、大厅和开敞式平面布局的办公室。职员餐厅、会议室包括服务室和发货部安排在第一层，大礼堂可用作会议和研究的地方。

The office building consists of two nine-storey projections and a seven-storey trapezoid middle section with a one-storey projection on the street side. Two main circulation cores give access to the offices. Standard floor plans based on a grid of 2.7 m offer 14.1 m deep spaces that can be partitioned as cellular, combination or open-plan offices. The staff cafeteria and conference area including service spaces and dispatch department are on the ground floor. The one-storey front projection contains a large auditorium for meetings and seminars.

在与业主——海塞事故保险公司的合作中，建筑师开发行政大楼以节能、便捷和运作投资费用低作为显著特征进行设计。建筑的设计理念不仅集中在外墙、遮阳和照明的细节，且更注重将建筑耗能减到最低。高品质的外墙的一半面积安装玻璃，这样做的目的是为了把通过热传导所失的能量减小到最低限度。在节能理念中，这幢建筑占有重要地位。

In co-operation with the client, the Hesse Accident Insurance, the architects developed an administration building marked by energy efficiency, convenience as well as low operational and investment costs. The architectural design not only focused on the façade, sun-shading and lighting details, but above all on minimizing the energy consumption of the building. Half of the high-quality façade is glazed and therefore reduces energy losses through heat transmission to a minimum. This was an important 'building block' of the energy concept.

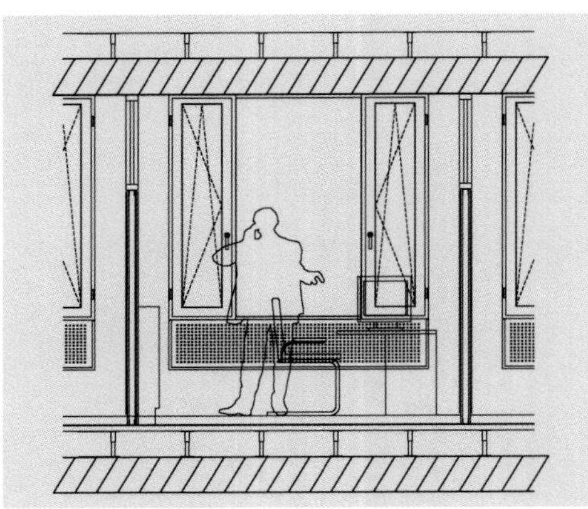

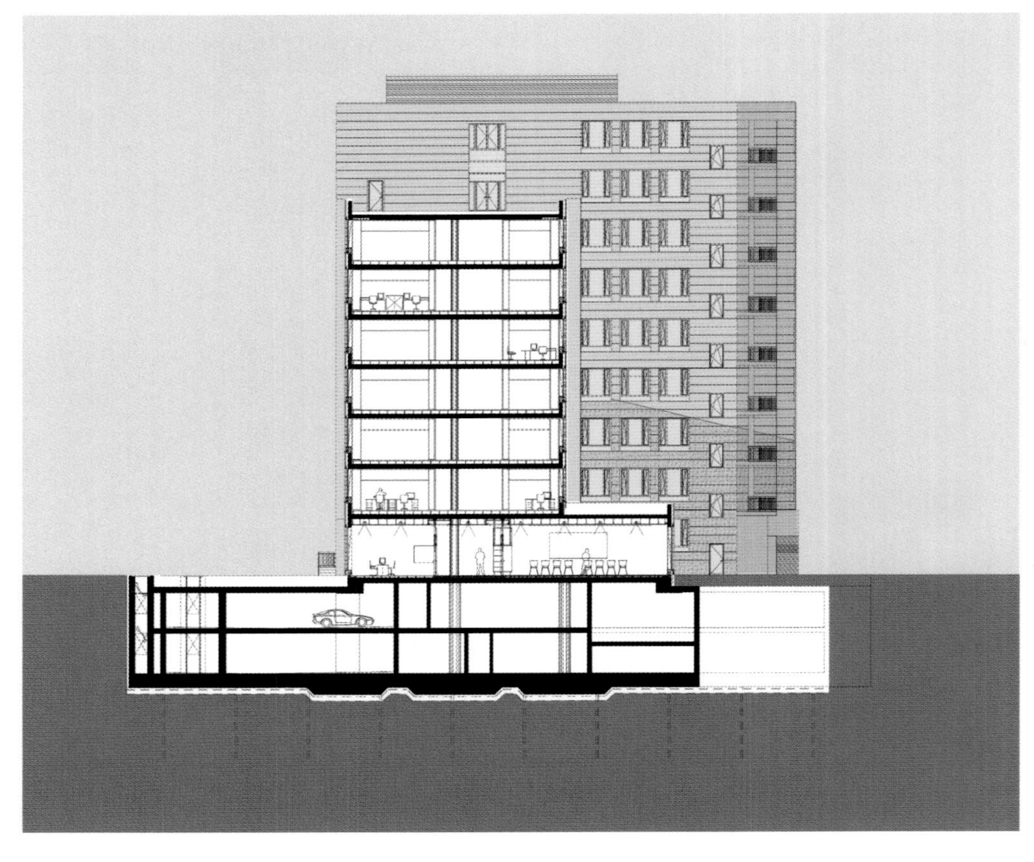

热能由热泵产生，热泵通过 68 个 50 米传感器吸收建筑余热和地热，大约 4 500 平方米的混凝土地面起到热交换器的作用。除此之外，吸入的空气采用 28℃暖水系统加热，每层楼的暖水系统都会并入到疏散供应的风箱中。

Heating energy is generated via a heat pump (heat recovery system) which uses both exhaust heat from the building and geothermal energy through sixty-eight 50-metre probes. The concrete floors (approx. 4,500 m^2) serve as heat exchangers. In addition, ingoing air is heated by a warm water system (28°C) integrated in the decentral supply air units (air boxes) along every office floor.

博朗与沃格特
美茵河畔法兰克福

海塞事故保险公司
美茵河畔法兰克福

HESSE ACCIDENT INSURANCE FRANKFURT/MAIN

节能建筑

ENERGY-EFFICIENT BUILDING

297 个这样的空气出孔，在负压力的作用下为办公室提供适度的新鲜空气，室内气温也可利用通风系统进行调节，通过可激活建筑储藏区的热泵来利用地热能量，所以室内夏季温度不会高于26℃，依此类推，吸入的空气经风箱可以降至 22℃。

297 such units, working with negative pressure, supply the offices with well-tempered fresh air. Interior climates can also be controlled through ventilation and by utilizing the geothermal energy from the heat pump (which activates the structural storage mass) so that interior summer temperatures do not rise above 26° C. In analogy to the heating process, ingoing air is cooled down to 22° C inside and via the air boxes.

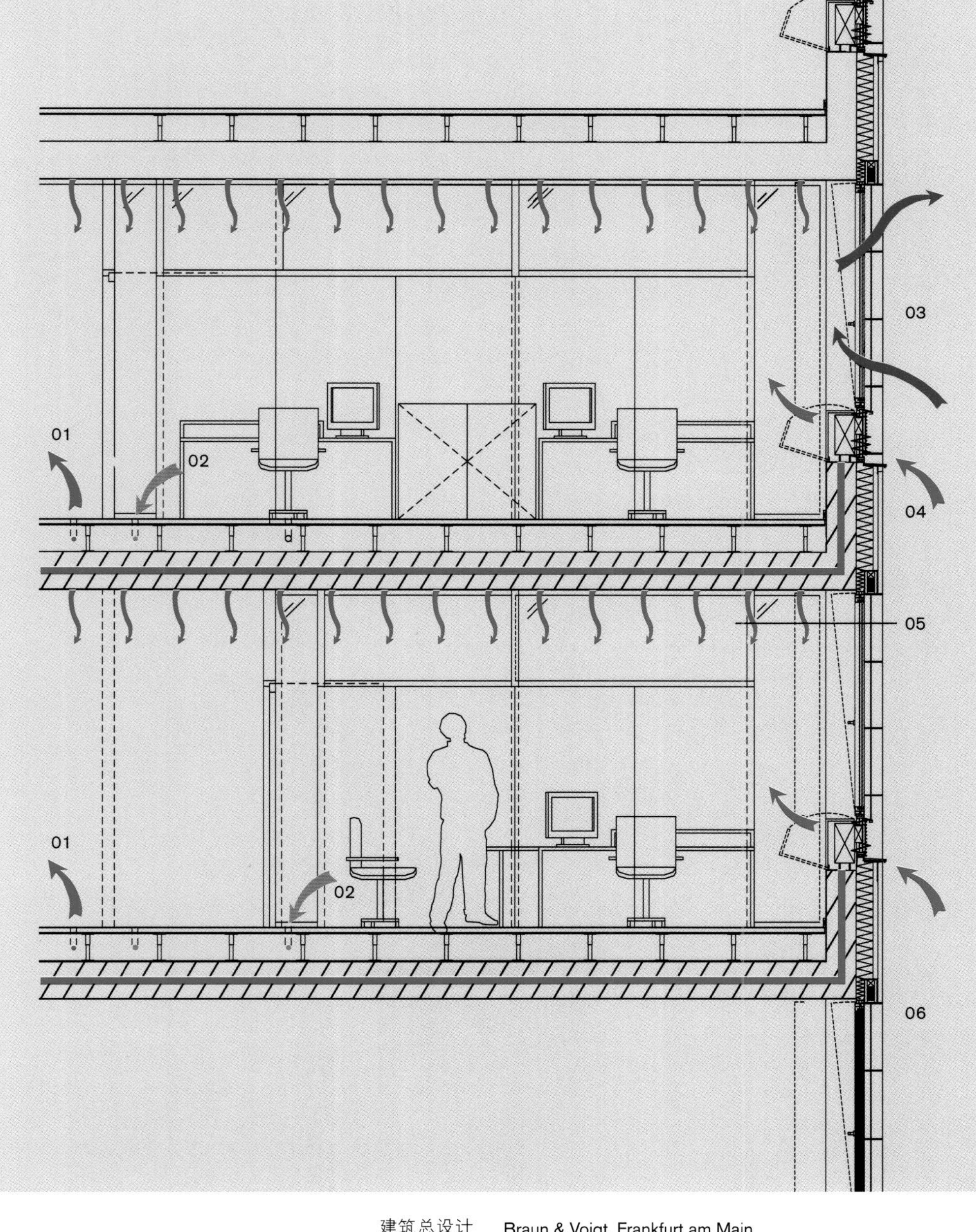

夏季 / 降温
Summer / Cooling
01 吸入空气
ingoing air
02 排出空气
outgoing air
03 开放式通风活板
openable ventilation flaps
04 手动式通风设备
manual ventilation device
05 最低温度控制下升温
heating under minimum temperatures
cooling under high temperatures
direct use of internal heat gains (offices, DP) by feeding 04 heat into thermally active structural parts
06 通过分散通风子设备（风箱）内 21℃的热水产生热能环境
creating thermal environment through 21° C warm water in decentral supply air units (air boxes)

室内温度数据：
热输出 < 10 W/m²
冷输出 < 40 W/m²
冬季温度：21 ~ 24℃
夏季温度：24 ~ 27℃
空气交换：40m³ / 人

Interior climate data:
heating output: < 10 W/m2
cooling output: < 40 W/m2
winter temperatures: 21°C – 24°C
summer temperatures: 24°C – 27°C
air exchange: 40 m^3 per person

建筑总设计	Braun & Voigt, Frankfurt am Main
业主	Unfallkasse Hessen, Frankfurt am Main
工程咨询	Bene Consulting, Frankfurt am Main
工程控制	VR Bauregie GmbH, Eschborn
三维图	Braun & Voigt, Piet Hohl, Frankfurt am Main
专业工程师	
楼体总装	Schindler Ingenieurgesellschaft mbH, Frankfurt am Main
塔吊安装	Bernhardt & Mertens Ingenieurgesellschaft mbH, Darmstadt
能源咨询	DS Plan, Frankfurt am Main
楼体模拟	Ingenieurbüro P. Jung, Köln
气象和风模拟	Ingenieurbüro P. Jung, Köln
建筑物理	ITA, Ingenieurgesellschaft für techn. Akustik mbH, Wiesbaden
生物工程师	GeoIngenieure Früchtenicht GmbH, Babenhausen
外部设计	Sommerlad, Haase, Kuhli, Gießen
电子数据处理	Weicom, EDV-Dienstleister, Mainz
防火安全	Brandschutz-Ingenieurbüro Consult GmbH, Darmstadt
大厦所有者	Albert Bau, Neunkirchen

保罗－吕博大楼和玛丽－伊丽莎白－吕德尔斯大楼是德国联邦政府新办公楼群，横跨柏林施普雷河湾，是著名的“联邦纽带区”，因为它象征着两德的统一和柏林的重建。为联邦议院新建的两幢联体楼，前后长 102 米，紧邻国会大厦，在河东西两岸延伸约 400 米。

The Paul-Löbe-House and the Marie-Elisabeth-Lüders-House, that form an appendage to the Federal Chancellery and extend beyond the River Spree, have come to be known as the „Federation Ribbon“, because they symbolize the reunification of Germany and of the city of Berlin. The Bundestag deputies' new twin buildings, 102 meters from front to rear, directly adjoin the Reichstag and extend for 400 meters on both sides of the river.

横跨施普雷河东西架桥的象征意义在于它是统一前联邦德国和德意志民主共和国的边界。遮盖河岸周围的无载体天棚和两个并列的小瀑布的设置为其增添了光彩。

The symbolism of this East-West bridging of the River Spree, the former border between Federal Germany and the German Democratic Republic, is enhanced by unsupported canopies that echo the contours of the river bank, and by two juxtaposed cascades.

斯蒂芬·布劳恩菲尔斯建筑设计事务所
慕尼黑 / 柏林

保罗 – 吕博大楼
玛丽 – 伊丽莎白 – 吕德尔斯大楼
柏林德国联邦议院

PAUL-LÖBE-HAUS
MARIE-ELISABETH-LÜDERS-HAUS
DEUTSCHER BUNDESTAG BERLIN

透明与民主

TRANSPARENCY AND DEMOCRACY

建筑成本	5 亿欧元
规划和建筑时间	1995–2001年，1996–2003年
总体积	750 000m³
楼层总面积	146 000m²
总使用面积	65 000m²
建筑监控	HOAI 1–9 人总设计者

大厅由整体无接缝钢管混凝土荧光屏天棚照明，把整个开放式梳状结构聚合在大楼中央。位于玛丽－伊丽莎白－吕德尔斯大楼的议会图书馆内设有中央圆形大厅，它作为独立整体结构与中央大厅相互辉映，这正符合了西岸总理办公楼的建筑理念。

A large Hall, lit by a monolithic jointless cast-concrete screen ceiling, holds the open comb-like structure together in the center. The central rotunda of the Bundestag library in the Marie-Elisabeth-Lüders-House stands as a self-contained entity in line with the central halls, thus corresponding from the constructional point of view to the Chancellery administrative building to the West.

100 米× 100 米见方的施普雷广场位于保罗－吕博大楼和玛丽－伊丽莎白－吕德尔斯大楼之间，施普雷河从“断开”的市政府之间流过，两岸之间架起两座大桥，西岸的环形建筑包括欧洲商会大楼和旅行饭店，并与东岸的中央报告厅相互辉映。

The 100-meter by 100-meter Spree Square is located between the Paul-Löbe-House and Marie-Elisabeth-Lüders-House. The River Spree, which flows through this open „city hall", is crossed by two footbridges. Correspondingly, the circular buildings housing the Europe Assembly Hall and the visitors' restaurant on the West bank, and the hearings hall on the East bank stand opposite each other.

斯蒂芬·布劳恩菲尔斯建筑设计事务所
慕尼黑 / 柏林

保罗 – 吕博尔大楼
玛丽 – 伊丽莎白 – 吕德尔斯大楼
柏林德国联邦议院

PAUL-LÖBE-HAUS
MARIE-ELISABETH-LÜDERS-HAUS
DEUTSCHER BUNDESTAG BERLIN

透明与民主

TRANSPARENCY AND DEMOCRACY

"联邦纽带"延施普雷河一带，除联邦总理府外，还包括保罗–吕博大楼和玛丽–伊丽莎白–吕德尔斯大楼，保罗 – 吕博大楼设置 21 个圆形会议室和为联邦议员准备的设有访者席的办公区，沿着高 25 米、长 200 米的大厅排成列，可俯瞰国会大厦、动物园及施普雷河湾公园。

The "Federation Ribbon" in a bend of the River Spree includes, in addition to the Chancellery, the Paul-Löbe-House and the Marie-Elisabeth-Lüders-House. The Paul-Löbe-House contains 21 circular committee rooms with visitors' galleries as well as offices for the Bundestag deputies. These are ranged in rows along a 25-meter high, 200-meter long hall. All the committee rooms and offices open onto outside courtyards with a view of the Reichstag, the Zoo or Spreebogen Park.

玛丽 – 伊丽莎白 – 吕德尔斯大楼位于东岸，内设议会图书馆、中央报告厅、联邦科技服务设施及后勤部。

The Marie-Elisabeth-Lüders-House, located to the East of the river, houses the Bundestag library, a hearings room and offices for the Bundestag's scientific services and logistics departments.

建筑设计 Stephan Braunfels Architekten, München / Berlin
业主 Bundesrepublik Deutschland
vertreten durch Bundesbaugesellschaft Berlin mbH
摄影 Ulrich Schwarz, Berlin, S. 086/087, S. 088 mitte, S. 089 mitte
Linus Lintner, Berlin, S. 088 unten
Reinhard Görner, Berlin, S. 089 oben

专业工程师
塔吊安装 Sailer, Stepan & Partner GmbH, München
房屋技术 HL-Technik AG, München
Cronauer Beratung Planung, München
外墙设计 R + R Fuchs, München

布伦纳及其合伙人
斯图加特

德国国家同步辐射加速器　柏林

BESSY, BERLIN

建筑设计

ARCHITECTURAL DESIGN

建筑成本	6 300 万欧元
计划和施工期	1992–2001 年
总面积	28 900m²
总体积	142 000m³
竣工验收	2001 年
使用面积	15 200m²

振动工程使同步加速器和实验区从振动外源脱离开。例如，利用开口结合和厚度为60厘米的无支柱底板，墙面可负荷风力。

Vibration engineering has freed the synchrotron accelerator and the experimental areas from outer sources of oscillation, e.g. wind loads on the façades, by means of open joints and an uninterrupted floor plate 60 cm thick.

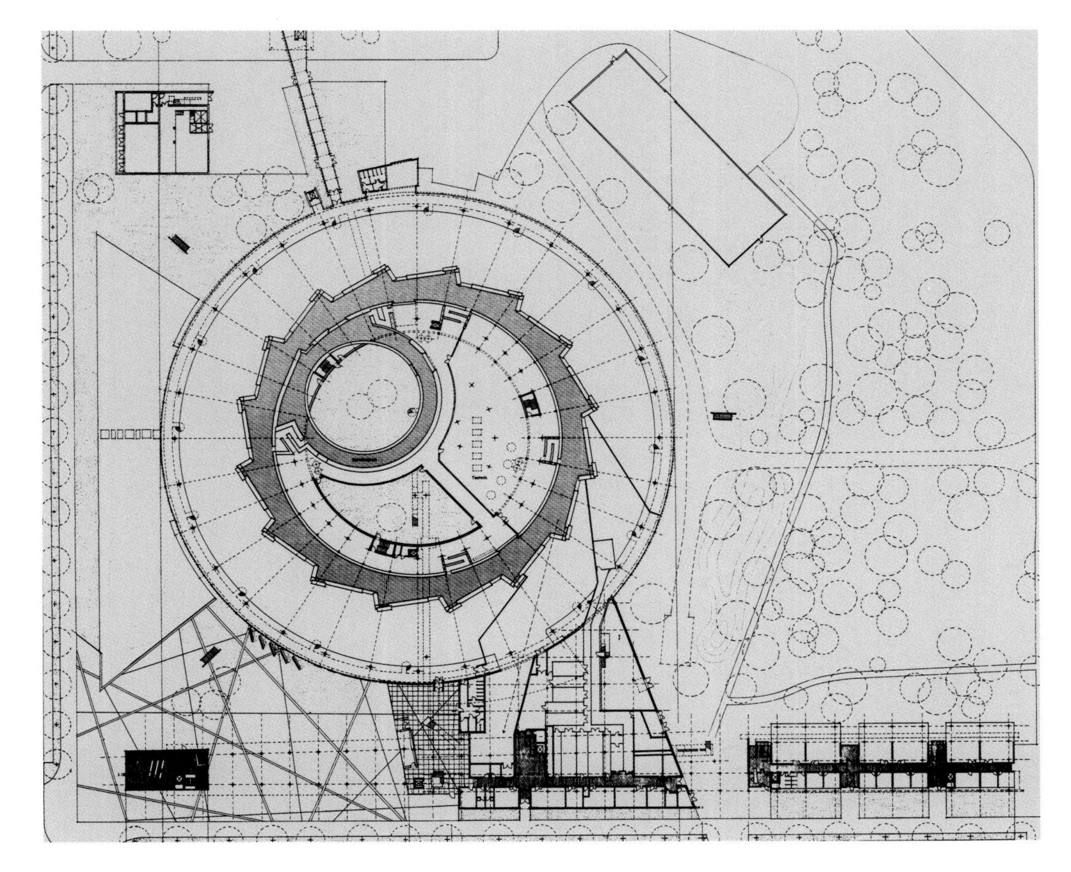

装置大厅平地而起，既可从内部欣赏外面的风景，来访者也可从外面看到楼内。在竣工之际，它代表了工程的一个全新的实施。

The storage hall rises from the ground to afford views (and to allow people to look inside). At the time of completion, it represented a new and innovative implementation of the programme.

BESSY 是德国国家同步辐射加速器，它是一个周长为 240 米的电子储存环，内含 40 米长阴极射线管，它应用于开展基础应用研究及发展计划。这就需要建立一个包括为“机器”提供有效服务系统的固定围场空间、形状及比例。装置大厅、实验室和办公楼的设计要符合特殊功能性要求，建筑要体现与众不同的重要性及影响力。

BESSY is a synchrotron accelerator, an electron-storing ring with a circumference of 240 m which contains 40 m long cathode-ray tubes used to carry out basic and applied research and development projects. This required a building enclosure including efficient supply systems for the ‘machine’, whose dimensions, shape and proportions were fixed. The design of the storage ring hall and the laboratory and office building followed their specific functional requirements and proposed different ‘signifying’ and ‘significant’ architectural bodies.

建筑主要由外包钢管的混凝土支柱构成，办公室及实验楼的钢筋混凝土层面采用无托梁平坦顶棚。装置大厅的屋顶檩条采用钢架大梁，横跨存储环隧道及高约 27 米的实验区。

The structure mainly consists of steel-tube-clad concrete supports. The RC floors in the office and laboratory building are flat ceilings without joists. Steel frame girders were used as roof purlins in the storage ring hall and span the storage ring tunnel and the experimental areas deep down over more than 27 m.

装置大厅、实验室和办公大楼由两层高玻璃门厅相连，使人在楼之间可自由出入，并备有公共服务区，还包括自助餐厅及形成楼区的交流中心。

The storage ring hall and the laboratory and office building are interconnected by the two-storey glazed main entrance hall which gives access to both. It also offers space for public functions and includes a cafeteria to form the communicative centre of the complex.

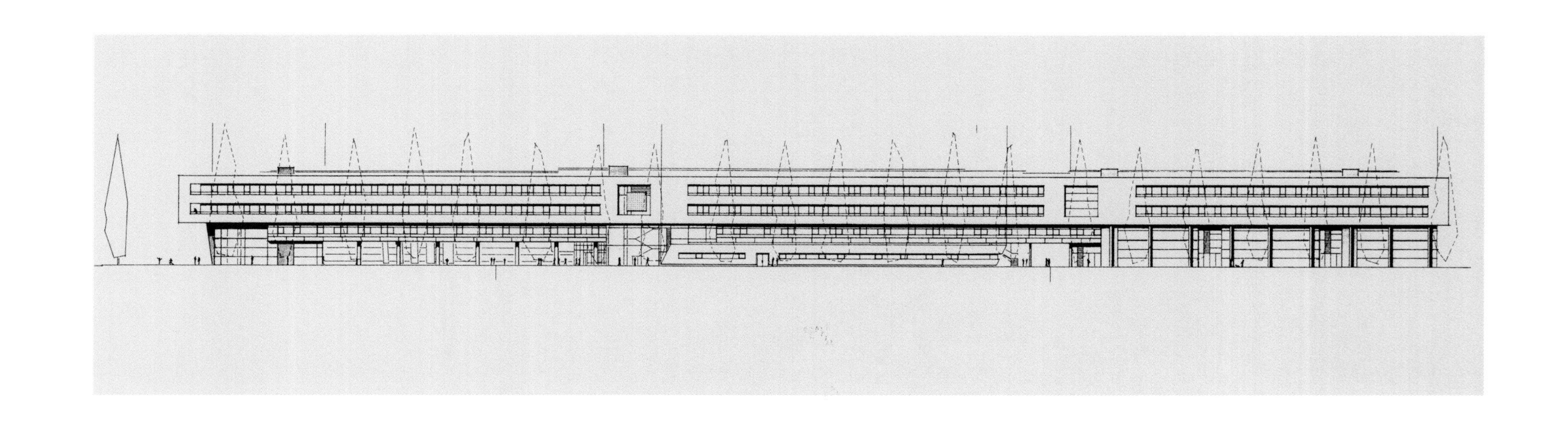

德国国家同步辐射加速器 柏林

BESSY, BERLIN

布伦纳及其合伙人
斯图加特

结构与建材

STRUCTURE AND MATERIALS

装置大厅外厚重的墙体由许多条形玻璃构成，使大厅内外直观地联系起来，这种设计使大厅更加明亮，也就是说大厅墙围顶部及底部的窄面窗户设计目的是防止科学家完全脱离自然环境的同时，高敏感装置和实验设备也不受到影响。同时，过路人可通过玻璃窗户看到内部的工作情景，从而对科学工作产生兴趣。

A number of strip windows in the otherwise massive walls of the storage ring hall visually connect the interior with the exterior. Lighting the hall from the sides, i.e. the bottom and top of the enclosure through narrow window bands means that the highly sensitive installations and test set-ups are not disturbed while the scientists do not entirely loose touch with the natural environment. At the same time, passers-by get glimpses of what happens inside so that they may become curious to learn more about this scientific work.

存储环位于实验区中央，实验区作为实验操作区及科学家工作区，所有装置可通过无振人行天桥进行观察。

The storage ring is surrounded by the experimental areas for test set-ups and the scientists' workstations. These installations are media-supplied via an outer vibration-free catwalk structure.

施工单位 BESSY GmbH, Berlin
设计单位 **B&P** Brenner & Partner
Architekten & Ingenieure, Stuttgart
摄影 H.G. Esch Photography, Hennef

专业工程师
TGA JMP Jaeger, Mornhinweg & Partner, Stuttgart
电工 KEP Klaus Engelhardt & Partner, Berlin
塔吊安装 Weiske & Partner GmbH, Stuttgart

布瑟和盖特纳尔建筑设计事务所
杜塞尔多夫

伏尔康工厂　科隆

VULKAN FACTORY COLOGNE

差异与现代化

CONVERSION AND MODERNIZATION

建筑成本	2 200 万欧元
计划和施工期	2002–2005 年
总面积	20 000m^2
总体积	101 800m^3
竣工验收	2005 年
占地面积	22 500m^2

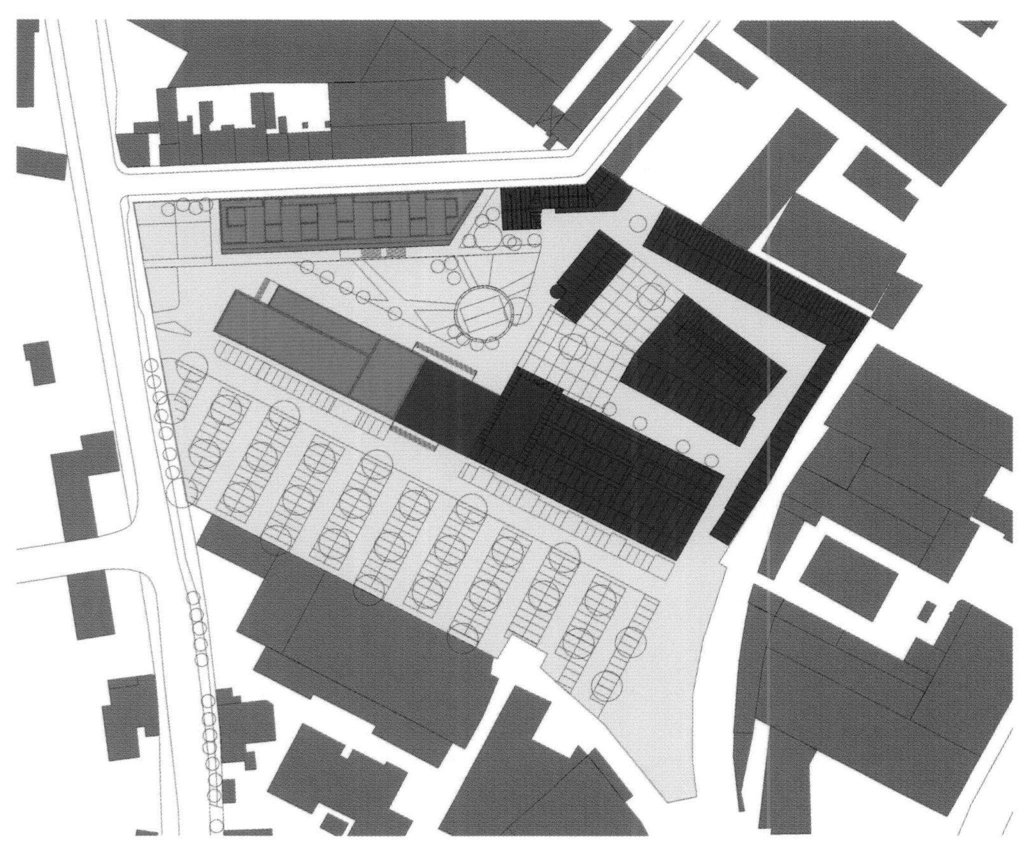

伏尔康工厂位于离市中心不远的科隆－艾伦菲尔德区，此地区正处于复兴阶段，经历了近 200 年的超负荷工业制造，工厂于 2001 年倒闭。转换观念的主要目的在于吸引寻求非传统工作环境的年轻现代企业家，结果在复兴有纪念意义的建筑区内建成了配有先进专业信息技术的服务网络中心。

The Vulkan factory site is in Cologne-Ehrenfeld, an urban area close to the city centre that is currently experiencing revitalization. Following almost 200 years of intensive industrial use, the factory was closed in 2001. The conversion concept aimed mainly to appeal to young modern entrepreneurs looking for an unconventional working environment. The result was a service network centre equipped with state-of-the-art professional communications technology in a revitalized industrial monument.

城市及工业区建筑重建的设计理念来源于“建筑的及时飞跃”。1898 年，原始建筑具有古典风格，1971 年扩建，在 2003 年直接并排建楼，现在的楼区与之前比较，具有自身的活力。

The design of the urban and architectural redevelopment of the industrial site developed from the idea of an 'architectural leap in time'. The historic styles of the original buildings of 1898 and the extensions of 1971 and finally 2003 are directly juxtaposed. The complex now draws its dynamism from these contrasts.

布瑟和盖特纳尔建筑设计事务所
杜塞尔多夫

伏尔康工厂　科隆

VULKAN FACTORY COLOGNE

多样性和统一性

DIVERSITY AND HOMOGENEITY

现存老建筑的差异性是可以接受的，并可以在此基础上进行扩建，在这块老的市区，每个建筑都有自己的风格，共同形成亮丽的效果，即大城市中的小城市。

The heterogeneity of the existing structures was accepted and deliberately translated into new extensions. In this old urban quarter, every building has its own character; together they form a vibrant urban ensemble – a small city within the larger city.

伏尔康的 6 座老建筑已重修，并建立了 3 座新建筑，这符合科隆市关于保存纪念性建筑的要求。这些开敞式平面的顶楼，按照承租人的要求个别划分及刷新。在对老楼房的部分建筑拆毁及扩建的过程中，建筑师创建了大片有活力的中心地带和公共设施，如 1 500 平方米的广场用于邻街咖啡馆和餐厅。

Six of the old Vulkan buildings were restored – in co-ordination with Cologne monument conservation – and complemented by three new structures. These offer spacious open-plan lofts which were individually partitioned and furbished according to tenant requirements. By demolishing some parts of the old production halls, and by extending them, the architects created a great number of dynamic intermediate spaces and open areas such as a 1,500 m^2 square for street cafés/restaurants, open-

建筑设计　Busse & Geitner, Düsseldorf
业主　Vulkan Grundstücksgesellschaft mbH & Co.KG.
摄影　Michael Reisch, Düsseldorf
大厦所有者　F.C. Trapp Baugesellschaft Wesel mbH

Sächsische Zeitung
Sächsische Zeitung

科耐尔森与西林格尔
达姆施塔特

德累斯顿印刷出版大楼

DRESDEN PRINTING AND PUBLISHING HOUSE

设计理念

DESIGN CONCEPT

建筑成本	540 万欧元
施工期	2003 年 8–11 月
总面积	14 300m²
总体积	48 000m³
竣工验收	2003 年
使用面积	12 200m²

建筑的设计在于运用巧妙的方式来改变现存的结构，经过仔细研究，从能源消耗、建筑美学及工艺学方面来看如果外部调节能有效地实现其功能性，就达到了改变现存建筑的目的，新设计把重点放在高层建筑上，从远处看去，大厦已成为城市的标志性建筑及出版社的形象载体。

The design was geared to transforming the existing structures in an intelligent way. Following detailed studies, they were altered only if and when the intervention improved functionality efficiently in terms of energy consumption, aesthetics and technology. As an urban landmark, visible from afar, and as image-carrier of the publishing house, the redesign focused predominantly on the high-rise.

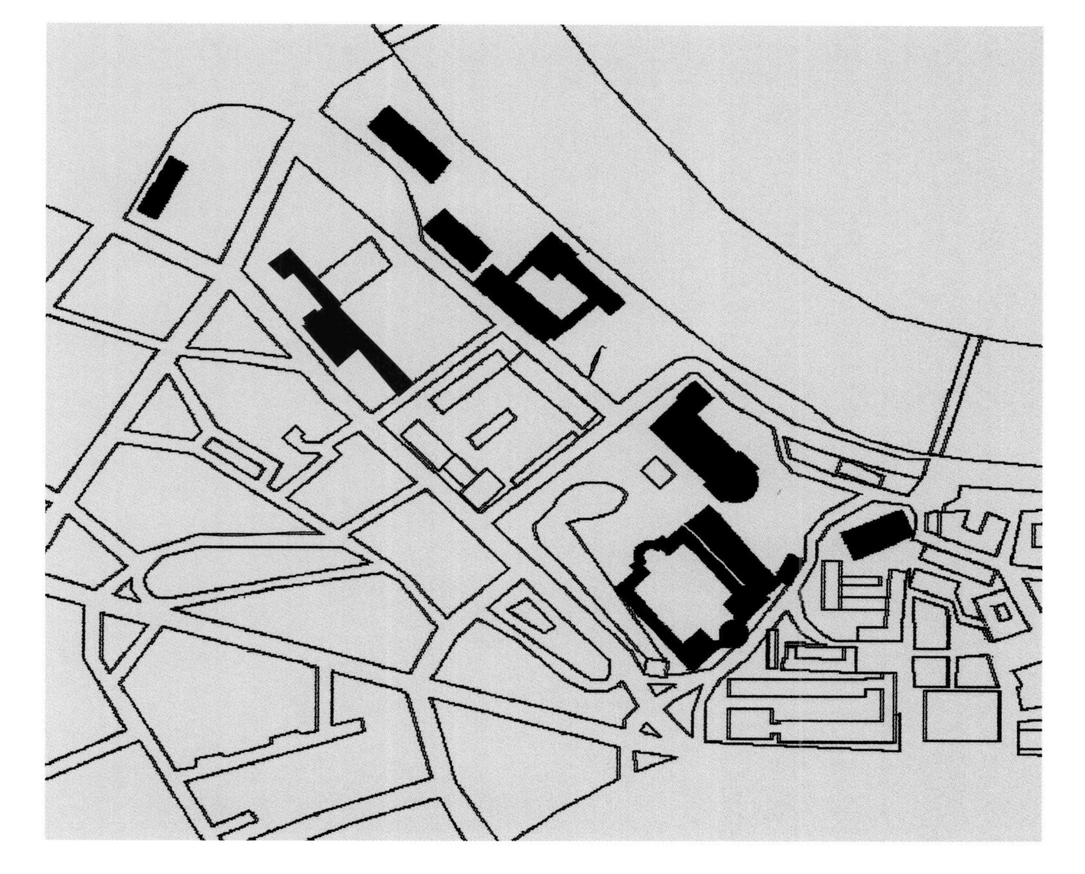

大厦外墙采用新式幕墙玻璃，墙面采用印刷技术，印上重叠、偏转及变形的字体，凭借日光及特殊印刷技术，外墙的字母装饰及日光反射，使建筑的玻璃外墙十分显眼。

The high-rise received a new curtain-wall glass facade screen-printed with lettering alienated through overlapping, reflection and distortion. Depending on the kind of daylight, and due to the special printing technique, the glass surface is either dominated by the printed décor or by reflections.

2002年夏天的易北洪水灾害及对建筑一层的最初修复后，德累斯顿印刷出版社对市中心的办公建筑又进行了全面整修，多数建筑（总房屋面积为 14 500 平方米）都是 20 世纪 60 年代建成的旧房屋，其办公环境的质量、能源消耗，尤其是建筑的外观对于现任出版社来说，不再符合其要求，所以 2003 年委任科耐尔森 + 希林格尔对建筑进行重新设计，使其具有现代化气息。

After the Elbe flood disaster in the summer of 2002 and after initial repairs to the ground floor, the Dresden Printing and Publishing House opted for a complete refurbishment of its office buildings in the city centre. Most of these buildings (total floor area: approx. 14,500 m^2) were erected in the 1960s and the quality of the work places, the energy consumption and, above all, their architectural expression, no longer matched the expectations of the present publisher owners who commissioned Cornelsen + Seelinger in March 2003 to redesign and modernize the buildings.

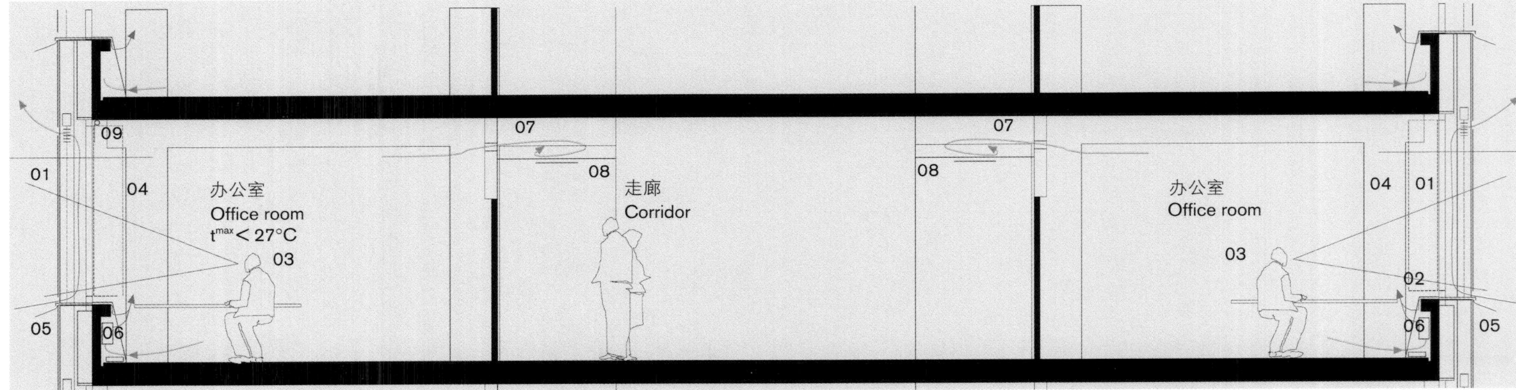

整修后状况（下）
Condition after Refurbishing（below）

01 木框窗户（外部遮阳设备设在开敞式后通风玻璃后面）
wood-framed window with exterior sun-shade behind open，back-ventilated glass pane
02 降低高度、配有电子、数字装置的轻质混凝土低墙
lightweight concrete parapet element of reduced height and electrical and digital installations
03 窗户低墙降低了 72 厘米高度后的改良视野
improved view through reduced window parapet height of 72cm
04 窗扇，宽 40 厘米，开启时不会遮挡视线
The window sash，40cm wide，is not in the way when opened
05 通风与夜间室内气温的降低，排气通风后外部空气进入，每层楼独立控制
ventilation and night-time reduction of room temperatures；intake of exterior air following exhaust ventilation，separately controllable on every floor
06 个人调控的冷热气流系统，冬日供热，夏日制冷（利用井水中的冷却能源）
individually controllable heating and cooling air-blast systems；heating in winter，additional cooling in summer（cooling energy from well water）
07 通过走廊地面排气
exhaust ventilation via corridor floors
08 防火天棚 F90，配有组合走廊照明
fire-proof ceiling F 90 with integrated corridor lighting
09 防眩屏（可选择）
anti-dazzle screens，optional

科耐森与西林格尔
达姆施塔特

德累斯顿印刷出版大楼

DRESDEN PRINTING AND PUBLISHING HOUSE

有效执行

EFFICIENT IMPLEMENTATION

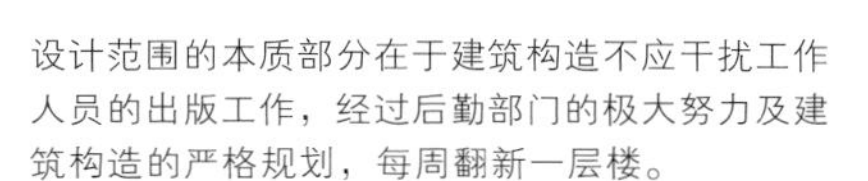

设计范围的本质部分在于建筑构造不应干扰工作人员的出版工作，经过后勤部门的极大努力及建筑构造的严格规划，每周翻新一层楼。

An essential part of the brief was that construction should not disturb the client's publishing operations. With great logistic effort and strict construction scheduling, one high-rise storey was refurbished every week.

设计及施工要在9个月内完成，并按照计划要求使工程费用控制在预算之内且准时完工，能源消耗降到15%以下，仲夏的办公室温度保持在27℃以下，建筑的下层构造设计精美，尤其是建筑本身已彻底改观了。

Planning and construction had to be completed in only nine months. The project was completed within budget and without any delay; every planning target was reached: energy costs were reduced to below 15%; in mid-summer office temperatures are kept below 27° C; the infrastructure is state-of-the-art and, above all, the architecture has been fundamentally changed.

这就是出版社所要求的现代化办公室，其成本低于建一座新大楼的费用。例如：经过印刷的双层墙面每平方米的建筑成本低于450欧元，原因在于独特的设计及简洁的细节构造。

Thus, the publishing house 'acquired' a modern office building which cost much less than a new structure. The printed double facade, for example, cost under 450 Euros per square metre, due to specially designed simple details.

建筑单位 Cornelsen + Seelinger, Darmstadt
业主 Dresdner Druck- und Verlagshaus, Dresden
摄影 Simone Rosenberg, München

专业工程师
塔吊安装 Bollinger + Grohmann, Frankfurt/Main
外墙设计 Cornelsen + Seelinger, Berlin
能源理念 Solares Bauen, Freiburg i. Br.
TGA Solares Bauen, Freiburg i. Br.
电工 Teamplan, Dresden
防火安全 Wackwitz, Dresden

汽车坡道的入口位于大厦前方的空地，可通往房顶平台停车场，大厦的高层形成跨越两地的桥梁。

The ramp leading to the car park on the roof deck is suspended from an opening in the front of the high-rise, with the top floors forming a bridge spanning this gap.

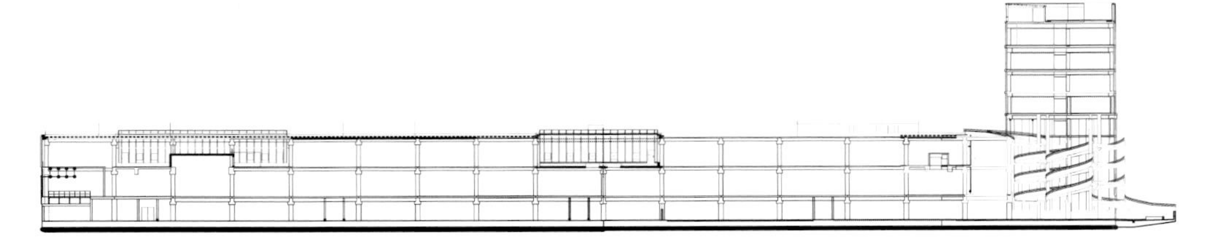

迪斯·乔比恩建筑设计事务所
美茵河畔法兰克福

耐尔曼售后服务部
美茵河畔法兰克福

NECKERMANN RETURNED GOODS DEPARTMENT, FRANKFURT/MAIN

城市位置

URBAN SITUATION

建筑开工时间	2000 年 6 月
建筑成本	4 700 万欧元
竣工验收	2002 年 6 月
总面积	45 450m^2
总体积	255 000m^3
净面积	53 800m^2
使用面积	32 000m^2

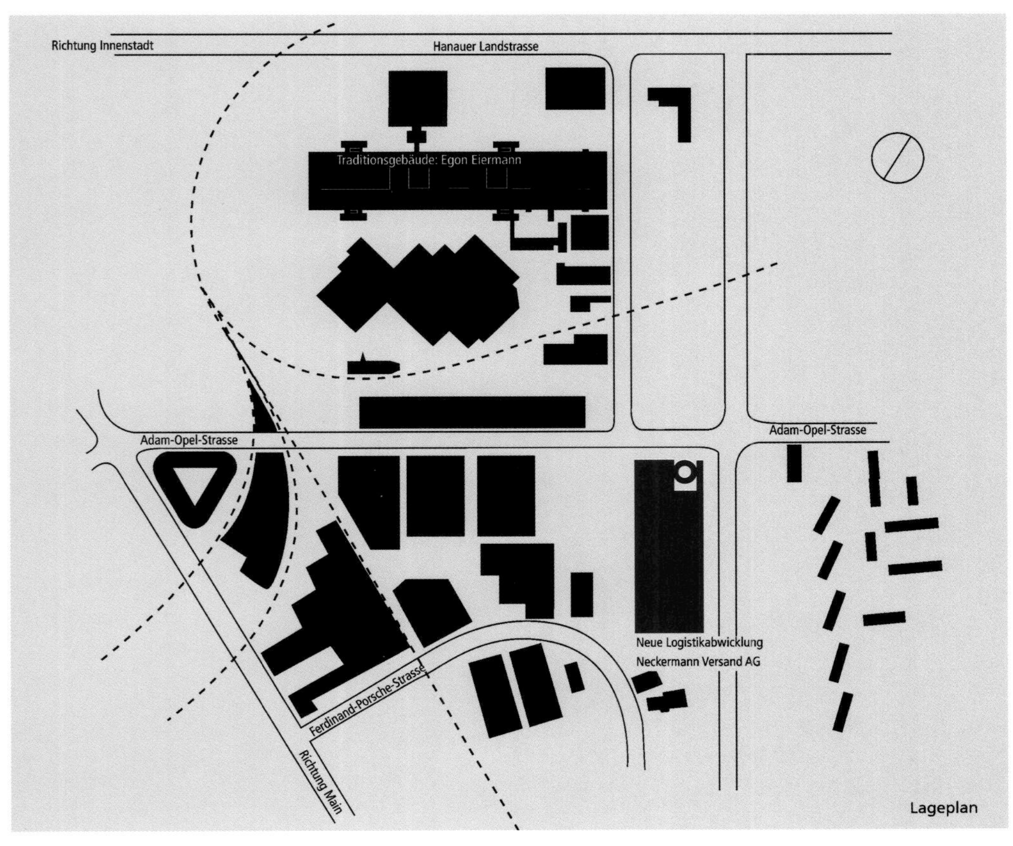

这种果断的设计理念在于借助埃贡·埃尔曼的设计，使公司世界闻名的建筑与新楼联系起来，公司对建筑的要求有些是相互矛盾的，考虑到这些矛盾，通过压缩与结合的办法建成具有高利润位置的简洁式大型建筑，通过布局等级划分，使不同种类的开放式平面布置环境得到有效利用。

The decisive design concept was to relate the new structure to the firm's world-famous building by Egon Eiermann. This was achieved by 'compacting' and combining partly contradictory user requests into a succinct large building volume which offers 'high-profile addresses' and organizes the available spaces of the heterogeneous, open-plan environment by means of layout hierarchies.

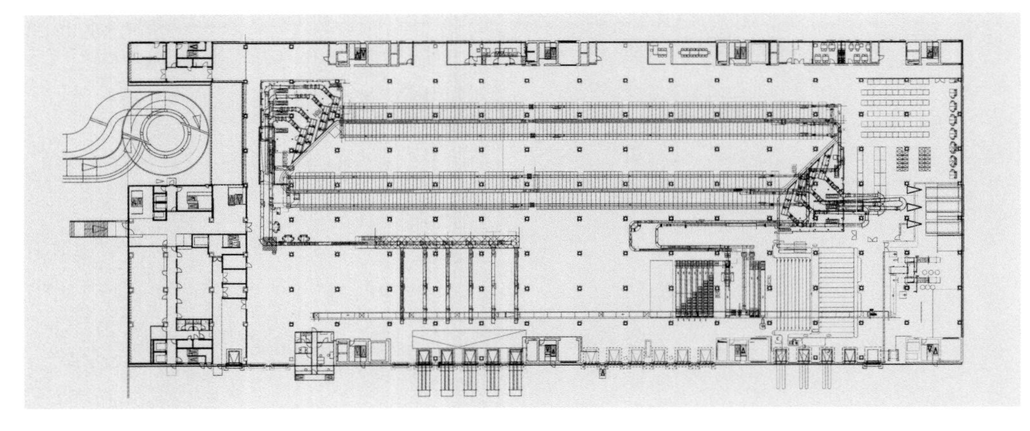

一层平面图：右侧是以绿岛状地带庭院为分隔的开敞式平面工作区，这样不会干扰运作程序的效率与灵活性。

Ground floor plan: The right proportions are created by means of green 'islands' in the form of courtyards which divide the large open-plan working area without disturbing the flexibility or efficiency of operational processes.

老板对于员工的积极态度体现在对工作环境的高质量要求——大面积自然采光、自然通风，按人员比例设计的建筑结构及环境优雅的休息大厅，智能、节能技术的利用及环境的保护装置，体现出建筑师传统与现代相结合的设计理念，使建筑显示出趣味性。

The employer's positive attitude to its staff is expressed in the creation of high-quality workplaces requiring largely natural lighting and aeration, architectural structures at a human scale and access to pleasant open areas. The architects created an interesting building which, while following tradition, adheres to contemporary principles – such as the use of intelligent, resource-saving technology – and the protection of the environment.

迪斯·乔比恩建筑设计事务所
美茵河畔法兰克福

耐尔曼售后服务部
美茵河畔法兰克福

NECKERMANN RETURNED GOODS DEPARTMENT, FRANKFURT/MAIN

内部流通与技术服务

INTERIOR CIRCULATION AND TECHNICAL SERVICES

大厅内部：外部安装可绕轴旋转的遮阳板条可以使能源得到补充，而且人们可以在楼内欣赏到位于阴影处的室外景色。

Inside the hall, the energy concept was complemented by exterior pivoting sunshading slats so that views outside can still be enjoyed even when they are in their shading position.

考虑到安全问题，私人汽车通道、发货运作区及后勤中心的运输设置在不同区域，汽车停车场设在后勤楼的屋顶，通过两条狭窄坡道直通停车场，可容纳260辆汽车，既解决了安全问题，又美化了楼前释放的空间。

For security reasons, private car traffic and operations in the delivery and transport yard of the logistics centre should not share the same areas. The solution to this problem consisted in placing the car park on the roof of the logistics hall, accessible via a two-lane ramp and offering parking space for 260 cars. The site area thus freed has been landscaped.

建筑设计	Dietz Joppien Architekten, Frankfurt am Main
业主	Neckermann Versand AG
工程管理	Kahl Hassdenteufel + Partner, Frankfurt am Main
摄影	Eibe Soenecken, Frankfurt
项目控制	Kahl Hassdenteufel + Partner, Frankfurt am Main
专业工程师	
房屋技术	Reuter + Rührgartner, Roßbach
固定	BGS Ingenieursozietät, Frankfurt am Main
物流运输	Pierau Planung, Hamburg
建筑企业	
主体建筑	Dressler Bau GmbH, Aschaffenburg
物流运输	van der Lande, Mönchengladbach

精心设计的大楼外墙避免了楼面的不和谐因素，并连接大楼的不同单元，窗户布局的逐步变化及中央门厅的构造，清晰明了地显示出大楼功能的划分。

This apparent contradiction was resolved by a façade design which made it possible, in a subtle way, to interlock the different sections of the building. Gradual variations in window formats and the central entrance structure clearly, yet unobtrusively, reveal its functional subdivision.

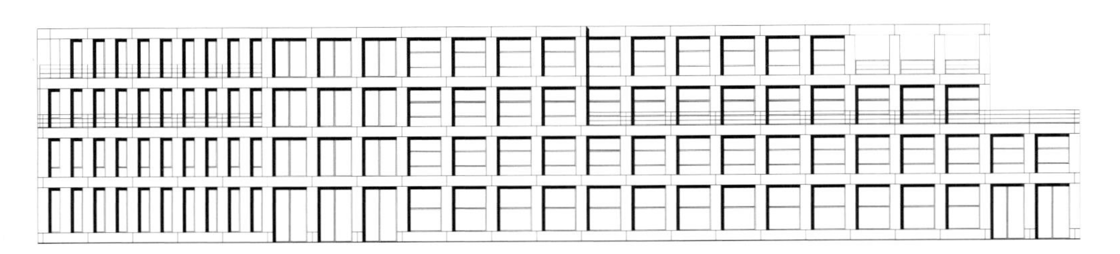

迈克斯·杜德勒建筑设计事务所
柏林

大学主任会议 / 德国公民研究基金会
波恩

CONFERENCE OF UNIVERSITY DEANS / STUDY FOUNDATION OF THE GERMAN PEOPLE, BONN

结构与建材

STRUCTURE AND MATERIALS

建筑成本	9 000 万欧元
施工期	2000–2002 年
总面积	7 237m^2
总体积	22 106m^3
竣工验收	2002 年
占地面积	2 776m^2
使用面积	1 814m^2

为大学主任会议及德国公民研究基金会而设计的多层大楼是一个高难度挑战。当然从成本与收益方面来看，大楼的双重功能保证了使用的高效性，尤其重要的是委托人要求大楼的设计要美观且醒目，并要突出两个协会的独立性，又不会削弱楼体的和谐外观。

Designing a multi-storey building for the Conference of University Deans (CUD) and the Study Foundation of the German People (SFGP) proved a fascinating challenge. Of course, the dual function of the building guarantees a high measure of efficiency as regards costs and benefits. But above all, the client required an aesthetically compelling design which would underline the independence of each of the two institutions without impairing the harmonious appearance of the complex.

两个协会的独立图书馆位于大楼内的大学主任会议区，而德国公民研究基金会的会议大厅设有休息室，一楼可通向花园。

The separate libraries of both institutions are accommodated in the CUD section of the building, while the large conference halls with a foyer and ground-floor access to the gardens were assigned to that of the SFGP.

迈克斯·杜德勒建筑设计事务所
柏林

大学主任会议 / 德国公民研究基金
波恩

CONFERENCE OF UNIVERSITY DEANS / STUDY FOUNDATION OF THE GERMAN PEOPLE, BONN

利用与效率

ACCESS AND EFFICIENCY

考虑到最大限度地利用建筑的灵活性，所有的服务设施（包括供热、通风、卫生、照明、供电及数据处理系统）都安装在栅格中，并延伸至建筑内各用户的角落，所以大楼拥有最大限度的可行性内部布局及用途。

With a view to maximum flexibility, all the service systems (heating, ventilation, sanitary, lighting and power supplies, DP media) were distributed on a grid, extendable in all directions, so that the building can accommodate the maximum number of possible interior sub-divisions and uses.

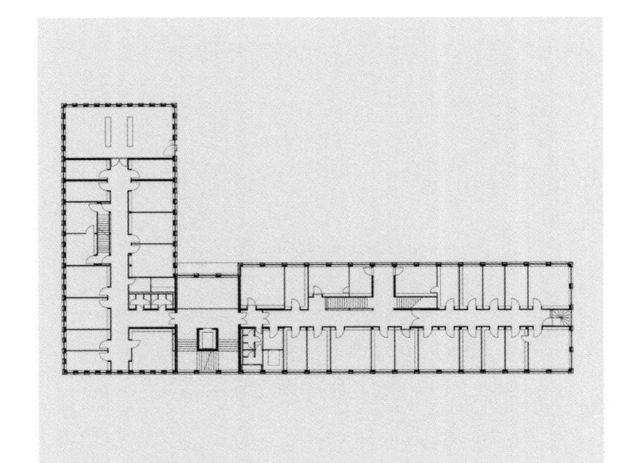

建筑的外墙设计特点体现在不同规格与尺寸的窗户，突出建筑用途的双重性，使来访者易于分辨出两个部门的方位，又体现了部门的个性。

The façades were designed with windows of different formats and sizes so that its dual institutional character is immediately revealed to approaching visitors and expresses the individual identity of each of the two clients.

经过中央主大厅可通往大楼内各单元，通向楼梯的休息室及接待区为两个协会的工作人员提供优雅的楼层间通道。从功能方面来看，两个协会相互联系，同时又有各自的划分界线。

The building 'sections' are accessed via the central main entrance hall. The lobby and reception area leading to the staircase offer users from both institutions a prestigious vertical access route which connects them in functional terms, yet also forms a vertical border between the two 'institutional halves' of the building.

建筑设计 Max Dudler Architekt, Berlin
业主 Hochschulrektorenkonferenz, Bonn
Studienstiftung des Deutschen Volkes, Bonn
园艺建筑 Lützow 7 Garten- und Landschaftsarchitektur, Berlin
摄影 Stefan Müller, Berlin

专业工程师
工程管理 Heberger Bau GmbH, Schifferstadt
塔吊安装 WMM – Walther Mory Maier Bauingenieure AG
房屋技术 FRANK Haustechnik
外墙 Ingenieurbüro Ludwig + Mayer

企业
大厦所有者 Heberger Bau GmbH, Schifferstadt

区域性发展区位于精心安排的功能性市区，它的布局是为了迎合整幢新建大楼南部的一层贸易博览会而设计的。从某种意义上讲，它是欣赏苏黎世城市景色的最佳位置，双子楼及其侧楼形成蒙太奇式图片效果。

This place-shaping development is a well-thought-out and functional urban area. Its arrangement points and leads to the trade fair ground south of the new high-rise ensemble which, in a certain sense, forms a gate, or vantage point that opens the view over the city of Zurich. Twin towers and extensions in a photo montage.

ARGE
艾特里尔.WW.苏黎世
罗夫・乌斯特
尤尔斯，乌斯特
瓦尔特・沃斯克勒
迈克斯・杜德勒建筑设计事务所　苏黎世

高层发展大楼
苏黎世

HIGH-RISE DEVELOPMENT,
HAGENHOLZSTRASSE, ZURICH

城市位置

URBAN SITUATION

建筑成本	14 200 万欧元
施工期	1999–2004 年双塔 扩建工程计划中
总面积	97 605m²
总体积	403 177m³
竣工验收	2004 年双塔 预计 2007 年扩建
占地面积	14 580m²
使用面积	42 060m²
可出租面积	供 4 860 人

利用了依次变化的窗户格局及节奏性接合技术，市区广场双子大楼的建造突出了阶梯式的建筑特色。

The design enhanced the high-rise character of the stepped new buildings by successively changing window formats and by applying the principle of rhythmical articulation to create twin towers around an urban square.

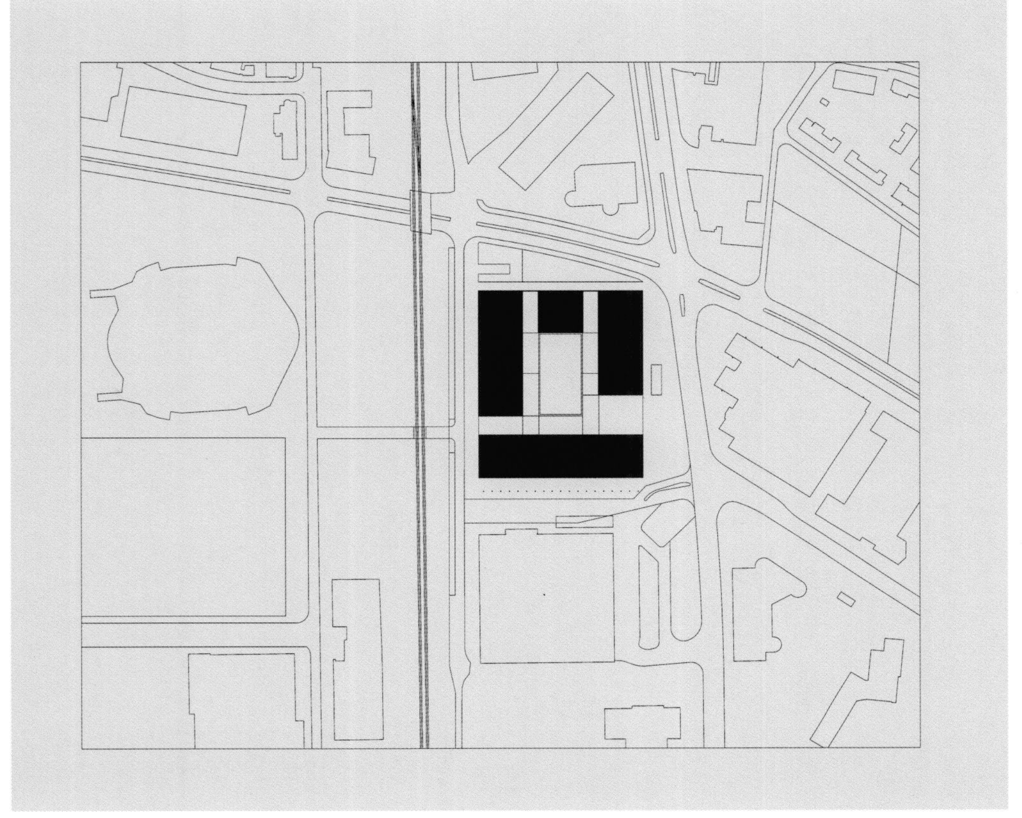

在平稳扩张的市区，大楼作为建筑实体，只有当延伸至市郊时，市区与市郊的矛盾便突显出来。市区的建筑要求紧凑，而市郊建筑要求功能多样化，并具有理想的建筑品质。苏黎世北部的劳尔森巴赫 / 奥立康区的开发面临同样的问题，考验设计者的地方在于要支撑且改善区域结构、社会条件及城市布局。

In places where a city expands to a moderate extent, it will always be recognizable as an architectural entity. It is only when it sprawls out into the countryside that problems will arise between the centre and the periphery. The first will become more compact, while the latter will develop as an agglomerate of anything but desirable architectural quality. This happened to Zurich North where a large area in the district of Leutschenbach/Oerlikon was to be developed. The challenge to the planners consisted in supporting and improving the area's structure, social conditions and urban layout.

根据市区与市郊的完全不同点，新建筑与城市环境相分离，使其具有不同的建筑规模。新建筑根据高度和规模而不是中间的空地，与周边环境相互联系起来。

The development remains separate from its urban environment by starkly differentiating between interior and exterior spaces so that it is the volumetric arrangement of the different buildings which links and relates it to the neighbourhood as regards heights and scale, and not the intermediate open space.

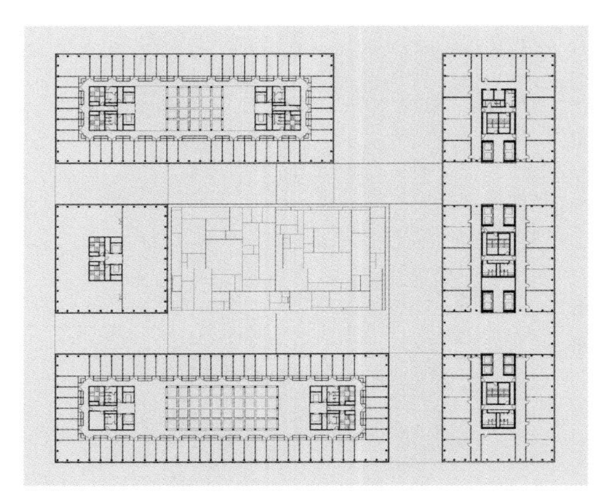

ARGE
艾特里尔.WW.苏黎世
罗夫・乌斯特
尤尔斯・乌斯特
瓦尔特・沃斯克勒
迈克斯—杜德勒建筑设计事务所　苏黎世

高层发展大楼
苏黎世

HIGH-RISE DEVELOPMENT, HAGENHOLZSTRASSE, ZURICH

利用与流通

ACCESS AND CIRCULATION

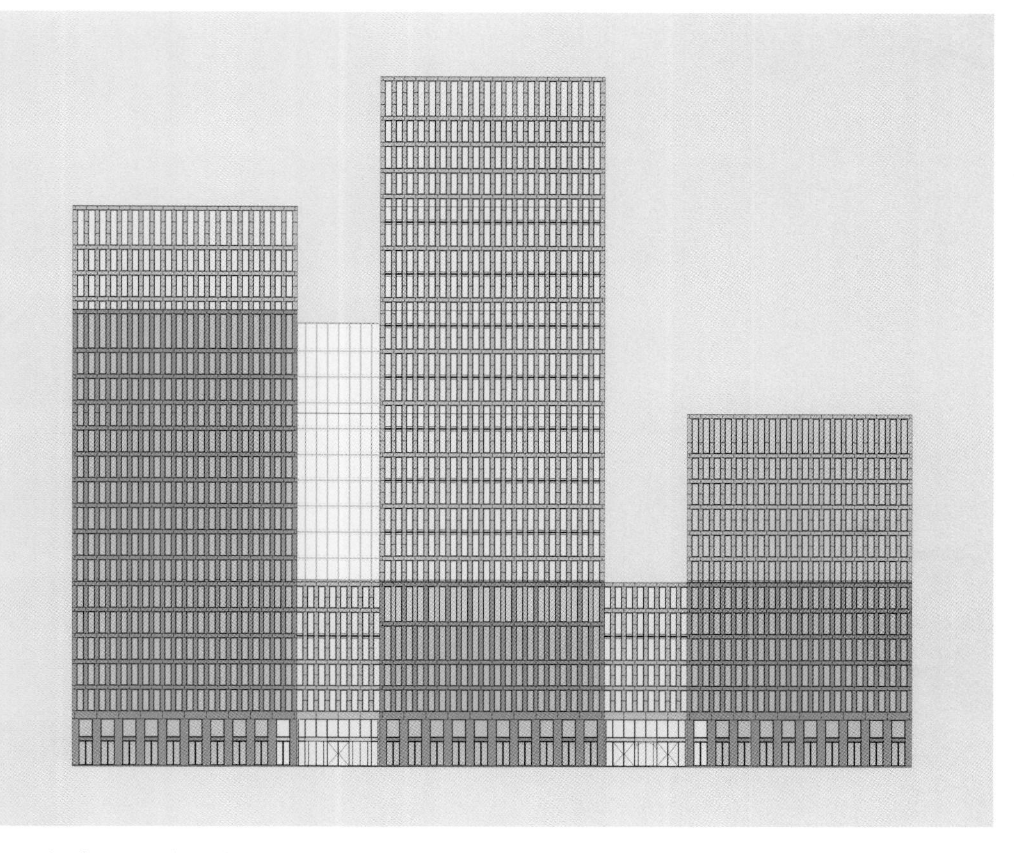

新建筑区的广场略高于周围水平面，目的是将城市外墙水平面延伸成为新建筑的垂直部分，连接塔楼与基础建筑的阶梯式建筑结构的出入口处，可通往大厦的广场区，从街道可以进入其西侧的拱廊式市中心区，这突出了公共场合的特点。

The square in the development area is slightly raised above the surrounding ground level and is to be understood as the horizontal urban façade which extends the vertical ones of the new buildings overlooking it. The entrances to the stepped architectural composition of interconnected towers and base buildings are oriented to the square of the estate, while the arcaded civic centre on its west side is accessed from the street which makes it a more public place.

建筑设计　Arge
Atelier WW, Zürich –
Rolf Wüst,
Urs Wüst,
Walter Wäschle
mit Max Dudler Architekt, Zürich

业主　Finanzdirektion des Kantons Zürich und TDC Switzerland AG, Kantonales Hochbauamt (BVK), AMAG

摄影　Stefan Müller, Berlin, S. 111–114
Hanne-Birgit Wiederhold, Eisingen, S. 110

专业工程师
园艺设计　Planetage GmbH, Zürich
房屋技术　Gruenberg & Partner AG, Zürich
外墙　Stäger & Nägeli AG, Zürich

迈克斯·杜德勒建筑设计事务所
柏林

办公大楼　达姆施塔特

OFFICE TOWER, DARMSTADT

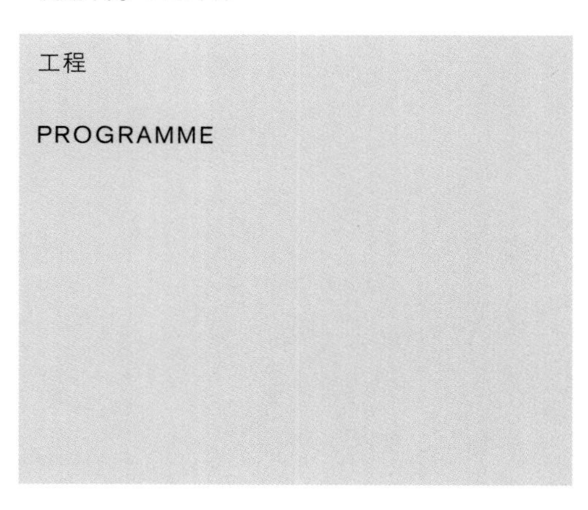

建筑成本	1 000 万欧元
施工期	2001–2003 年
总面积	5 145m² (+ 外围建筑)
总体积	18 240m³ (+ 外围建筑)
竣工验收	2003 年
占地面积	2 210m²

突显建筑物表面的垂直部分使其线条分明，即从莱茵河大街的角度看上去更像一座塔。

Stressing the verticals in the façade structure aimed to make its typology clearer, i.e. to make it appear a bit more like a tower which is now the case seen from Rheinstrasse.

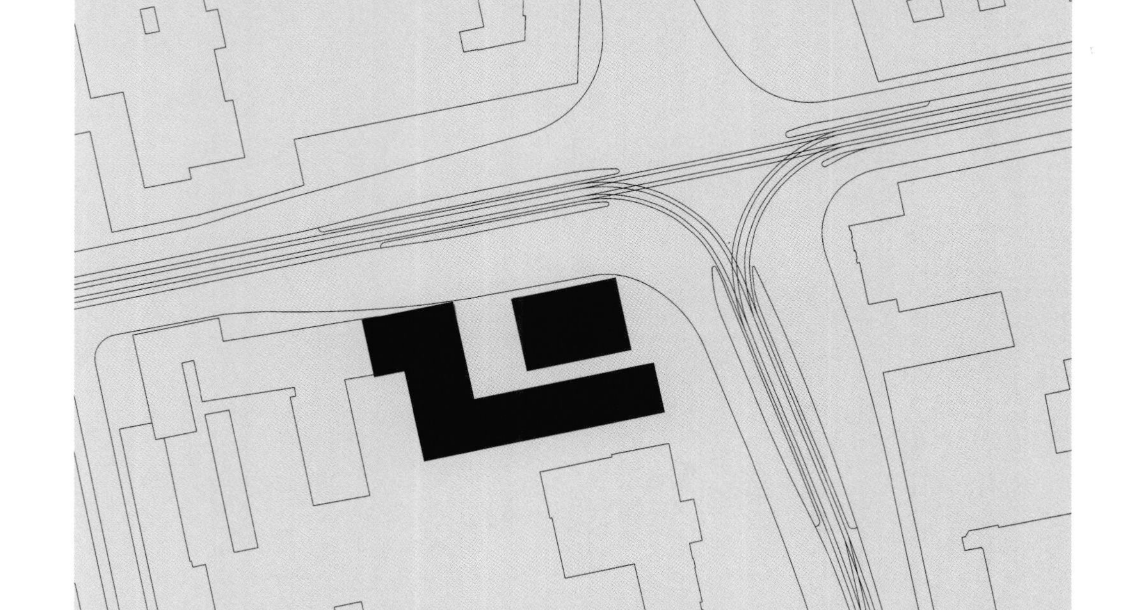

20 世纪 60 年代的建筑是黑森州达姆施塔特镇的标志性建筑。坐落于莱茵河大街的 13 层的办公楼也同属于这一时期。陈旧的窗子、古老的栏杆不再符合当代的审美观，更不能抵御严寒酷暑和噪音的干扰。因此，为了改变这种结构，在建筑物的表面贴上两层立面。外层由大格石板构成，内层注重现有的承重结构。当然，这种砖石立面决定了建筑物的共同特征。

Buildings from the 1960s have marked the town of Darmstadt, Hesse. The thirteen-storey office building on Rheinstrasse is also of this period. The old façade, structured in window and parapet bands, no longer met contemporary aesthetic expectations and was also insufficiently insulated against noise and cold or heat. Thus the building was 'pared down' to the structure and then re-enclosed with a two-leaf façade, the outer one of large-format stone slabs and the inner one followed the existing load-bearing structure. Of course, the stone cladding now determines the public character of the building.

迈克斯·杜德勒建筑设计事务所
柏林

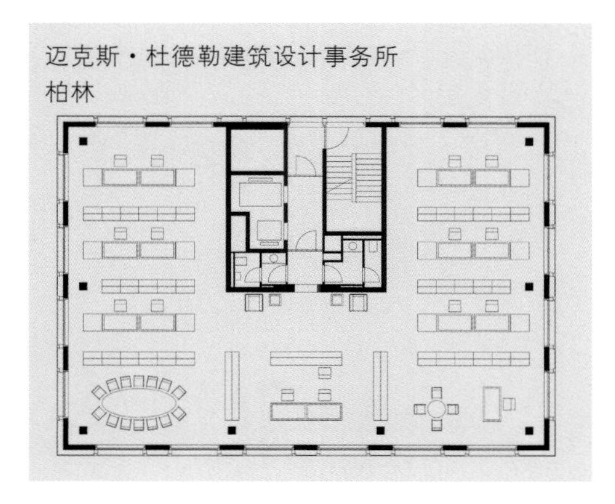

办公大楼
达姆施塔特

OFFICE TOWER, DARMSTADT

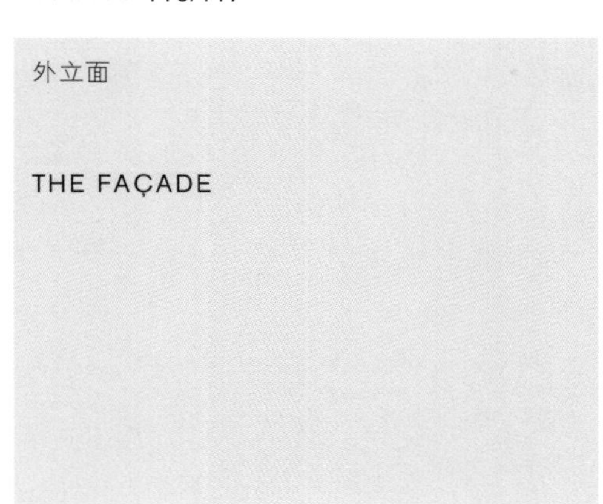

外立面

THE FAÇADE

被毁坏的内部结构提供了设计不同办公室楼面布置图的机会。

The gutted interior offered the chance to design different office floor plans.

外立面采用海角绿色花岗岩窗户规格，从楼底向楼顶逐步扩大。

The façade of Cape Green granite is structured rhythmically by the window for-mats increasing in size towards the top of the building.

一楼购物拱廊以上的楼外立面由海角绿色花岗岩构成，并布满窗格，窗格规格从底到高逐步扩大。早先粗短的建筑外观已被淘汰，相反，创作出这种纤细垂直的视觉差，看似一座高层大楼。

The building skin of Cape Green granite, generously perforated by windows, now rises from a ground-floor shopping arcade. Its rhythmic pattern was created by gradually broadening the windows towards the top. This eliminates the previous rather stumpy appearance of the building and, instead, produces the optical illusion of slender verticality, so that it really looks like a 'high-rise'.

建筑单位 Max Dudler Architekt, Berlin
施工单位 Newcom Property Projekt Darmstadt GmbH
摄影 Stefan Müller, Berlin

基层的绿岩用来效仿周围历史建筑物，而其上方的灰白砾石突出了新式建筑的现代风格。办公室楼层设置在中央露台附近。

The green stone of the base course emulates the historic buildings in the neighbourhood, while the pale sandstone above it emphasizes the modernity of the new construction. The office floors are arranged around the central patio.

迈克斯·杜德勒建筑设计事务所
柏林

新建办公大楼与翻新过的旅馆
弗里德里希大街　柏林

OFFICE BUILDING (NEW) AND
HOTEL (RESTORATION),
FRIEDRICHSTRASSE, BERLIN

位置与料材

LOCATION AND MATERIALS

建筑成本	1 200 万欧元
施工期	2000–2004 年
总楼层面积	11 000m²
总体积	36 000m³
竣工验收	2004 年
占地面积	1 320m²

当位于弗里德里希大街著名的商务贸易大厦的门面恢复了自身的原始设计，并且整个大厦经过改造、扩建形成一面横墙和两座侧楼的时候，一座新式的7层办公室楼也与其毗邻而建。上两层的倾斜式屋顶已失去了其存在的意义（转作旅馆使用）。新式建筑的另一个显著特点是两种不同颜色的石面形成了新旧的鲜明对比。

While the façade of the historic business and commercial building on Friedrichstrasse was restored to its original design and the building converted and extended by one transverse and two side wings, a new seven-storey office block was built on the adjacent site. The two top floors of the new block step back following the pitched-roof gradient of the adjacent building (which was converted into a hotel). Another special feature of the new office building is that it plays on the contrast between old and new by using façade stones in two different colours.

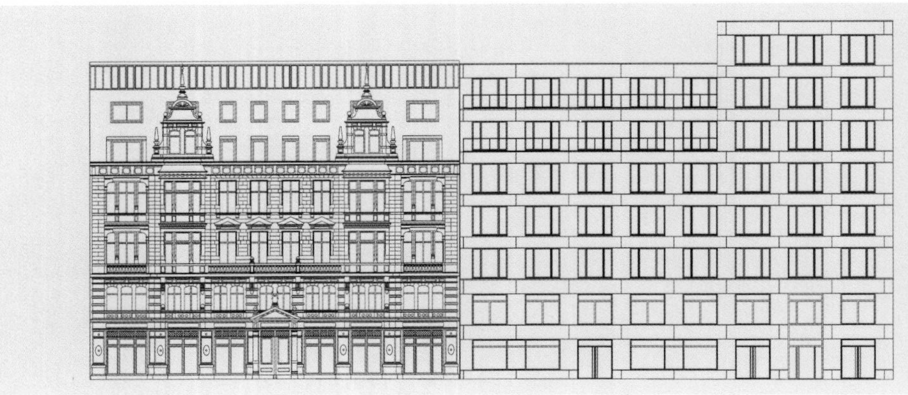

坐落在街角的整个办公大楼将传统的柏林式周边街区与附近的在国际建筑展期间建造的楼群有机结合起来。其地基框架连接的有效性、楼层地面的标准性以及顶层设计的独特性，将垂直建筑的本色与水平特质的柏林式建筑浑然一体。

In its corner situation the office building mediates between the traditional Berlin perimeter street block and the nearby free-standing buildings erected at the time of the International Building Exhibition IBA. Its articulation – basement, standard floors and tower storeys – is a hybrid of a vertical point building on the one hand and a horizontal Berlin block on the other.

迈克斯·杜德勒建筑设计事务所
柏林

新建办公大楼与翻新过的旅馆
弗里德里希大街 柏林

OFFICE BUILDING (NEW) AND
HOTEL (RESTORATION),
FRIEDRICHSTRASSE, BERLIN

建筑设计

ARCHITECTURAL DESIGN

这个设计考虑到具有悠久历史的建筑门面的修复。

The design took into account the restoration of the historic front building on the neighbouring property.

建筑单位 Max Dudler Architekt, Berlin
业主 FVG Familienverwaltungsgesellschaft mbH/ Berlinhaus
摄影 Stefan Müller, Berlin
Claus Graubner, Berlin, S. 108 unten

专业工程师
工程管理 B + S-Consulting
塔吊安装 Schwarzbart & Partner, Berlin
能源技术 HHS, Stuttgart Planungsbüro für techn. Gebäudeausrüstung

费斯切建筑设计事务所
德国建筑师联合会
曼海姆　科隆

瑞特赛斯律师事务所
曼海姆

RITTERSHAUS LAWYERS' OFFICES
MANNHEIM

艺术与建筑

ART AND ARCHITECTURE

建筑成本	约 920 万欧元
施工期	约 10 个月
总面积	6 500m²
总体积	21 500m³
计划草案	2001 年
竣工验收	2002 年

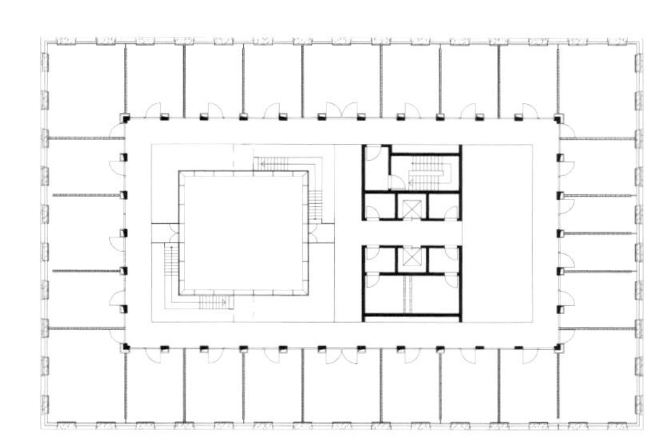

安德里亚斯・冯・魏兹佐克教授的书中描述装饰性的中楣，客人们以出版文字形式来表达他们自己的想象。因此，艺术便融入“移动图书馆”这样一个储藏了丰富知识的建筑物之中。在预制混凝土制成的中楣装饰下，内部空间的容量与类型变得更加清晰。

A decorative frieze of books by Professor Andreas von Weizsäcker, Munich, describes the clients' self-image: the printed word as the expression of their work. Art therefore enters into a close relationship with a 'floating library', i.e. the store of knowledge inside the building. The content and typology of the interior space becomes legible from the frieze of prefabricated concrete façade elements.

先进的办公室设计过程不单是个人的想法，而是业主、建筑师和专业工程师集体智慧的结晶。设计、功能和财政效率三者是可以调和的。

新建的律师办公楼与整体建筑设计的一致性是值得注意的。

Progressive office designs are not just the result of an idea, but of a planning process in which the client, the architects and specialized engineers are involved. Design, function and financial efficiency are by no means incompatible contradictions.

The new office building of Ritterhaus lawyers is remarkable for the consistency with which such an integral planning and building process was implemented.

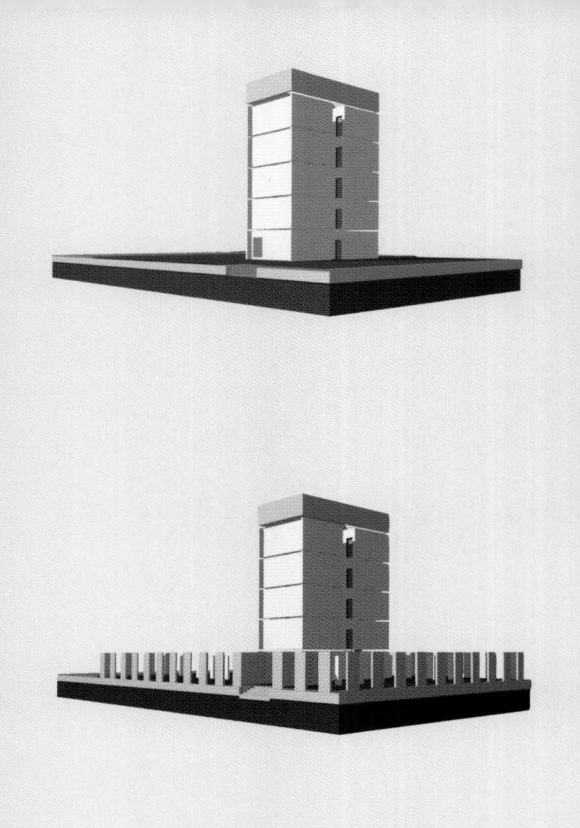

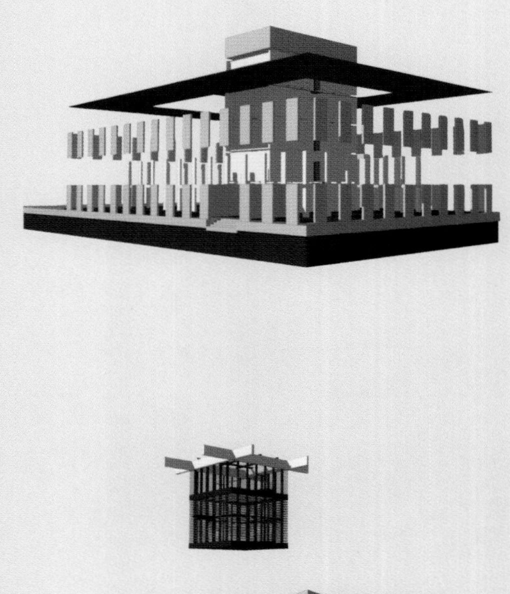

预制混凝土成分的承重结构，拥有综合技术装置功能的弯曲天花板，玻璃钢筋制成的“移动图书馆”。

Load-bearing structure of prefabricated concrete components; curving ceiling elements with integrated technical installations; ‘floating library’ of glass and steel.

针对外墙与内壁设计的可承重的预制剪力墙网格结构，采用了大量内部支柱，使内部空间易于划分。预制地板、天花板及综合设备（暖通空调系统和喷淋器）形成曲线环境，体现出建筑学、结构设计与分析、维护技术、建筑物理学之间相互贯通的模式规划。

The design is based on a load-bearing prefabricated shear-wall grid for both the exterior enclosure and the interior façade. This structure made interior columns superfluous and offered the chance to partition the interior in a flexible way. The curving elements of the prefab floors/ceilings with integrated services (HVAC and sprinklers) underline the modular principle of the interaction and ‘cross-over’ between architecture, structural design and analyses, services technology and the physics of construction.

费斯切建筑设计事务所
德国建筑师联合会
曼海姆　科隆

瑞特赛斯律师事务所
曼海姆

RITTERSHAUS LAWYERS' OFFICES MANNHEIM

设计与结构

DESIGN AND STRUCTURE

01 总水管 / 洒水器
water mains / wide-range sprinkler
02 热坑（供热 / 冷却循环补偿结构质量惯性；集成入墙壁模块用来单独控制内部温度）
hypocausts (heating / cooling loops compensating the inertia of structural mass; integrated into wall modules for individual interitor climate control)
03 空气对流
air circulation caused by convection
04 通风口
air intake opening
05 带有消声器的空气排气口
air exhaust with silencer
06 用于调和中庭气候的凹缝通风口
air outlet in recessed joint for tempering the atrium climate
07 通过电缆和基底凹槽进行能量分布
power distribution via cable ducts and floor cavity
08 通风管
ventilation duct
09 通过弯曲的天花板进行优化日光控制
optimized daylight control via curving ceiling
10 通过天花板模数的冷热循环调整结构部分
coops in wiling modules tempering structural sections by means of heating/cooling
11 配电
electrics
12 热交换机
heat exchanger
13 土地探测器
soil probes

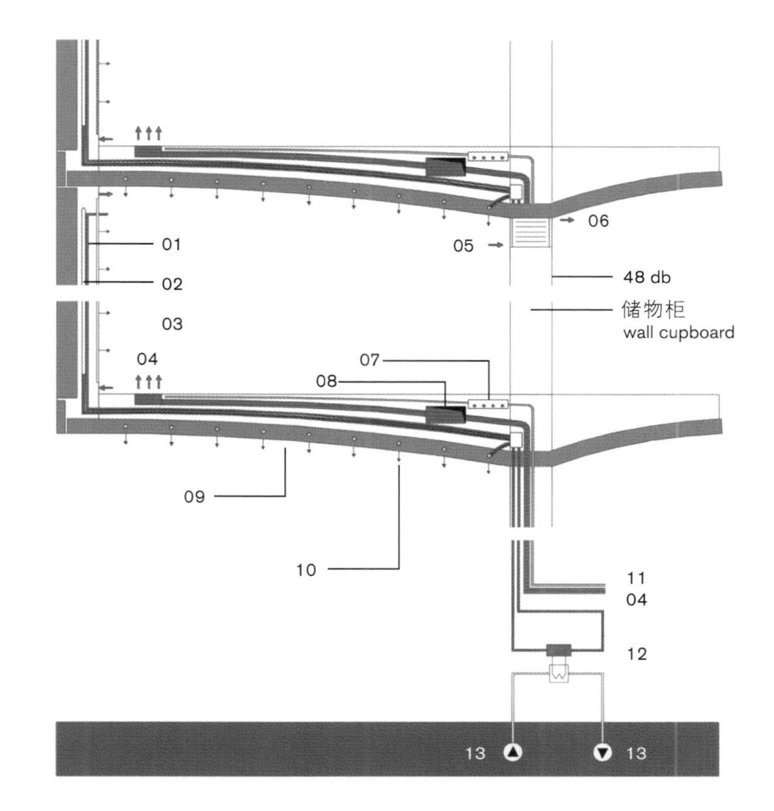

建筑物描绘了一个五层门廊连接中庭环绕“移动图书馆”通过楼梯和电梯中心。阳光通过通风窗和拱形天窗照射到前庭。人工照明是建筑现场的主要组成部分。

Typologically, the building represents a 'twin atrium' with the five-storey entrance hall connected to the atrium enclosing the 'floating library' via the stairway and lift core. The atria are daylit via clerestory bands and domed lanterns. Artificial lighting is an integral part of the architectural scenario.

地下水采暖和制冷的能量通过贯穿于混凝土制成的地板与天花板之间的细状管道，为建筑物提供了能量。特别开发的一项工程是每间房通过热网可以分别控制温度。使用土地探测器得知地热能可以供给高效环保的能量， 因此，每平方米的成本可以减少到不超过 0.5 欧元。

The environmental heating and cooling energy of groundwater is utilized for heating and cooling the building via capillar pipes embedded in the concrete floors/ceilings. Every room can be climate-controlled separately with a 'hypocaust wall element', specially developed for this project. The use of geothermal energy via soil probes provides cost-efficient and environmentally friendly energy supplies, the costs of which are therefore reduced to less than € 0.50 per square metre of office space.

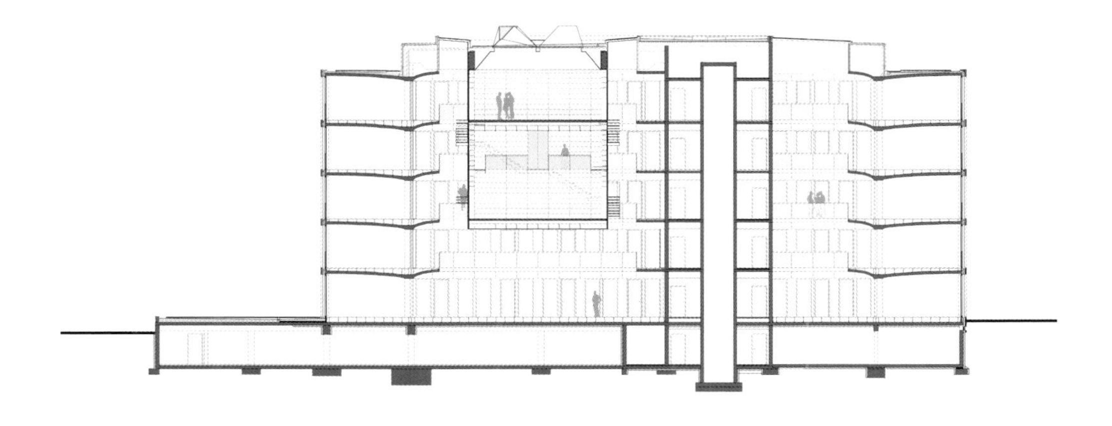

建筑物的一个最显著的特征就是“移动的”即悬浮的图书馆，位于中庭并由玻璃钢筋制成。更重要的一点是业主的自我想象和理解在他们的作品中得以体现。这座图书馆体现了人们别出心裁的想法，即悬浮式的而不是固定不变的。

The most prominent feature of the building is the ‘floating’, i.e. suspended, library of glass and steel in the atrium, and an important representation of the clients’ self-image and notion that knowledge in print is at the heart of their work. The library suggests the reverse of habitual views – suspending instead of standing; book ‘bellies’ instead of book spines. Andreas von Weizsäcker’s book frieze further expounds this idea.

图书馆内 3 层高悬浮空间面积为 9 米 × 9 米，悬浮室重 40 吨，利用 24 根扁钢条（400/200 毫米）悬挂在中庭屋顶架的钢盘大梁上。通过馆内各楼梯与通道可进入悬浮空间。

The library glass cube has a floor area of 9 x 9 m and three storeys. It weighs 40 tons and is freely suspended from the RC girders of the atrium roof structure by means of 24 flat-steel bars (400/200 mm). The three-storey cube is accessible from all sides via stairs and bridges with balconies.

费斯切建筑设计事务所
德国建筑师联合会
曼海姆　科隆

瑞特赛斯律师事务所
曼海姆

RITTERSHAUS LAWYERS' OFFICES
MANNHEIM

结构

STRUCTURE

业主　LEG – Landesentwicklungsgesellschaft Baden-Württemberg mbH, Stuttgart
建筑设计单位　Fischer Architekten BDA, Mannheim – Köln
摄影　Thomas Ott, Mühltal
艺术设计　Prof. A. v. Weizsäcker, München

塔吊安装　Bollinger + Grohmann, Frankfurt am Main
能量理念　köhler beraten und planen, Wiesbaden
楼体装备　köhler beraten und planen, Wiesbaden

乔・福兰斯克建筑设计事务所
美茵河畔法兰克福

A × A 保险公司住宅及办公区
勃肯海姆大街
美茵河畔法兰克福

RESIDENTIAL AND OFFICE BLOCK FOR AXA VERSICHERUNGS-AG, BOCKENHEIMER LANDSTRASSE FRANKFURT/MAIN

设计方法

DESIGN APPROACH

建筑成本	3 840 万欧元
施工期	2001–2003 年
总楼层面积	25 000m²
总体积	90 000m³
草案设计	2001 年
出租面积	10 000m²
居住面积	3 300m²
停车位	70

可以说建筑分为：地基、柱身和柱头三个部分，各部分的结构与窗户设计都是不同的。

The building is zoned, so to speak, in three parts: 'base, shaft and capital' with different structures and window arrangements.

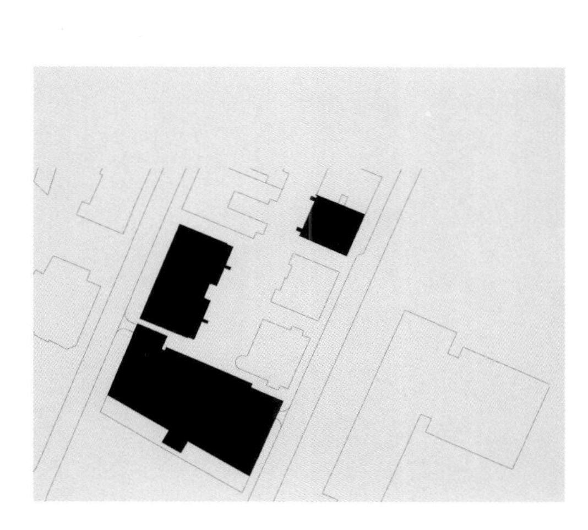

平面图显示，新建筑邻近街道，住宅的发展形成19 世纪末 20 世纪初法兰克福的西边市政厅又一个特色的主题。

The site plan shows that the new structure closes the street front. The residential development re-enacts the motif of the turn-of-the-century townhouses in Frankfurt's Westend.

位于法兰克福西区的原 5 层建筑已被拆毁，并新建一座 11 层高的办公大楼，楼后建了 3 座 4 层别墅。建筑师的设计十分周全，适合居民、客户和使用者。

The five-storey predecessor structure in Frankfurt's Westend area was demolished and replaced by an eleven-storey office building. Three four-storey urban villas were erected on the rear site area. The architects developed a well thought-through design aiming for urban density which suits residents, client and users alike.

正如这里的情况，最大限度和最适宜的城市密度显示出建筑群之间的和谐共存，其中包括3座列名市政厅及一些新的建筑。新旧建筑并不孤立看待，这就是新与旧、传统与现代和谐统一的建筑理念。

In a situation like this one, maximum and optimal urban density presupposes a balanced juxtaposition of existing buildings, here three listed townhouses, and the new structures. The desired harmony of old and new was achieved by an architectural philosophy which does not regard either old and new or tradition and modernity (or stone and glass as the symbols of these concepts) as opposite poles.

乔・福兰斯克建筑设计事务所
美茵河畔法兰克福

A × A 保险公司住宅及办公区
勃肯海姆大街
美茵河畔法兰克福

RESIDENTIAL AND OFFICE BLOCK
FOR AXA VERSICHERUNGS-AG,
BOCKENHEIMER LANDSTRASSE
FRANKFURT/MAIN

城市密度

URBAN DENSITY

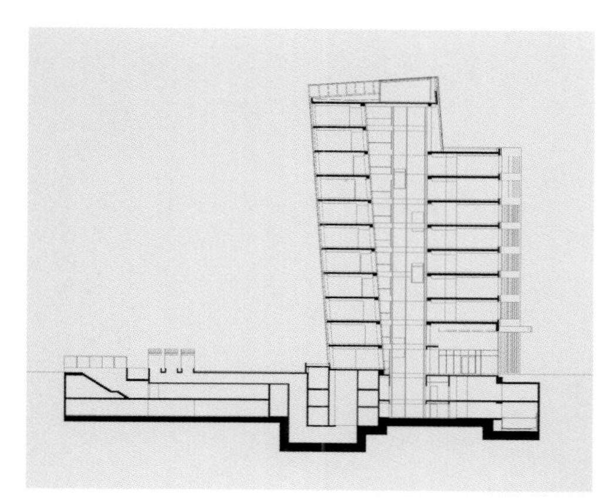

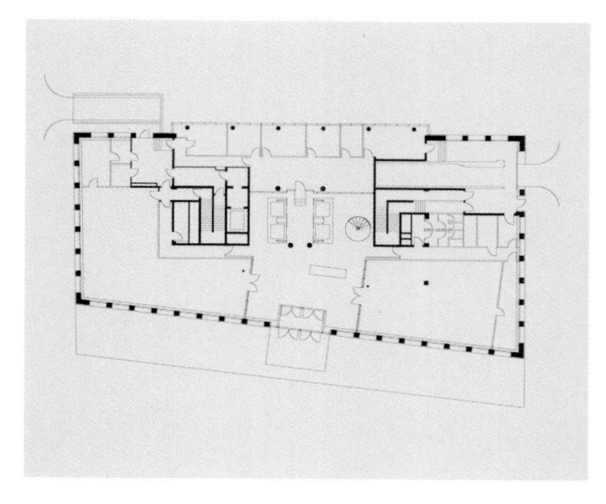

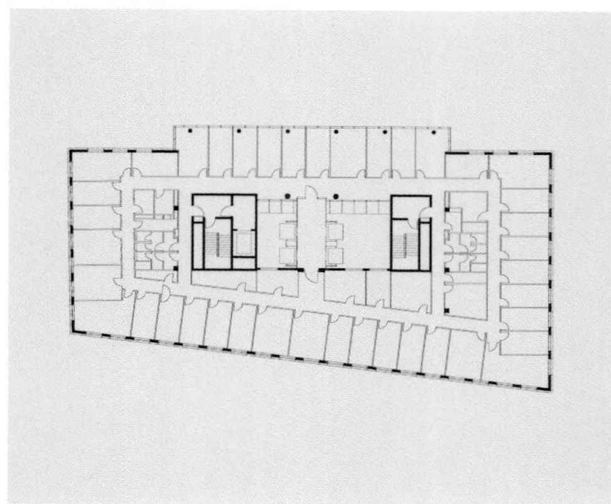

工程的设计方案是在勃肯海姆大街建成紧凑的建筑群。在梯形平面图的周边建一圈楼区，中间建有一座玻璃斜塔。

The project proposed a compact building complex on Bockenheimer Landstrasse, consisting of a massive block-edge structure on a trapezoid plan with a 'leaning' glass tower rising out of it.

在周边的楼区中插入新建筑的设计理念，也大大地扩宽了勃肯海姆大街的街边区，并在街区中间建有花园、绿带区，用于间隔地区。

The idea of carefully inserting the new building into its surroundings, also meant that the large front site area on Bockenheimer Landstrasse was generously planted and gardens created in the middle of the block. The landscaping design was based on the fact that in contrast to natural greenery, inner-city green areas are always planned and used to define spaces.

由于高楼的倾斜设计，日光可穿过一楼大厅至楼顶的空隙，使三开间基础结构所形成的中央空间自然采光。

Due to the high-rise 'gradient', the central spaces of the three-bay base structure spanning are naturally lit through a void reaching from the ground-floor lobby to the top of the building.

乔・福兰斯克建筑设计事务所
美茵河畔法兰克福

A × A 保险公司住宅及办公区
勃肯海姆大街
美茵河畔法兰克福

RESIDENTIAL AND OFFICE BLOCK
FOR AXA VERSICHERUNGS-AG,
BOCKENHEIMER LANDSTRASSE
FRANKFURT/MAIN

位置

LOCATION

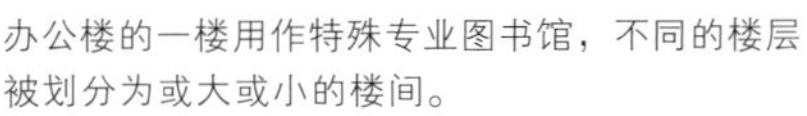

办公楼的一楼用作特殊专业图书馆，不同的楼层被划分为或大或小的楼间。

The ground floor of the office building is used as a specialized professional library. The different floors may be partitioned as smaller or larger suites.

工程参与者

业主	AXA Versicherungs AG, vertreten durch AXA REM, Köln
总担保人	ABG GmbH, Köln
主要企业	Hochtief AG, Niederlassung Frankfurt am Main
工程控制	Pockrandt Management, Wiesbaden
设计单位	Jo. Franzke Architekten, Frankfurt am Main
园艺设计	Dr. Gabriele Schultheiß, Berlin
灯光设计	ag Licht, Bonn
塔吊安装	Pirlet & Partner, Köln
大楼技术	P&A, Petterson & Ahrens, Obermörlen
电工设计	SHI, Schad Hölzel und Partner, Mörfelden
图书馆	Deutsche Werkstätten Hellerau, Dresden
摄影	Jean-Luc Valentin, Frankfurt am Main

湿盐库的外观从远处望去也很突出。夜晚，当从库内透出光来就越发显明。

The shape of the damp salt store is impressive also from a distance, even more so at night when it is illuminated from inside.

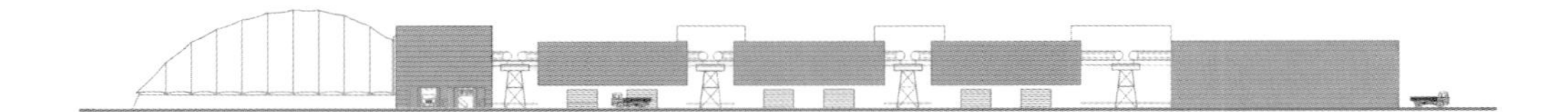

布莱谢罗德化学工业园中的湿盐库

DAMP SALT STORE IN THE CHEMICAL INDUSTRIAL PARK OF BLEICHERODE

弗瑞克·雷查特建筑设计事务所
美茵河畔法兰克福

建筑设计

ARCHITECTURAL DESIGN

建筑成本	386 万欧元
建筑施工期	2003 年
总面积	3 750m²
建筑标准	52m × 30m × 20m
使用面积	2 900m²
用途	仓库和生产车间

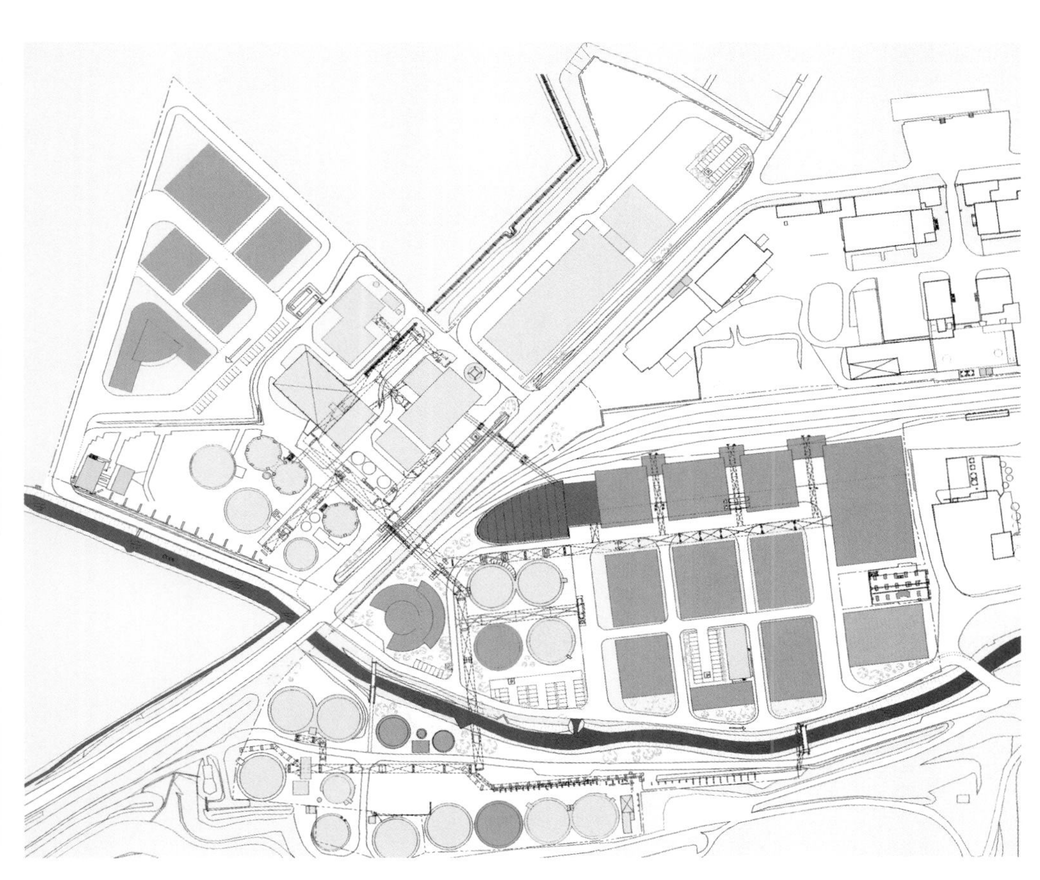

湿盐库位于布莱谢罗德市碳酸钾矿业杜刹国际有限公司，占地 140 000 平方米的工业区，是弗里克·莱舍特建筑设计室的总体规划中的第一建筑阶段。盐库位于化学工业园入口处，从远处看，它外型简洁、设计明朗，使杜刹国际有限公司的新形象更具特色。

The damp salt store is on the 14-hectare industrial site of the potash-mining company Deusa International Ltd. in Bleicherode. It represents the first building stage of the masterplan designed by Frick.Reichert Architects. With its concise shape and clear architectural design, it stands at the entrance of the ‘Chemie-Industrie-Park’ – visible from afar – and characterizes the new image of Deusa International Ltd.

湿盐库的形状像一个倾倒的锥体，建筑包括一个3米高的露天混凝土圆环及用于支撑的几个间隔5米的微拱形木梁，外裹聚酯材料用于抵抗天气变化。因为这种材料是透明的，日光可进入库内。相对减少了因照明而消耗的能量。

The form of the damp salt store resembles a dumping cone. The construction consists of an exposed concrete ring, 3 m high, supporting several correspondingly arched wooden girders at intervals of 5 m that are covered with polyester fabric for weather protection. As this material is diaphanous, it lets in daylight which considerably reduces the energy costs for additional artificial lighting.

布莱谢罗德化学工业园中的湿盐库

DAMP SALT STORE IN THE CHEMICAL INDUSTRIAL PARK OF BLEICHERODE

弗瑞克·雷查特建筑设计事务所
美茵河畔法兰克福

结构与材料

STRUCTURE AND MATERIAL

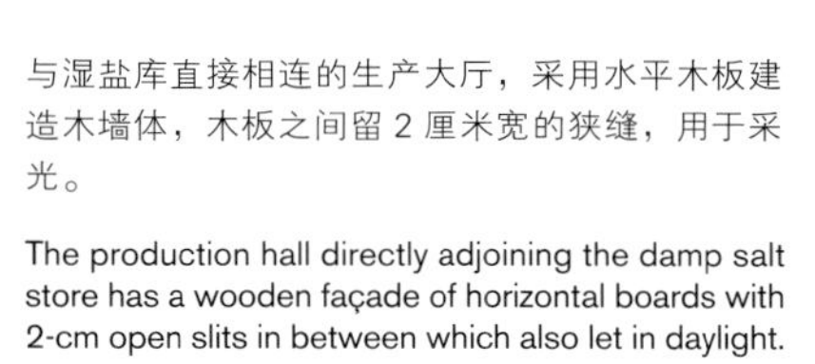

与湿盐库直接相连的生产大厅，采用水平木板建造木墙体，木板之间留 2 厘米宽的狭缝，用于采光。

The production hall directly adjoining the damp salt store has a wooden façade of horizontal boards with 2-cm open slits in between which also let in daylight.

建筑单位	Frick.Reichert Architekten, Frankfurt am Main
业主	Deusa International GmbH, Bleicherode
摄影	Philippa Phaler, Frankfurt am Main
专业工程师 支撑装置设计，设备技术 结构设计，建筑工程管理	Ercosplan, Erfurt
整体承包公司 焊装	HTI Bau Greußen, Greußen
钢结构	Stahl- und Anlagenbau Grüßing GmbH, Kambachsmühle
木结构	Nolte Systembau GmbH, Bischofferode
隔膜	Cenotec GmbH, Greven
窗口	Glasbau Kempf, Bleicherode

这座两层高的办公楼是在原有两座汽车车间基础上后加盖的。它的突出特点是其隔板式玻璃窗外墙设计，墙面设有可外翻小窗，内设新颖独特的遮光装置和整体墙面灯光系统，用于内部照明。自然通风装置与地面冷热系统相连，通过可外翻小窗可达到简单而高效的气候构想。

The two-storey office building is an extension to two older car workshop halls. Its special feature is the cell-glazing façade with hatches opening out and upwards; inlying novel sun-shading (retro technology) mechanism, and a façade-integrated lighting system for illuminating the interior. In conjunction with the floor heating and cooling system, the natural ventilation via open hatches produces a simple, yet highly effective climatic concept.

凯特曼与斯科西格
科隆

摩羯宫　门兴格拉德巴赫

CAPRICORN, MÖNCHENGLADBACH

能源概念

ENERGY CONCEPT

竣工验收	2001 年 11 月
开工时间	2001 年 3 月
总面积	359m²
总体积	1 247m³
关键字	正面墙体上光复旧技术，分别开启盖——外墙体连接，内部照明
制图	维斯特普基金会建筑，2002 年优秀奖

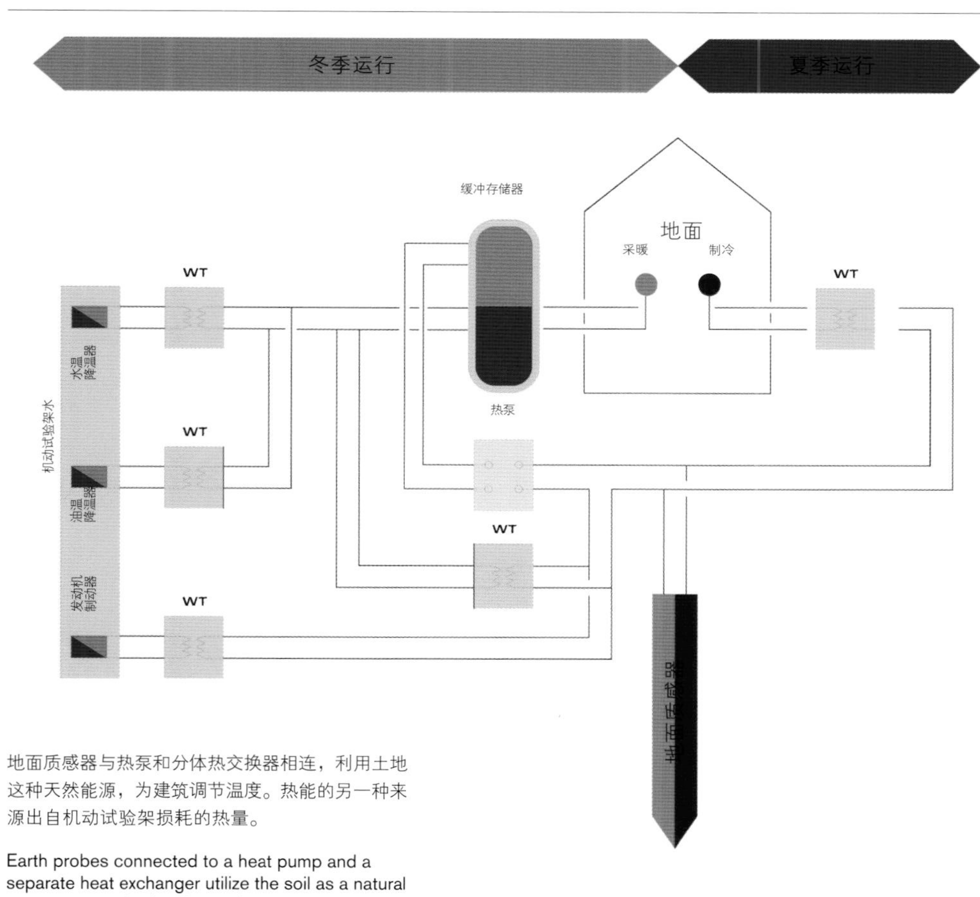

地面质感器与热泵和分体热交换器相连，利用土地这种天然能源，为建筑调节温度。热能的另一种来源出自机动试验架损耗的热量。

Earth probes connected to a heat pump and a separate heat exchanger utilize the soil as a natural energy source for heating and cooling the building. An additional source of heating energy is the waste heat from the motor test rig.

建筑单位	GATERMANN + SCHOSSIG, Köln
业主	Capricorn Engineering GmbH & Co. KG, Mönchengladbach
摄影	M. Milde
技术工程师	
固定	IGH-Ingenieurgesellschaft Höpfner
TGA	H. Schalm GmbH mit Ingenieurbüro Zingsem & Partner

凯特曼与斯科西格
科隆

波鸿市政厅

MUNICIPAL SERVICES BOCHUM

城市位置

URBAN SITUATION

竣工验收	2004 年 12 月
开工时间	2003 年 3 月
总面积	22 073m²
总体积	69 476m³
能量消耗	4 256kWh/m²a
关键字	集成墙面 IntegralFassade
职位	总设计师

市政服务厅的新楼位于波鸿市中心，形成了都市的新标志。6 层高的底座构造的中央，设有中庭，由 60 米高的塔状建筑构成，成为紧邻波鸿环路的显著标志。

This new building of the city's municipal services in the centre of Bochum forms a new urban landmark. The six-storey base structure with an atrium in the middle is complemented by a 60-metre tower which sets a significant sign right next to the Bochumer Ring (road).

建筑蕴涵了多种技术创新。建筑师为了这个工程项目特别设计了新的集成墙面，即 60% 为透明玻璃，40%采用深灰遮光玻璃并装有新式集成遮阳装置（逆向薄片）。除此之外，在墙面形成一体的光源，则体现办公室照明系统的现代化设计的复杂及精致。

The architecture includes a number of technical innovations. The architects developed a new 'integral façade' specially for this project, which consists of 60 % transparent and 40 % dark grey opaque glass with a novel integrated sun-shading mechanism (retro lamellae). In addition, the luminaries integrated in the façade form a contemporary office lighting system with a very intricate and sophisticated design.

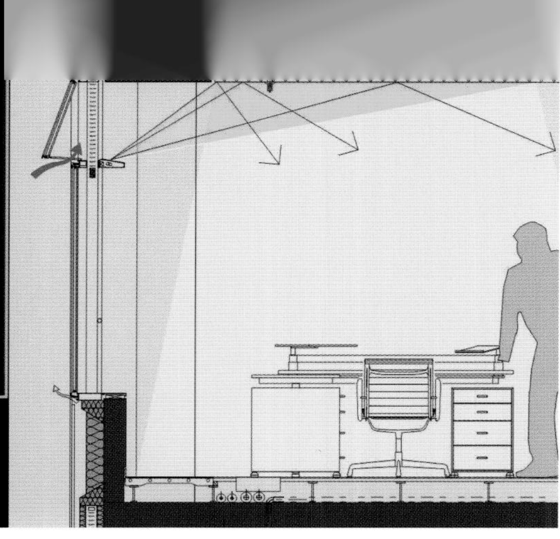

外墙横截面示意图：有效传音建材的供热/制冷面通过新鲜空气墙面活板进入室内——为室内照明的整体光源——光控百叶窗式内部逆光遮阳板——背透气式玻璃女儿墙板。

Schematic cross-section of the façade: cooling/heating floors of acoustically effective material – fresh-air intake via facade flap – integrated luminary for interior lighting – inlying retro sunshade in the form of a light-controlling blind – back-aerated glass parapet panel.

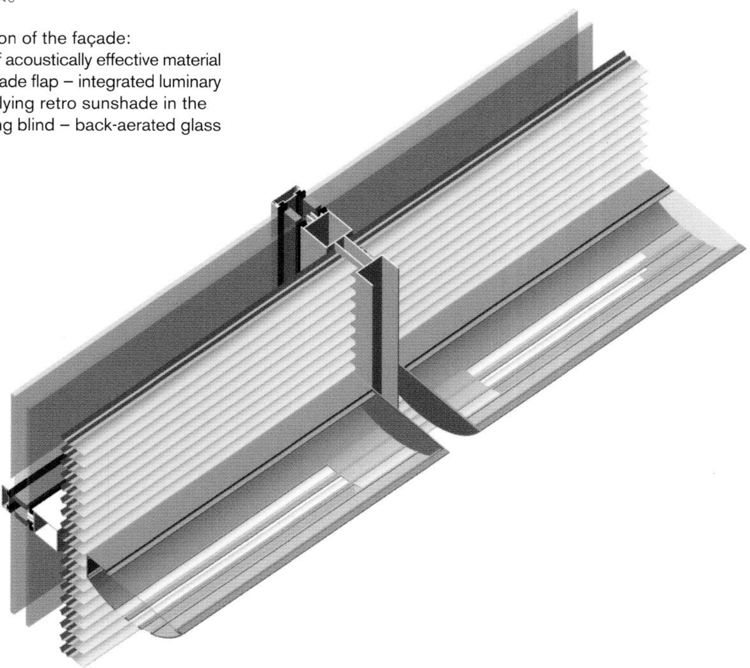

集成墙面：遮阳—日光控制—内部照明—供热—制冷—通风—供能（选择）

IntegralFaçade: sun-shading – daylight control – interior lighting – heating – cooling – ventilation – energy generation (optional).

当遮光系统开动，即使是夏天被动吸热量最低的炎炎烈日，墙面玻璃的透明度也要求很高，鉴于降低制冷能源的消耗，冬天不反光玻璃经曝晒而被动吸热，其采用免维修或低维修率保养技术。

Adequate transparency of the façade, even when sun-shading system is activated – minimal passive heat gains, even under summer sunlight incidence, with a view to reducing cooling energy consumption – passive heat gains from insolation in winter – dazzle-free insolation – maintenance-free/low-maintenance technologies.

凯特曼与斯科西格
科隆

波鸿市政厅

MUNICIPAL SERVICES BOCHUM

外墙设计

FAÇADE DESIGN

从节能标准来看，新式墙体取得很好成效，利用探测器预先采用供热／制冷技术，所以能源消耗可有效地控制在客户所要求的范围内。

Innovative façades achieve very good results in terms of energy savings. In connection with heating and cooling technologies preconditioned by probe operarations, energy consumption can be kept significantly below the figures requested by the client.

建筑单位 GATERMANN + SCHOSSIG, Köln
业主 Stadtwerke Bochum GmbH
摄影 Barbara Staubach, Wiesbaden

技术工程师
固定 Horz + Ladewig
热量分析 Weber & Partner Ingenieurgesellschaft mbH
空闲房间规划 Schümmelfeder Stöcker Partner
照明 Köster + Köster

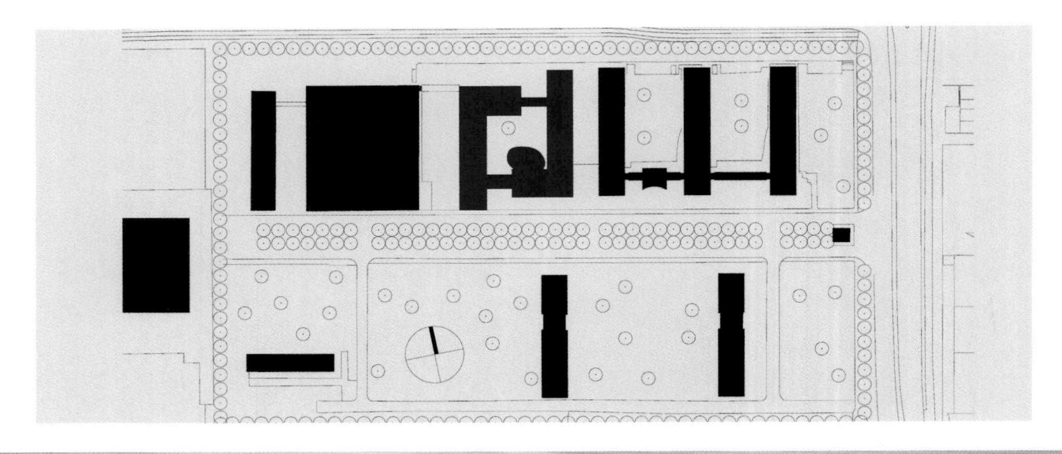

凯特曼与斯科西格
科隆

北莱茵 – 威斯特法伦州杜塞尔多夫市为学校教育及少年儿童所新建的部门

NEW MINISTRY FOR SCHOOL EDUCATION, YOUTH AND CHILDREN OF THE STATE OF NORTH-RHINE WESTPHALIA DÜSSELDORF

结构

STRUCTURE

获奖情况	一等奖
竣工验收	2005 年 5 月
开工时间	2003 年 11 月
总面积	11 117m²
总体积	38 007m³
关键字	集成墙面 通过夜间中心降温器操控 横向通风 联合办公室
职位	总设计师

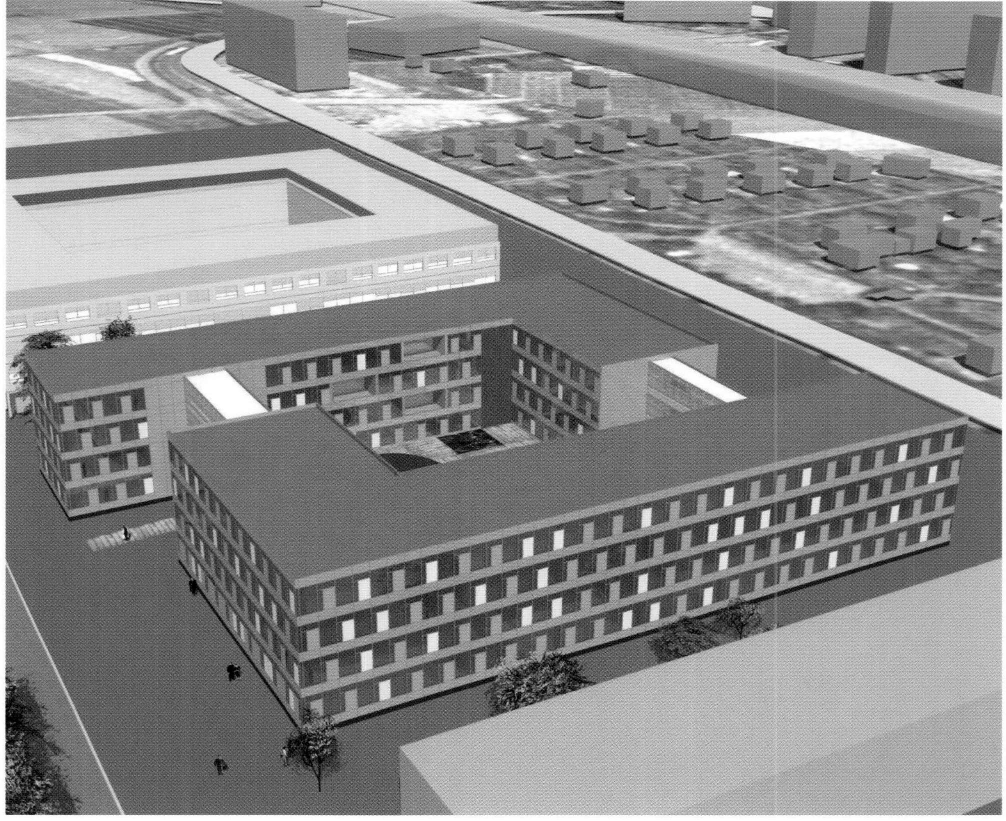

两座 4 层楼高的 L 形建筑，以玻璃通道和内部环形构造错位式相连。中间形成内院。建筑设计目的在于，在中心走廊两边建造办公室开敞式平面办公区或两者的结合。一楼设有中庭会议区，可对外出租，连接两座楼的透明的通道与走廊交汇处，形成多种不同的环形区域，便于欣赏周边景色。

Two four-storey L-shaped buildings, offset against each other, interconnected by glazed access and interior circulation structures to enclose a patio. The buildings are designed to provide offices on both sides of central corridors or open-plan group offices, or combinations of the two. The ground floor accommodates a central conference area which may also be used by external groups. The transparent connecting structures, meeting points extending from the corridors, and loggia terraces all combine to create varied circulation areas offering views of the surroundings.

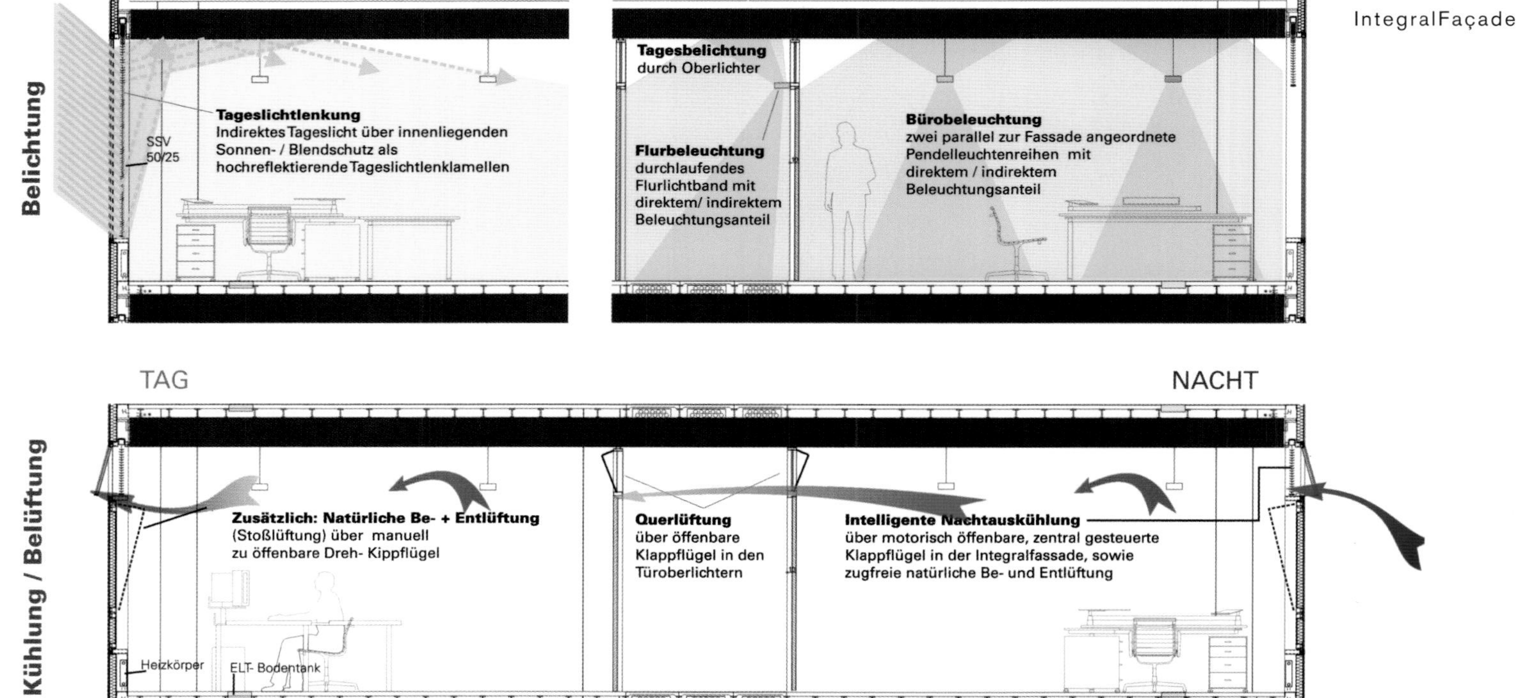

IntegralFaçade

凯特曼与斯科西格
科隆

北莱茵－威斯特法伦州杜塞尔多夫市为
学校教育及少年儿童所新建的部门

NEW MINISTRY FOR
SCHOOL EDUCATION, YOUTH AND
CHILDREN OF THE STATE OF
NORTH-RHINE WESTPHALIA
DÜSSELDORF

外立面

FAÇADE

所有带外墙面的房间通过斜面气窗和上下开关的窗户采用自然通风。夜晚暴露的混凝土顶棚冷却下来，室内日间温度在炎热的夏天可降低 1 ～ 3℃。

All the rooms with outer walls are naturally aerated, through tilting transom windows and sashs. The exposed concrete ceilings cool down over night and therefore reduce interior daytime temperatures in hot summers by 1 to 3° C.

外墙的透明部分、玻璃和光控式遮光系统经过优化可在夏天提供舒适的室内温度，同时保持明亮的室内环境，室外的风景一览无余。

The transparent parts of the façades, the glazing and the light-controlling sun-shades have been optimized so that they provide the most pleasant interior climate possible in the summer while maintaining a light-flooded interior atmosphere and unhindered views.

没有外墙面的楼内办公室配备了具有吸热回收功能的通风系统：提供必要的自然通风。利用现场局部操作网络总线系统，可控制遮光装置。机动开启式窗户可由局部操作网络总线控制，夜晚自动制冷。

The interiors without outer walls are equipped with ventilation systems with heat recovery; these provide the necessary fresh air. The sun-shading mechanism is controlled by means of a LON bus system. For automatic night-time cooling, mechanically openable windows are also controlled by LON bus.

建筑单位
业主 GATERMANN + SCHOSSIG, Köln
摄影 Bau- und Liegenschaftsbetrieb NRW, Niederlassung Düsseldorf

专业工程师
固定和 TGA Brandi IGH
空闲房间规划 club L 94
照明 Brandi IGH
Kress & Adams

戈瓦斯・昆・昆
柏林

玻璃塔　美茵河畔法兰克福

GLASS TOWER, FRANKFURT/MAIN

建筑设计

ARCHITECTURAL DESIGN

建筑成本	约 6 700 万欧元
计划和施工期	2002–2004 年
总面积	59 000m²
总体积	206 500m³
竣工验收	2004 年
占地面积	13 211m²
使用面积	21 040m²（表面）
可出租面积	约 34 450m²（表面）

细长的塔楼高约 80 米，从顶楼向远处望去，法兰克福西区迷人的景色尽收眼底。塔楼采用双层墙面，可起到隔音的作用，且背面的窗户也可敞开。U 形中庭采用可视设计，把楼区的三个单元结合为一体。

From its top at a height of almost 80 m, the slender tower affords a fantastic view over Frankfurt's west. It was clad with a double façade designed to function as a noise barrier, but also to make it possible to open the windows at the rear side. The U-shaped atrium is clearly visible and interconnects all three building sections to form a compact complex.

工程的设计与执行由一些指导思想决定。线性与宽敞空间界限外墙与中庭。背墙面由大块玻璃构成，与优质板条相结合直至楼顶，使内部环境宽敞明亮，通风良好。

Just a few guiding ideas determined the design and its implementation. The façades – and the atrium 'table' are marked by linearity and generous dimensions. The north façade consists of large glazings combined with fine slats up to the top of the tower so that the interior is full of air and light.

新建透明塔楼的整个房屋面积达 35 000 平方米，作为德累斯顿投资银行的总部大楼，它与市中心法兰克福展览中心相对。80 米高的塔楼中央为带有两个活门的空心墙，两侧各建一个低层的楼房，由中庭贯穿 3 个单元楼。明确的设计理念使建筑清晰大方。

New transparent tower with a gross floor area of 35,000 m², located opposite the Frankfurt Fair grounds in the city centre, as the bank headquarters of Dresdner Kleinwort Wasserstein. 80-metre glass tower with two-leaf cavity façade (high-rise) flanked by two lower buildings, with an atrium to connect the three. The determining design ideas were clarity and spaciousness.

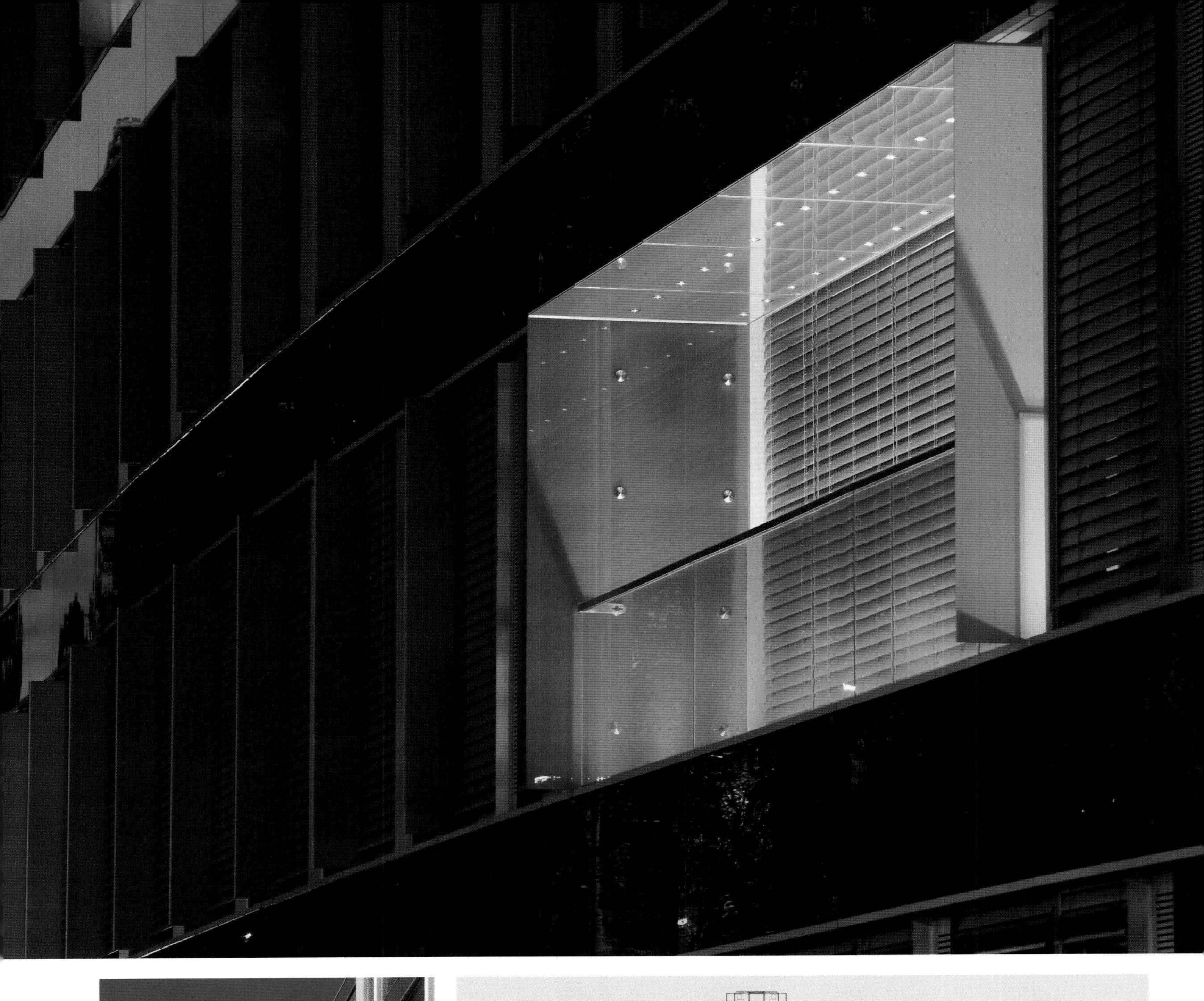

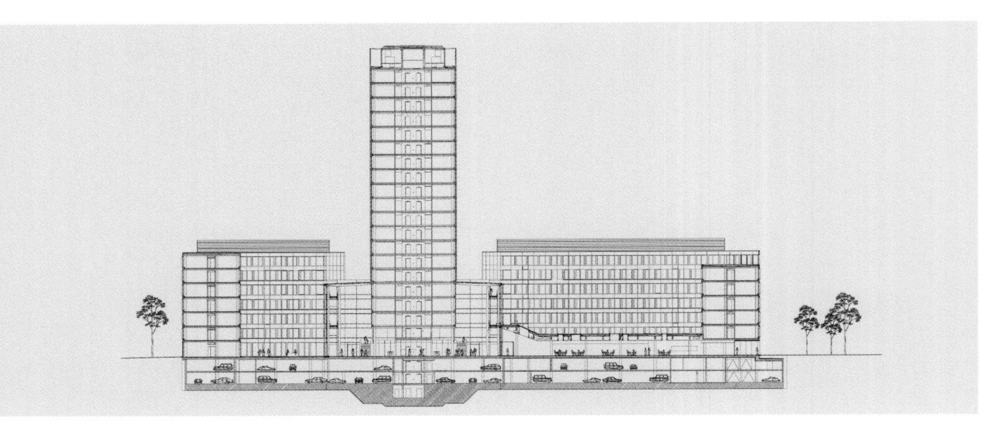

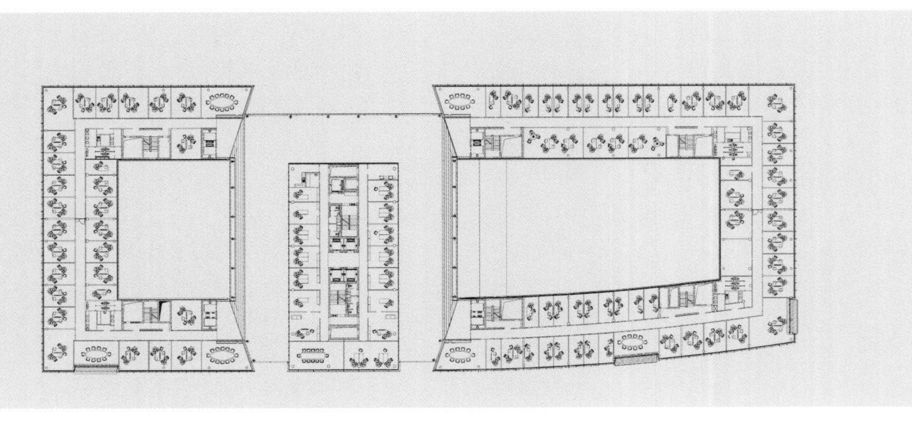

玻璃塔　美茵河畔法兰克福

GLASS TOWER, FRANKFURT/MAIN

戈瓦斯·昆·昆
柏林

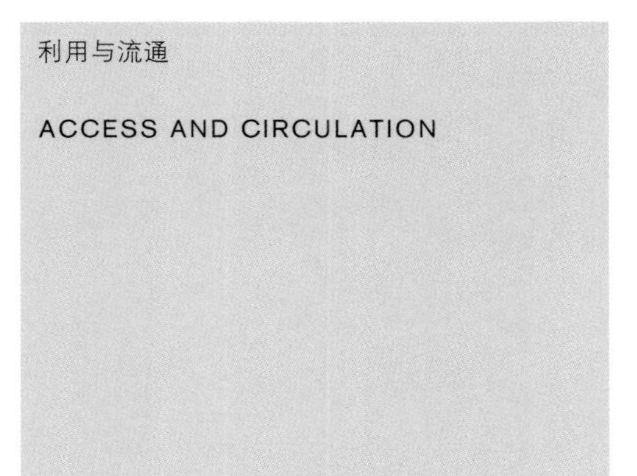

利用与流通

ACCESS AND CIRCULATION

两侧低层楼房内设有娱乐活动室（内部景观以运动为主题）。前方为朝南的凉廊，夜晚在灯光的照耀下十分明亮。

At night the south-facing loggias of the lower buildings, in front of the rec-reational facilities (whose 'inside panorama' designs were based on themes like sports) are brightly illuminated.

中庭的举架很高，光线充足，人们可经此进入大楼。在大楼侧面可经过中庭通向后院和员工餐厅。这是早期建筑师与艺术家共同创作的结果。餐厅明亮的天棚与室内装饰是建筑师与柏林苏普集团共同设计，而北面染色玻璃墙则是与伦敦亚历山大贝雷斯科恩科合作的。

The building is accessed via the high, light-flooded atrium which forms the hallway. On its sides the atrium opens into planted patios and the staff restaurant. The architects collaborated with artists from early on. The illuminated ceiling of the restaurant and the 'inside panoramas' were designed in co-operation with the Berlin group 'Soup', while the northern stained-glass wall was co-designed with Alexander Beleschenko from London.

制图	Gewers Kühn und Kühn Gesellschaft von Architekten mbH, Berlin
艺术	soup berlin, Alexander Beleschenko, London
业主	Dreyer Brettel & Kollegen Management GmbH
接管方	HochTief Construction AG, Frankfurt am Main
摄影	Claus Graubner, Frankfurt am Main, S. 148, 151 mitte soup berlin, S. 151 unten Jens Willebrand, Köln, S. 149, 150, 151 oben
专业工程师 房屋技术	EB Ebert Ingenieure, Nürnberg
景观设计	ST raum a Garten und Landschaftsarchitektur, Berlin
固定	Hartwich / Mertens / Ingenieure, Planungsgesellschaft für Bauwesen mbH, Berlin
电路设计	PBE Beljuli Planunsgesellschaft mbH, Pulheim

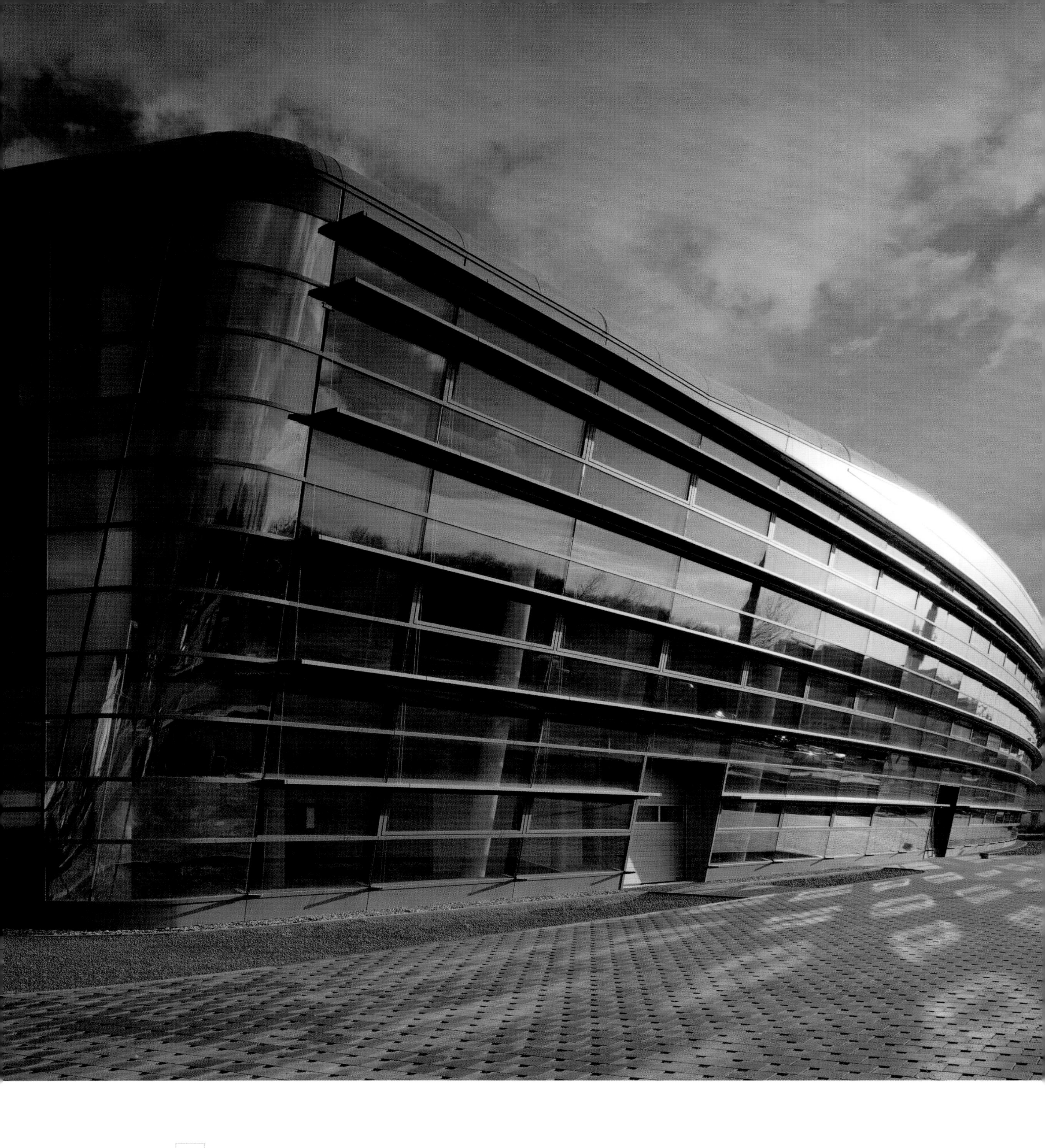

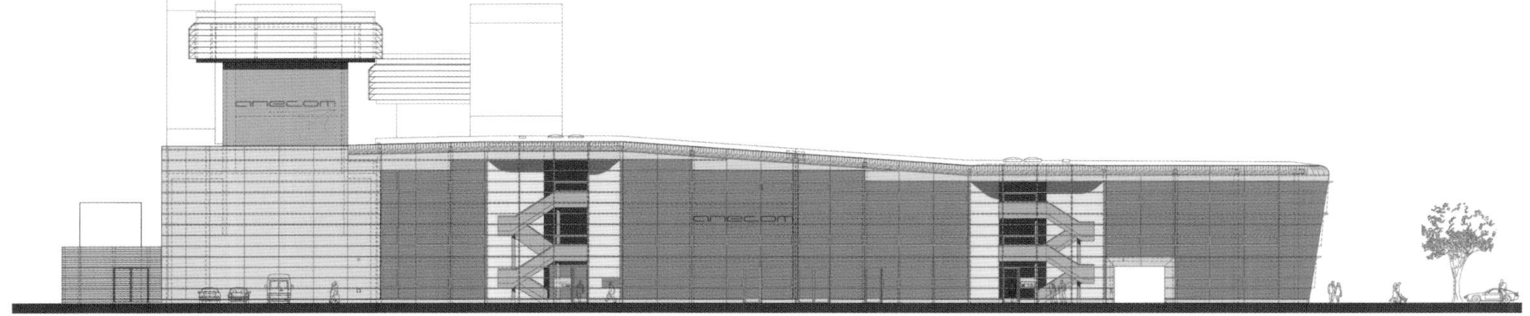
cinecom
cinecom

戈瓦斯·昆·昆
柏林

柏林邻区威尔道航空航天中心

AEROSPACE AND AVIATION CENTRE, WILDAU NEAR BERLIN

工程方案

PROJECT

建筑成本	1 500 万欧元
施工期	18 个月
总面积	8 200m^2
总体积	36 900m^3
竣工验收	2003 年
占地面积	12 790m^2

在过去的百年中，经过不断的拆毁及重建，这个地区已发展成为重要的工业区。这里是新建的职能中心，体现出当今高科技水平的航天工业区。

Over the past century, the area developed into an important industrial centre marked by alternate demolition and reconstruction. Here, the new competence centre sets a sign for today's high technological standards of aircraft engineering.

建筑设计把航空工业及飞行事业以楼形和墙面的形式体现出来。玻璃及铝材构成的外形似乎是模仿了风道的轮廓。为航空技术发展新建的职能中心将使威尔道老工业区复兴繁荣。

The architectural design translates the language of the aircraft industry and aviation into form and façade. The contours of the building of glass and aluminium seem to have been taken from a wind tunnel. The new competence centre for aviation technology will contribute further to revitalizing the former industrial location of Wildau.

位于柏林南区的新职能中心用于航空航天技术的研究与发展，为此设计出涡轮实验台、生产设备办公区及培训中心。东面，楼区与铁路线相邻，所以具有噪音的实验台和生产设备安置在此。为楼区西侧的科学大学工程专业学生所建的主要通道、门厅到明亮的办公室和培训区都可俯瞰公园。

The new centre for aviation and aerospace technological research and development south of Berlin is equipped with turbine test rigs, production facilities, offices and a training centre. To the east, the complex borders on a city railway line so that all noisy testing and production facilities were placed here. The main access and entrance, light-filled offices and the training facilities for the engineering students of the University of Applied Sciences in the western section of the complex overlook a public park.

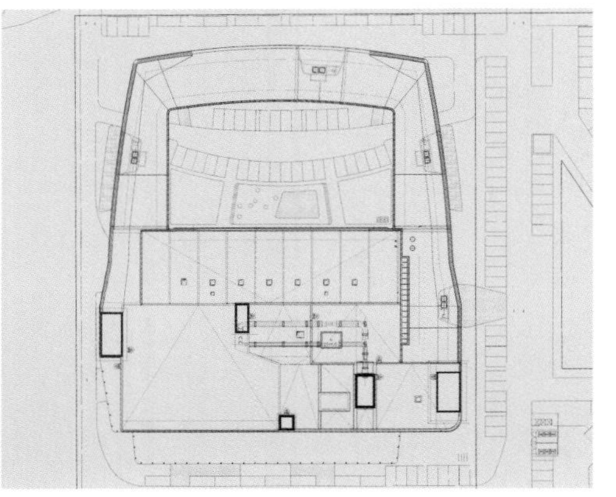

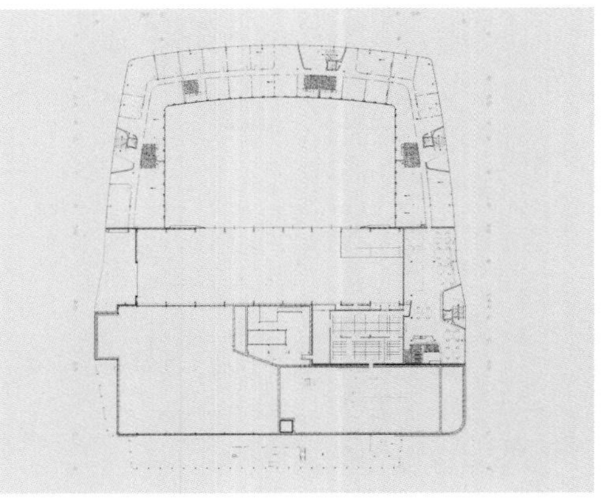

办公楼与工厂围成一个大面积内院，内部空间与环庭相连，以促进工作的合作交流及效率，按照楼体的形状、类型及建材来设计内院，使整个建筑更具活力。融合科技与自然，为航天技术而新建成的职能中心在面向未来发展的方向上可作为代表性建筑。

The offices and workshops surround a spacious interior courtyard. All the interior spaces are connected by circular passages which facilitate efficient, co-operative and communicative work processes. The courtyard design was based on the same shape, style and materials as those of the building mass and geared to enlivening the structure. With its combination of technology and nature, the new competence centre for aviation technology presents itself as an example of a future-oriented architecture.

戈瓦斯·昆·昆
柏林

柏林邻区威尔道航空航天中心

AEROSPACE AND AVIATION CENTRE, WILDAU NEAR BERLIN

展望未来的建筑

FUTURE-ORIENTED ARCHITECTURE

建筑的设计以统一方式提供了特殊功能及所需的内部构造，目的在于建出高质量的并具有影响力的楼房，从建筑的外形玻璃和铝材构成的外墙以至楼房的各个细节来看，无不体现航空技术及空气动力学的高科技特点。

The design accommodates the specific functions and provides the required interior organization in a unified gesture, a building of high architectural quality and expressive power. The high-tech character of aviation technology and aerodynamics is echoed by the shape of the building mass, the glass-and-aluminium façade as well as by every detail.

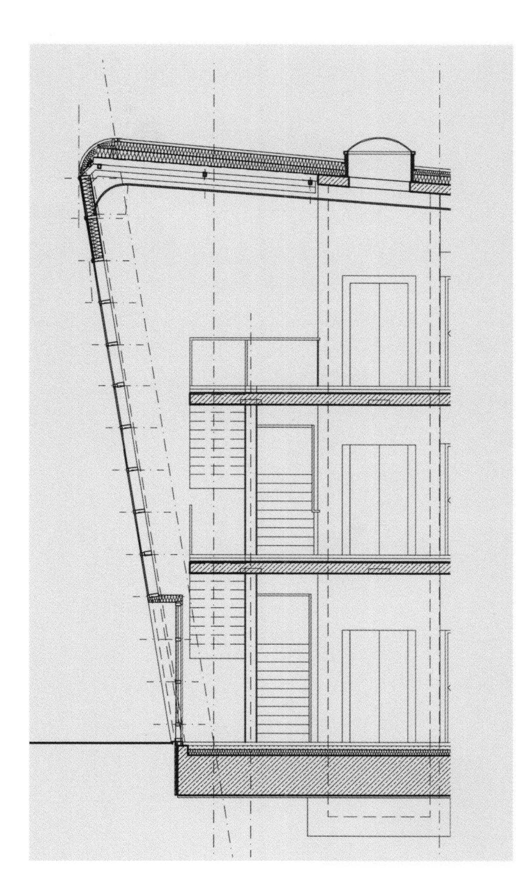

建筑设计 Gewers Kühn und Kühn
Gesellschaft von Architekten mbH, Berlin
施工单位 WFG Regionale Wirtschaftsförderungsgesellschaft
Dahme-Spreewald mbH
用户 Anecom Aero Test GmbH
摄影 Ralph Baiker, Berlin, S. 152, 153
Eberle & Eisfeld, Berlin, S. 154 oben, 155
GKK, S. 154 unten

新建大楼选址在威斯巴登的郊区，先前是机器制造工厂，现在是一片荒地，同样地经历了动态变化。从城市重建的意义来说，整个市区需要再次发展。城市规划采用花园式城市理念，按此要求设计了新储蓄银行保险综合楼。建筑包括不同的区域，区域之间由楼梯、电梯、通道及商业街相连。

The site chosen for the new building on the outskirts of Wiesbaden was a wasteland left by a former machine-building factory and as such subject to dynamic changes. The entire urban area required redevelopment in the sense of urban repair. The urban project followed the garden city idea and the new savings bank insurance complex was designed accordingly. The building consists of different modules interlinked by stairways and lifts, and by a passage or mall.

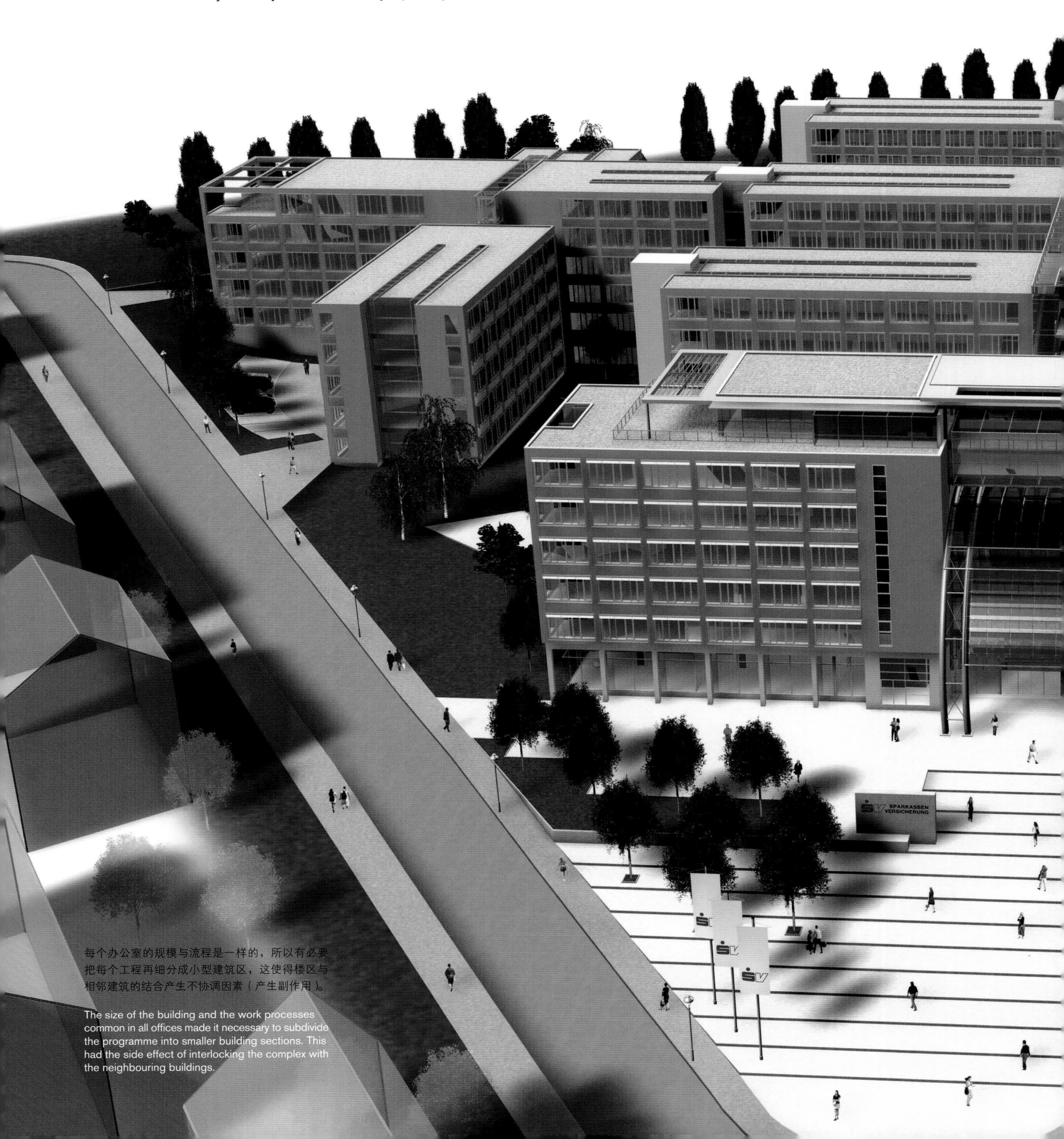

每个办公室的规模与流程是一样的，所以有必要把每个工程再细分成小型建筑区，这使得楼区与相邻建筑的结合产生不协调因素（产生副作用）。

The size of the building and the work processes common in all offices made it necessary to subdivide the programme into smaller building sections. This had the side effect of interlocking the complex with the neighbouring buildings.

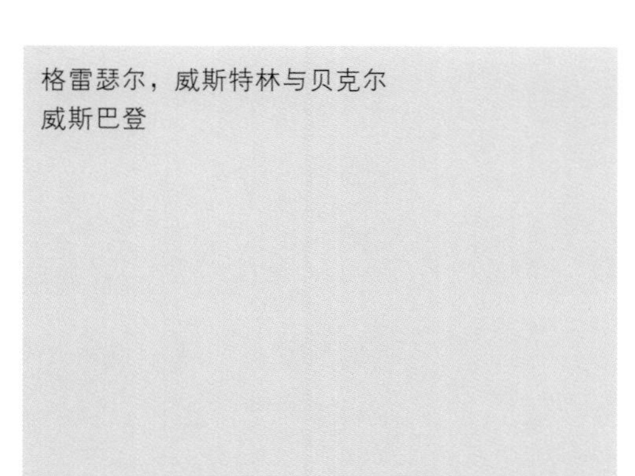

格雷瑟尔，威斯特林与贝克尔
威斯巴登

储蓄银行保险总部
威斯巴登

SAVINGS BANK INSURANCE
HEADQUARTERS, WIESBADEN

建筑成本	8 000 万欧元
总面积	60 419m^2
总体积	238 365m^3
占地面积	28 470m^2
使用面积	31 612m^2

位于街道交叉口与广场的建筑及其出入口常常依靠凸肚窗、塔楼和曲线形街角来装饰。

Buildings at street crossings and squares – and similarly the entrances – are often marked by special features like oriel windows, turrets or canted corners.

为了使建筑、设施与城市环境、地形形成一体，建筑群要符合基地的最高群，而在鲁道夫大道与卡尔—温—林得大道以对角建楼也是处于这样的考虑。建筑群的末端只能从倾斜的角度望见，所以从街道望去，楼群看起来比实际规模要小，且跃过中间的绿化带，与邻边的居民区形成一体。

To integrate it with its urban context and topography, the complex was adapted to the contours of the site. The same purpose was pursued by arranging them on a diagonal along Rudolfstrasse and Carl-von-Linde-Strasse. The ends of the building sections can only be perceived from an oblique perspective so that, seen from the street, the complex appears smaller than it really is and seems to enter into dialogue with the adjacent residential buildings across intermediate green areas.

储蓄银行保险总部
威斯巴登

SAVINGS BANK INSURANCE HEADQUARTERS, WIESBADEN

格雷瑟尔，威斯特林与贝克尔
威斯巴登

除了街道的十字路口作为门口的路标，西面的小型公园也成为楼区的又一个突出的特点。

Apart from the street crossing as a 'sign-posting' entrance situation, a 'pocket park' on the west side forms a second identifying feature of the complex.

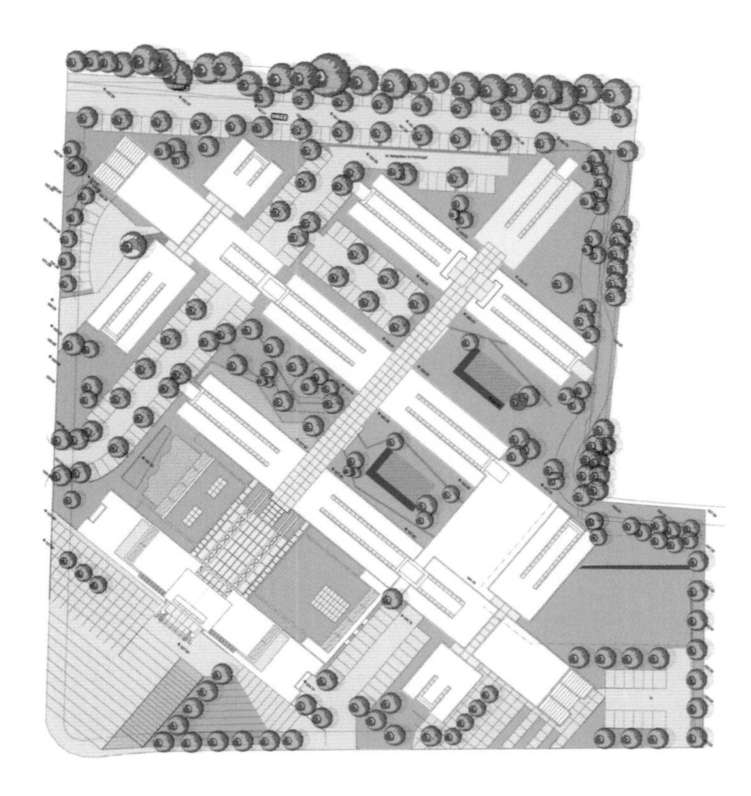

施工单位	Gresser, Vesterling + Becker Planungsgesellschaft, Wiesbaden
业主	SparkassenVersicherung, Wiesbaden
三维图像	pinx. Design-Büro, Wiesbaden
技术工程师	
静力学	Dipl.-Ing. Hartwig Euler, Hanau
通风 / 采暖	Ingenieurgesellschaft Abels · Habicht, Mühltal/Traisa
金属外墙	a + f – Architektur und Fassadenplanung, Frankfurt am Main

格鲁勃与克雷尼·柯恩勃格
美茵河畔法兰克福

工程工会 IG Metall 的大楼主体
美茵河畔法兰克福

MAIN FORUM – IG METALL TRADE UNION HIGH-RISE, FRANKFURT/MAIN

设计与结构的概念

DESIGN AND STRUCTURAL CONCEPT

施工期	2000–2003 年
建筑成本	12 500 万欧元
竣工验收	2003 年
高度	80m
总面积	60 430m^2
面积	35 000m^2
总体积	244 630m^3

法兰克福市区外围的建筑与周边环境的相互作用表现在高层垂直板层，水平穿插进整个建筑。

The interaction between this building at the periphery of Frankfurt's inner city and its urban surroundings finds expression in a high-rise slab – the vertical element – inserted into the horizontally marked environment of stone.

建筑区的规划（低层建筑环绕高层建筑），使外广场与内中庭显出其重要性。坛式建筑是工会与公众联系的重要场所，而中庭的作用主要在于为楼内提供各项功能。建筑内部，可直观地感受到平行与垂直结构。

The arrangement of the building masses (the high-rise being 'set' in a 'ring' of lower buildings) made it possible to create the outer 'Forum' and the inner 'atrium' as the most important spaces. The Forum is an essential instrument of the trade union's public relations while the atrium is oriented to and focuses on the interior functions of the building. In this place, the transition between the horizontal and the vertical can be experienced visually.

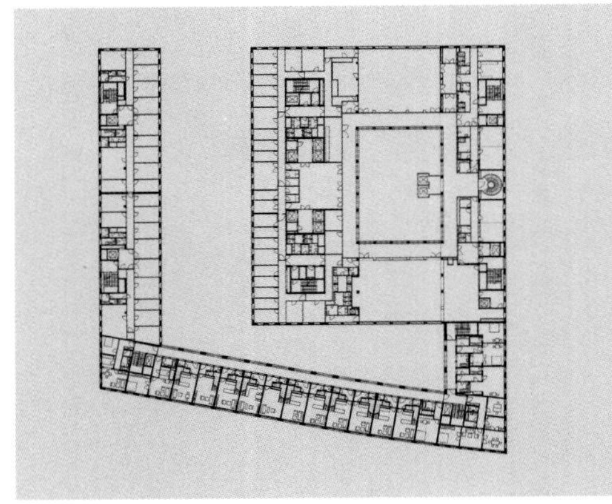

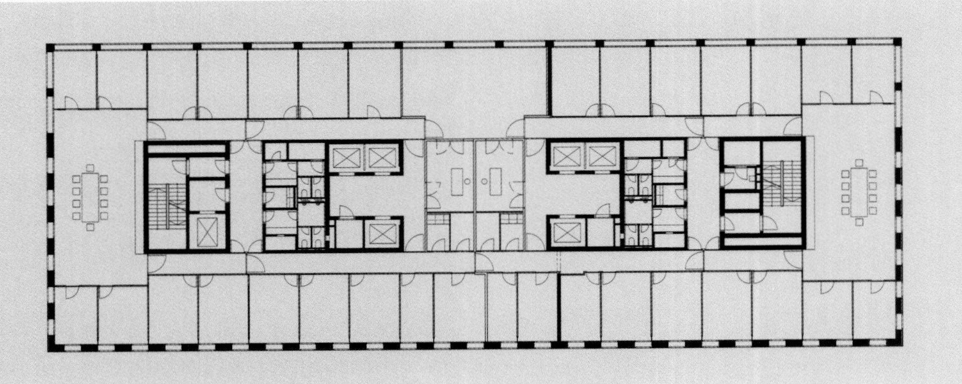

坛状主楼附有明显的生态学特点，因为它利用主楼水源来替代对吸热的机械冷却，节省大量原始能量。除此之外，土地深层的冷冻能力经基础桩也可用来制冷，所有这些都代表着建筑生态质量的潜能。

The Main Forum building sets a sign to an ecological future: as it utilizes water from the Main as a substitute for the mechanical cooling of heat gains, it saves a considerable amount of primary energy. In addition, the refrigeration power of deeper soil stratas is also utilized for cooling via the foundation piles. All these measures represent a substantial contribution to the overall ecological quality of the building.

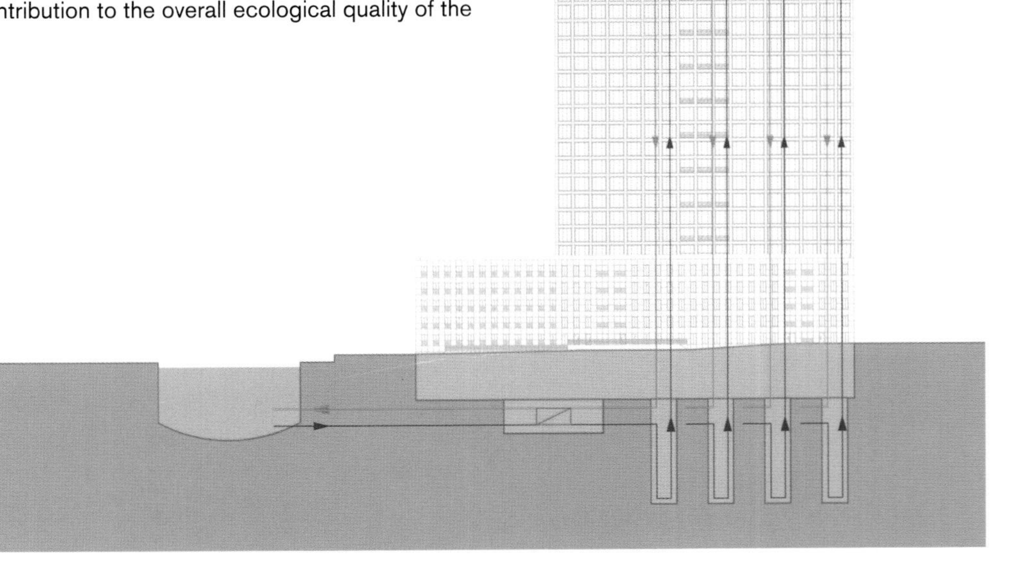

地基桩起着能源输送管道的作用。在炎热的夏天，存在地下的冷能可以导出。

The foundation piles serve as 'energy ducts': in hot summer months, the cooling energy 'parked' in the sub-soil, so to speak, can be tapped.

工程工会 IG Metall 的大楼主体
美茵河畔法兰克福

MAIN FORUM – IG METALL TRADE UNION HIGH-RISE, FRANKFURT/MAIN

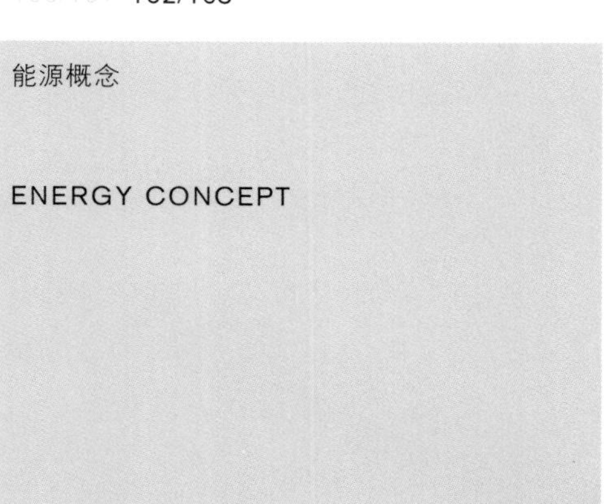

能源概念

ENERGY CONCEPT

巨大的混凝土建筑也可在夜间储存冷气，这样在白天就可用作制冷肋材，为室内提供舒适温度。

The enormous concrete masses also store the coolness of the night air and thus, during the day, function as 'cooling ribs' which create a pleasant interior climate.

坛式楼的风景：工会延展区的重要构成。

View into the 'Forum' – an important element in the trade union's public outreach.

建筑设计	Gruber + Kleine-Kraneburg, Frankfurt/Main
施工单位	Treuhandverwaltung IGEMET GmbH, Frankfurt/Main
工程控制	Gründer und Partner, Ingenieurges. GbR, Hamburg, Projektbüro Frankfurt/Main
摄影	Stefan Müller, Berlin
能源方案草图第 162 页	schaafhausen grafik-design, Frankfurt/Main
技术工程师	
建筑管理行列中心论坛	Gruber + Kleine-Kraneburg, Frankfurt/Main
塔吊	BGS Ingenieurgesellschaft mbH, Frankfurt/Main
HLS	Schmidt Reuter Partner, Offenbach
电工	Reuter Rührgartner, Rosbach v.d.H.
建筑物理	Graner + Partner, Bergisch Gladbach
外墙	IFFT Institut für Fassadentechnik
整体承包公司	
建筑施工	HOCHTIEF CONSTRUCTION AG, Frankfurt/Main

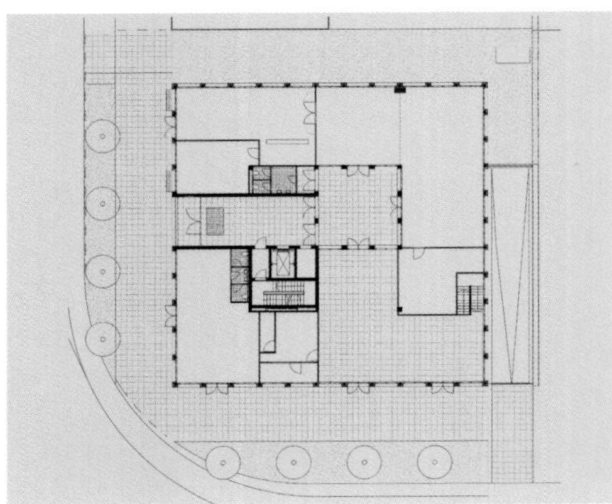

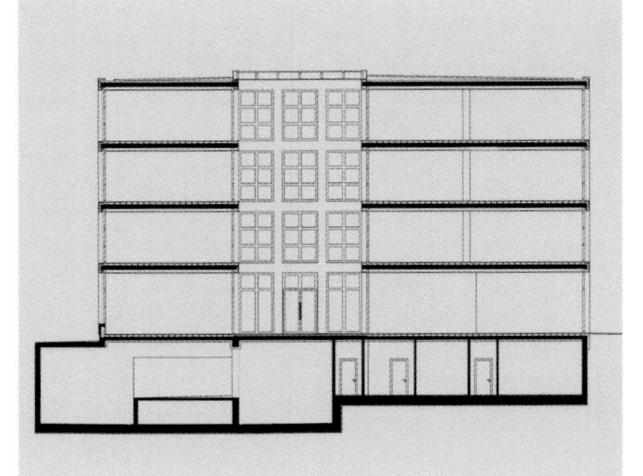

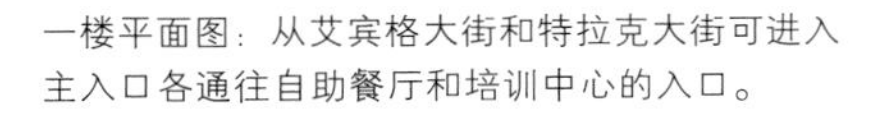

一楼平面图：从艾宾格大街和特拉克大街可进入主入口各通往自助餐厅和培训中心的入口。

Ground floor plan: The main entrances as well as the entrances to the cafeteria and the training centre are accessed from Elbinger Strasse and Trakehner Strasse.

林克大楼是一座4层独立方块建筑，位于法兰克福市豪森区居住区与工业区过渡地段。在节能与土地的有效利用方面，它是建筑的典范。其高质量且灵活的可进行自然采光及通风空间围绕着一个带有玻璃棚顶的中庭，中庭担当气候缓冲区的作用。因为外墙具有热存储能力，又使其成为建筑突出的生态要素。

'Rinck House' is a free-standing four-storey cube in the Frankfurt borough of Hausen and marks the transition from residential to industrial urban areas. In terms of efficient use of energy and area, it represents an important architectural contribution. Its high-quality and flexible, naturally lit and ventilated spaces sur-round an atrium with a glass roof which acts as a climatic buffer zone. The façade is another ecological element as it has a large thermal storage capacity.

林克大厦
美茵河畔法兰克福

RINCK HOUSE
FRANKFURT/MAIN

格鲁勃与克雷尼・柯恩勃格
美茵河畔法兰克福

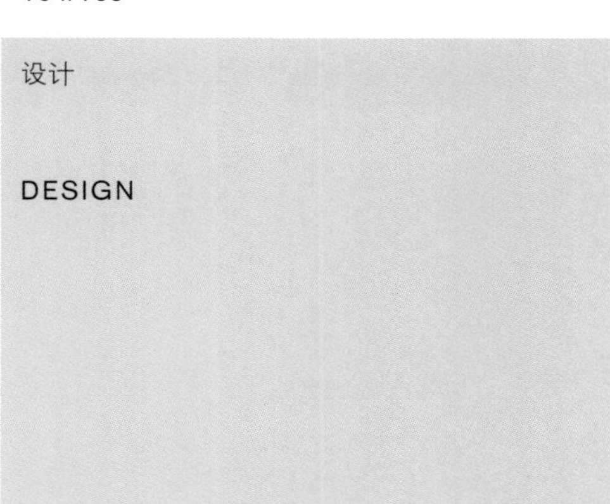

施工期 2001–2002 年
建筑成本 600 万欧元 / 年
竣工验收 2002 年
高度 15m
总面积 2 989m²
净地表面积 580m²
使用面积 2 270m²

明亮的外墙使独立方形建筑的特点尤为突出，它由黑石的墙面支撑，木框落地窗被漆成深色。

The clear-cut façade enhances the signature character of the free-standing building cube. This is supported by the dark stone cladding of the façade and the articulation of the floor-to-ceiling windows with wooden frames varnished in dark hues.

西北方向：一座两层楼的仓库从北墙延伸出来。

North-west view: A two-storey warehouse projects from the north façade.

建筑设计 Gruber + Kleine-Kraneburg, Frankfurt/Main
施工单位 Rinck + Co. KG, Frankfurt/Main
工程控制 Ing.-Büro für Hochbau Thomas Schleemilch, Weilrod
摄影 Stefan Müller, Berlin

专业工程师
塔吊 Fischer & Werle, Frankfurt/Main
HLS GfG H. Brandl, Seligenstadt
电子 IAS Gerhard F. Stefan, Södel/Wetterau
建筑物理 Fischer & Werle, Frankfurt/Main
外墙设计 Gruber + Kleine-Kraneburg, Frankfurt/Main

费迪南德海德建筑师事务所
美茵河畔法兰克福

BLFP–布瑞莫劳伦兹教授
弗里林豪斯建筑规化有限责任
公司佛莱德伯格

N49 办公区
美茵河畔法兰克福

OFFICE BLOCK N 49
FRANKFURT/MAIN

优化办公性能

OPTIMIZED OFFICE PROPERTY

建筑成本	1 930 万欧元（KGr. 300–500）
	893 万欧元 /m^2 总面积
总面积	17 400m^2 上部
	4 200m^2 下部
总体积	57 420m^3 上部
	12 600m^3 下部
使用面积	14 790m^2
楼体面积	60m × 50m
楼体高度	25.80m
竣工验收	2003 年
占地面积	7 763m^2

建筑的清晰外形给人留下深刻印象，楼体经过精心设计的结构，保证了建筑的持续性和耐久性。它的模块设置及直体结构都是从经济及功能方面考虑的。

The building impresses with its clear architectural profile. Its carefully designed details promise sustainability and longevity, and its modular configuration and straightforward structural system stand for economic and functional efficiency.

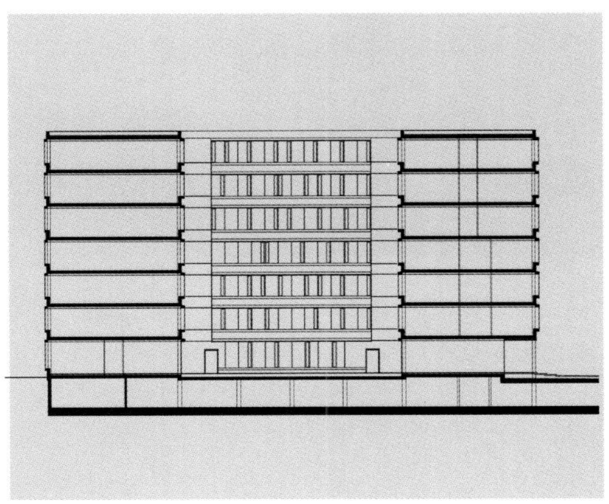

N49是优化办公区。从功能、资金及城市规划角度看，它符合全方位的需求。清晰的结构是城市规划的重要元素，并占据城市的枢纽位置。建筑紧凑的结构使它的地理位置更为突出。从资金角度看，它的整个计划用地的面积与可供出租区域、外墙及内部空间的比例都是合理的。所有平面布置、入口及交通流域的概念从经济和功能方面得到加强。

N 49 is an optimized office block which, in terms of function, finances, architecture and urban design meets a number of requirements all at once. Its clear-cut form is a significant urban design element and a pivotal point on the urban map. With its compactness, it pin-points and creates a 'good location'. In financial terms, its ratio of gross plan area and rentable floor space, as well as the ratio of outer façade area and interior volume, are sound. Both floor plan arrangements, and access and circulation concepts have all been optimized, economically and functionally.

四个入口和环形中心及两个门口大厅为出入提供极大的方便，其特点还在于可以把 190 多平方米的空间划分为可租用的小型办公单元或包括小间及大办公室的大单元或把两者结合起来的办公结构。建筑总面积的93% 用于租赁，说明了楼层利用的有效性。

Four access and circulation cores and two entrance halls offer utmost flexibility. It is possible to partition smaller rental office suites from 190 m^2 upwards or to create larger ones containing cellular and large open-plan offices, and combinations of the two. 93% of the gross floor area are rental spaces – and proof of the efficient floor plans.

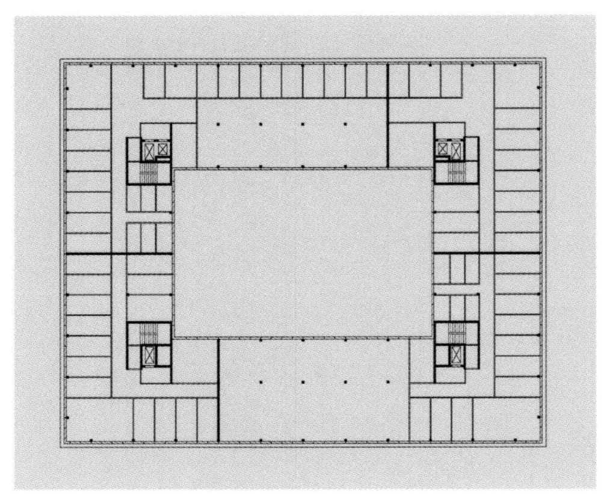

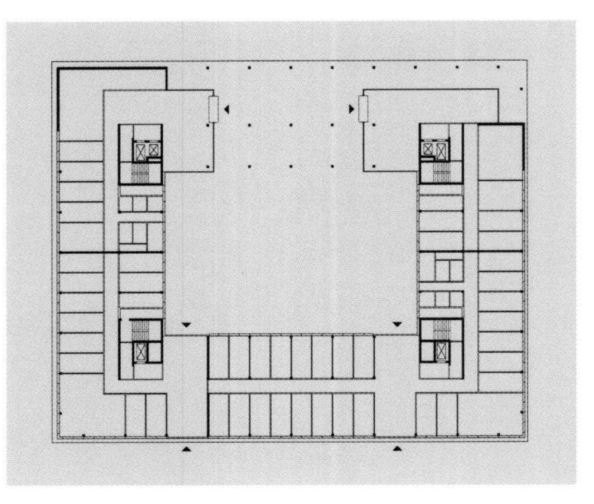

费迪南德海德建筑师事务所
美茵河畔法兰克福

BLFP- 布瑞莫劳伦兹教授
弗里林豪斯建筑规划有限责任公司
佛莱德佰格

N49 办公区
美茵河畔法兰克福

OFFICE BLOCK N 49
FRANKFURT/MAIN

结构效率

STRUCTURAL EFFICIENCY

施工单位 Niederrad N 49 Immobilien GmbH, Bad Hersfeld
建筑设计 Arbeitsgemeinschaft:
Ferdinand Heide Architekt BDA, Frankfurt am Main
Bremmer, Lorenz, Frielinghaus Planungsgesellschaft mbH, Friedberg
摄影 Christoph Kraneburg, Köln/Darmstadt

专业工程师
大厦所有者 HKI Bauunternehmung, Bad Hersfeld
塔吊安装 S.A.N. Stöffler Abraham Neujahr GmbH für Tragwerksplanung
房屋技术 IBK Ingenieurbüro H. & A. Klöffel, Bruchköbel

海恩建筑设计事务所
慕尼黑 柏林

建筑成本	约 1215 万欧元
计划和施工期	1999–2001 年
总面积	81 600m²
总体积	530 000m³
竣工验收	2001 年
使用面积	43 000m²

德国大众汽车的透明工厂
德累斯顿

VOLKSWAGEN'S 'GLÄSERNE MANUFAKTUR', OR SEE-THROUGH MANUFACTORY, DRESDEN

视觉与创新

VISION AND INNOVATION

Baukosten	121,5 Mio. Euro
Bauzeit	1999–2001
Bruttogeschossfläche (BGF)	81 600 m²
Bruttorauminhalt (BRI)	530 000 m³
Fertigstellung	2001
Hauptnutzfläche	43 000 m²
Leistungsphasen	HOAI 1–8 Generalplaner

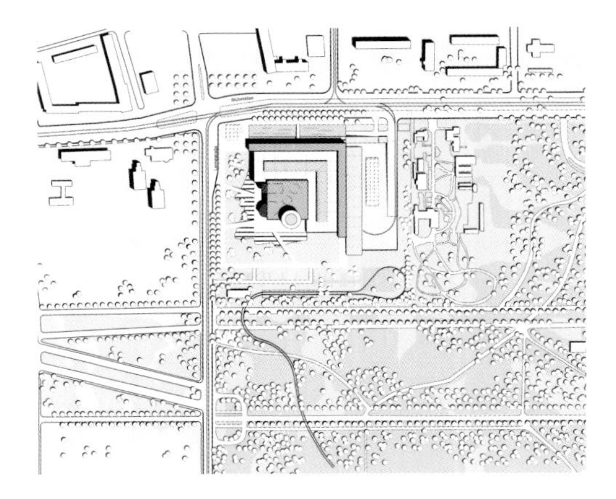

5 000 平方米的户外场地顺应了城市的示范式规划，工厂位于斯特拉斯堡广场，与城市植物园相距约 100 米，总共投资约 600 万欧元。从远处可看到工厂的标志性建筑——40 米高的玻璃塔，所有下线待取的汽车存放于此。

The 50,000 m^2 of open area were adapted to the city in exemplary fashion. These measures cost about 6 million euros for the factory site on Strassburger Platz, about 100 metres away from the Botanical Gardens. The landmark of the factory, visible from afar, is a glass tower 40 m high containing the cars waiting for their buyers to come and fetch them.

2001 年 12 月，位于德雷斯顿市中心边区的工业透明玻璃工业大楼投入运营，大众汽车新型豪华轿车辉腾在宽敞的大厅内进行组装，等待客户提货。厂址所在的位置——斯特拉斯堡广场，与城市植物园（大花园）相邻。广场原是展览及外贸交易场所，是极富幻想及创新的地方。

December 2001 saw the opening of this industrial glass building at the edge of Dresden's city centre. In its spacious halls, the new VW Phaeton luxury limousines are assembled and handed over to customers. The site on Strassburger Platz, in the immediate vicinity of the Grosser Garten (great garden), is a former exhibition and trade fair ground that has always been a place of visions and innovations.

与其他公司的传统销售渠道不同，玻璃工厂把汽车的生产变成一种体验和一道引人入胜的风景。外墙由透明玻璃构成。来访者及过路人可看到明亮的生产车间内世界知名汽车的手工装配过程。汽车零部件由特制的运货电车送往工厂装配，电车在生产区电车轨道上行驶。

Unlike other conventional sales outlets, the Gläserne Manufaktur turns not only the handing-over of the cars into an experience and attraction, but also their production. As the façades are made of clear glass, visitors and passers-by are able to watch the assembly by hand of the prestigious automobiles in brightly lit, studio-type production halls from outside. The parts are delivered to the assembly line by a custom-built CargoTram which uses the urban tramways.

海思建筑设计事务所
慕尼黑　柏林

德国大众汽车的透明工厂
德累斯顿

VOLKSWAGEN'S 'GLÄSERNE MANUFAKTUR', OR SEE-THROUGH MANUFACTORY, DRESDEN

透明

TRANSPARENCY

在运往生产线之前，汽车车体放置在车体仓库的橱窗内。完成组装的汽车在交货之前存放在同样透明的陈列窗式高楼内。

Before being moved to the assembly line, the car bodies are displayed in the 'shop window' of the bodywork storage hall. The finished cars are then displayed in the equally transparent high-bay showcase tower before they are handed over to their buyers.

生产区与销售区之间设有展览厅、技术模型厅、餐厅和快餐店。它是购车之行的最后一站，为买主及德累斯顿市民提供服务。整个建筑的设计成为全市的大事，也是特殊的公共市场。

Displays, technical simulations, restaurants and snackbars are arranged in the spaces between production and sales areas. They round off the 'car-purchasing experience', not only for the buyers, but also for the citizens of Dresden. The whole project was designed as an urban event and a specialized public market place.

透明工厂是集多种用途于一体的聚集地，它继承了历史古城德累斯顿的文化传统并且赋予其现代气息。这座新标志性建筑为德累斯顿市增添了光彩。

The Gläserne Manufaktur represents a venue for different events which uses Dresden's cultural heritage of historic monuments and extends it by a decidedly contemporary element. The city is thus enriched by a new identifying place.

2002年10月26日晚，乔治・比才的经典歌剧《卡门》在德累斯顿市的透明工厂进行首场演出。

Georges Bizet's opera 'Carmen' performed in Dresden's Gläserne Manufaktur; first night on 26 October 2002

有玻璃外墙的大众汽车生产厂商是第一个安装由传统的汽车制造技术和手工工艺相结合的生产线的厂家。

With its ‘see-through factory’, Volkswagen is the first car manufacturer to have installed an assembly line which combines the classical industrial production process with manufactory-type work.

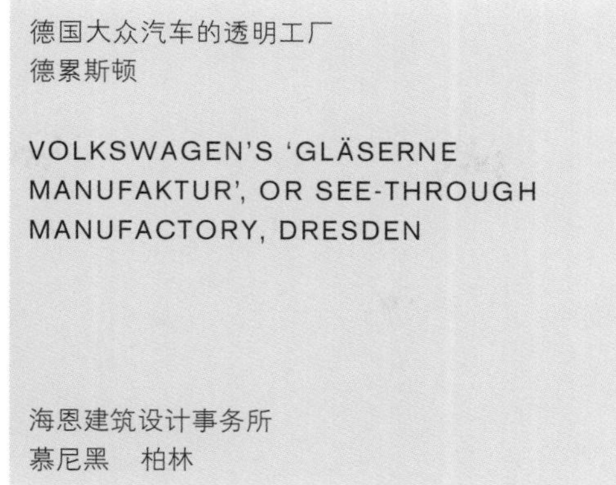

德国大众汽车的透明工厂
德累斯顿

VOLKSWAGEN’S ‘GLÄSERNE MANUFAKTUR’, OR SEE-THROUGH MANUFACTORY, DRESDEN

海恩建筑设计事务所
慕尼黑　柏林

工业与文化

INDUSTRY AND CULTURE

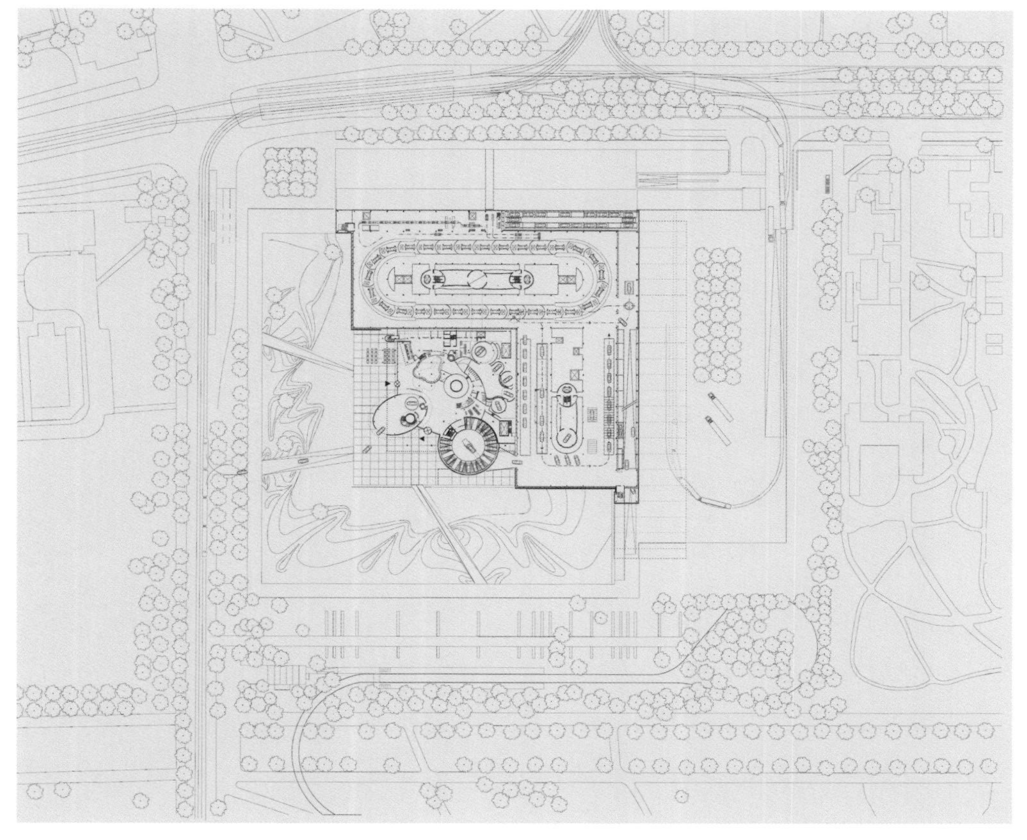

由于透明工厂将技术与文化相结合，它成为重新审视德累斯顿市的催化剂。同时，它展示了不同的城市特点。德累斯顿市在整个工业区所起的作用、富含文化气息的博物馆和展览馆及以公园和庭园为特点的花园式古城特点保持至今。

By combining technology and culture, the Gläserne Manufaktur may serve as a catalyst for a new way of perceiving the city of Dresden. At the same time, it addresses different aspects of the city’s identity: its role as an integrative industrial location; its rich culture of museums and exhibitions; its historic garden-city character with many parks and gardens still preserved today.

玻璃外墙和 24 000 平方米的拼花地板使内部环境宽敞明亮，反映出生产过程创新的基本理念，即采用精细的手工装配来弥补工业生产过程的不足。

The glass façades and the 24 000 m^2 of parquetry floors create light-flooded interior atmospheres which reflect the basic innovative philosophy of a manufacturing process that supplements industrial production processes with careful manual assembly.

建筑设计	Henn Architekten, München, Berlin
施工单位	Volkswagen AG, Wolfsburg
摄影	H. G. Esch, Hennef-Stadt Blankenberg, S. 173 unten, 174, 175 unten Werner Huthmacher, Berlin, S. 170/171, 172, 173 oben, 175 oben
专业工程师	
规划房屋技术 HLS	Heinze, Stockfisch, Grabis und Partner, Hamburg
塔吊	Leonhardt, Andrä und Partner, Berlin
园艺设计	Stötzer & Näher, Berlin

赫尔曼与博世
斯图加特

农业生产合作总部的整修与改进
斯图加特

REFURBISHMENT AND IMPROVEMENT OF LSV HEADQUARTERS, STUTTGART

计划方法

PLANNING APPROACH

建筑成本	约 2 200 万欧元
计划和施工期	2001–2003 年
总面积	11 500m²
总体积	约 34 000m³
竣工验收	2003 年
使用面积	约 5 500m²

因为老楼不符合国家关于建筑安全的有关规定。所以，在南外墙新增安全梯和消防电梯。为建筑安装外置纵向环形通道，减少老楼的内部楼梯，并重新设计楼内空间，形成阁楼式办公楼。为此，也要相应地增加辅助结构以加固建筑。

As the old building no longer complied with the official high-rise guidelines, two new emergency staircases and a fire-fighter lift were added on the south façade. Fitting the building with external vertical circulation routes made it possible to eliminate the old internal staircase and to redesign the interior as spacious, loft-type office floors, but it also made supplementary constructions necessary to stabilize the building.

混凝土结构外部用扁钢筋杆加固，可使危楼加强负荷支撑力。

The load-bearing capacity of the gutted structure was strengthened by gluing flat steel bars onto the existing concrete structure.

标准层平面图

Standard floor plan

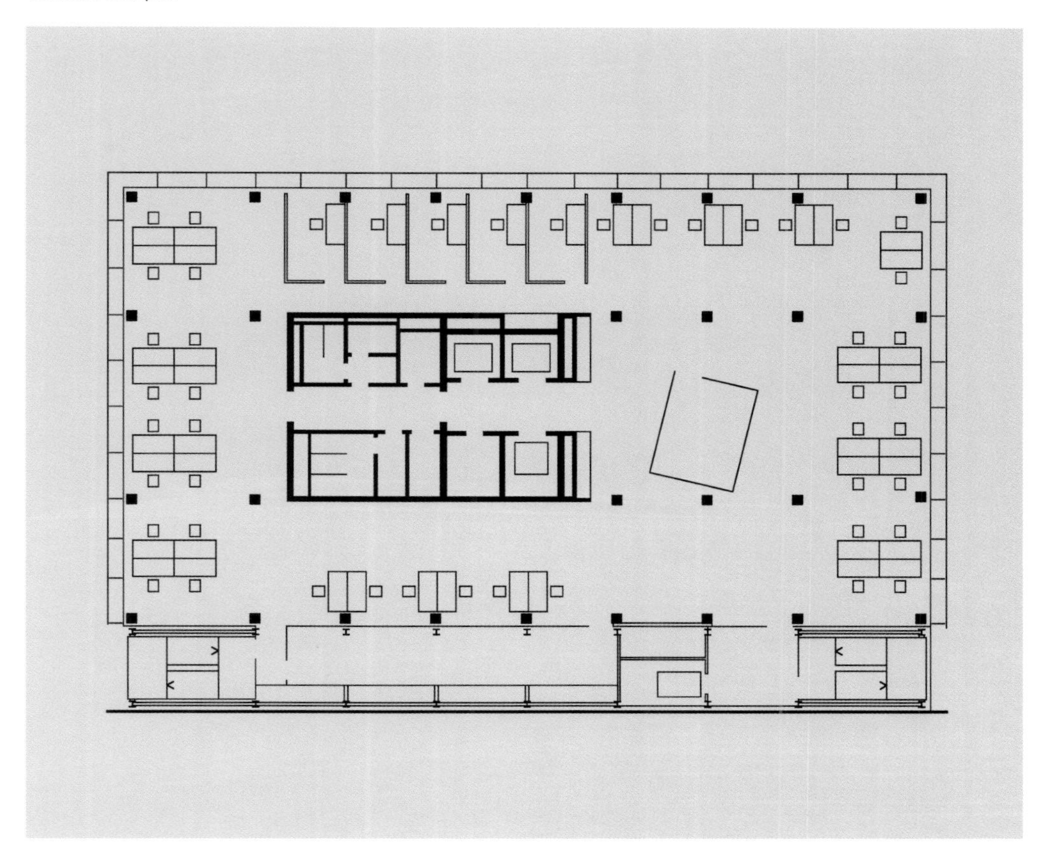

能量概念包括原有混凝土结构产生的温度适应。带有整体块毛细管的垫子铺满地面，并浇上沙浆，这种方法可使楼内冬暖夏凉，不必安装暖气。

The energy concept included climatization via the thermal mass of the existing concrete floors. Mats with integrated capillary tubes were flushed into the floors and then topped with screed. As this system cools the interiors in the summer and heats them in the winter, radiators could be omitted.

20 世纪 60 年代巴登符腾堡的农业合作社的总部不符合当代办公建筑的要求，委托 The Kubus Generalplaner 有限公司采用最新式的办公楼设计，对 12 层楼建筑进行改装和整修。委托书中还规定安装新型防火、保温防潮等设施和新增结构及服务功能。

The 1960s headquarters of the Landwirtschaftliche Sozialversicherung (agricultural social security) Baden-Württemberg no longer met the requirements of a modern office building. The Kubus Generalplaner GmbH (limited company) was therefore commissioned to modernize the twelve-storey building and to refurbish it with state-of-the-art office workplaces. The commission also included the installation of up-to-date fire protection, insulation and damp-proofing, new structures and service functions.

每层楼的外墙都有必要的水平防火玻璃墙，外墙内部设有结构支柱。11、12 层楼的墙面向内凹进，露出支柱。整个楼层从上到下十分明亮，在边缘形成了一个夹壁的空间走廊。

The necessary horizontal fire-proof 'bulkheading' of the façade cavity from floor to floor was done with glass. On the 11th and 12th upper levels, the façade steps back behind the structural piers (which are inside it on the lower levels) and is glazed from floor to ceiling. This has created a perimeter walk-in cavity space.

大楼坐落在直通斯图加特的南面主干道，由于新建内森巴赫塔尔桥的开放，LSV大厦门前交通拥挤。设计师设计了一个明亮的全玻璃大楼，为城市道路增加亮点。玻璃墙是两叶夹壁墙的外层，起到保护墙面内部木制结构及隔音的作用。这些结构被大窗户遮挡，总的来说，墙面还是很透明，可远观全方位景色。

The high-rise is located on a main thoroughfare leading into Stuttgart from the south. Following the opening of the new Nesenbachtal bridge, heavy motor traffic has been passing very close to the LSV high-rise. The designers erected a clear, fully glazed building to set a special accent in this urban gateway situation. The glazing is the outer skin of a two-leaf cavity wall, it functions both as a noise barrier and weather-protection for the inner wooden parts of the façade. As these are interrupted by large windows, the façade is on the whole very transparent and affords views in all directions.

赫尔曼与博世
斯图加特

农业生产合作总部的整修与改进
斯图加特

REFURBISHMENT AND IMPROVEMENT OF LSV HEADQUARTERS, STUTTGART

外墙结构

FAÇADE STRUCTURE

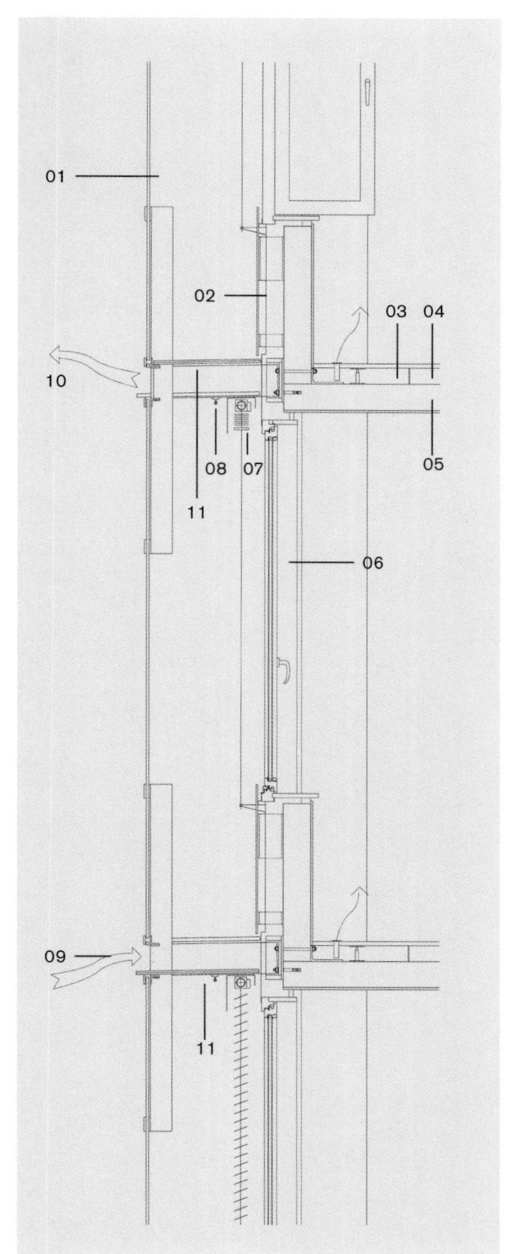

01 外墙面：玻璃窗，铝制点接头，支撑架（电镀面），悬臂（电镀面）
outer façade: glazing, aluminium point fixings, spiders (galvanized flats) and cantilevers (galvanized flats)
02 内墙面：镀层，后通风，绝热系统，钢筋混凝低墙（现存），底涂
interior façade: cladding, back-ventilation and thermally insulation systems, RC parapets (existing), rendering coat
03 空气输入管
air intake duct
04 供电管
electrical supplies duct
05 钢筋混凝土楼板（现存）Stahlbetondocke（Bestand）
RC floor (existing)
06 木框窗户（多层加固的云杉木）
wood-framed windows (spruce with multiplex stiffening)
07 铝板条上拉百叶窗
pull-up blind of aluminium slats
08 安全生气系统
work-safety system
09 新鲜空气入口
fresh-air inlet
10 废气出口
waste-air outlet
11 半透明薄片式安全玻璃窗
translucent laminate safety glazing

墙面的精心雕琢，甚至达到 1：1 的比例。采用多个公司合作的设计细节，设计师通过热能空气动力模拟及优化建筑结构和玻璃窗厚度，创造出一座具有成本效益的双层空间围合体系。它的几何结构设计，使建筑的透明度进一步增强。

By carefully detailing the façade, even at a scale of 1:1; by means of details developed in co-ordination with various firms; by thermal and aerodynamic simulations and by optimizing structural parts and glass-pane thicknesses, the designers were able to develop a cost-effective two-leaf cavity enclosure. Its geometric and structural design lends the building a high measure of transparency.

单层玻璃的外表符合隔音标准，可挡风遮雨，而且内部木制墙面为楼内提供隔热功能，从而产生舒适的内部温度。

The outer skin of single-glazing complies with noise-barrier standards and keeps out rains and winds, while the inner wooden façade provides the interior with thermal insulation and well-tempered interior climates.

施工单位 Landwirtschaftliche Berufsgenossenschaft, Landwirtschaftliche Krankenkasse Baden-Württemberg
总设计 KUBUS Generalplaner GmbH, Stuttgart
建筑设计 Herrmann + Bosch Freie Architekten BDA, Stuttgart
摄影 Roland Halbe, Stuttgart

专业工程师
塔吊安装 Pfefferkorn Ingenieure, Stuttgart
建筑技术 Trippe + Partner, Stuttgart

HPP 建筑设计事务所
汉特里奇・派斯奇尼格及其合作方 KG
杜塞尔多夫

阿连兹 – 凯
美茵河畔法兰克福

ALLIANZ-KAI, FRANKFURT/MAIN

城市规划

URBAN REGULATION PLAN

建筑成本	12 800 万欧元
计划和施工期	1997–2002 年
总体积	420 000m³
占地面积	100 000m²
使用面积	34 900m²
竣工验收	2002 年
楼层高度	60m

通过延长将弗歌尔魅德大街至厂区解决了车辆通行的问题，并设置了人行道。住在工厂后面生产区的人们可通过人行道直达河边。这种设置与西南边广场设计相结合，形成一种新的城市规划。

The problem of vehicular access was solved by extending Vogelweidstrasse into the site, and a footpath was laid out from this site road which gives those living in the residential area behind it direct access to the river. Combined with a newly designed square to the southwest, this created a new urban layout.

斯特莱斯曼大街建 6 层高围墙式建筑的设计构想获得比赛大奖。泰德 – 星 – 凯建的 16 层高楼为主入口，它标志着梅茵区从紧凑连体建筑向透明梳状结构转变的思路。与带有露台的建筑面朝江河低矮的码头将扩建成未来的梅茵堤岸公园。

This design, which won a competition prize, proposed a six-storey perimeter block structure on Stresemannallee, and a sixteen-storey high-rise on Theodor-Stern-Kai which marks the main entrance and the beginning of the transition from compact attached buildings to a more transparent comb-like structure along the Main. The building opens towards the river with planted patios, and the lower quay area is extended as another building block of the future Main embankment park.

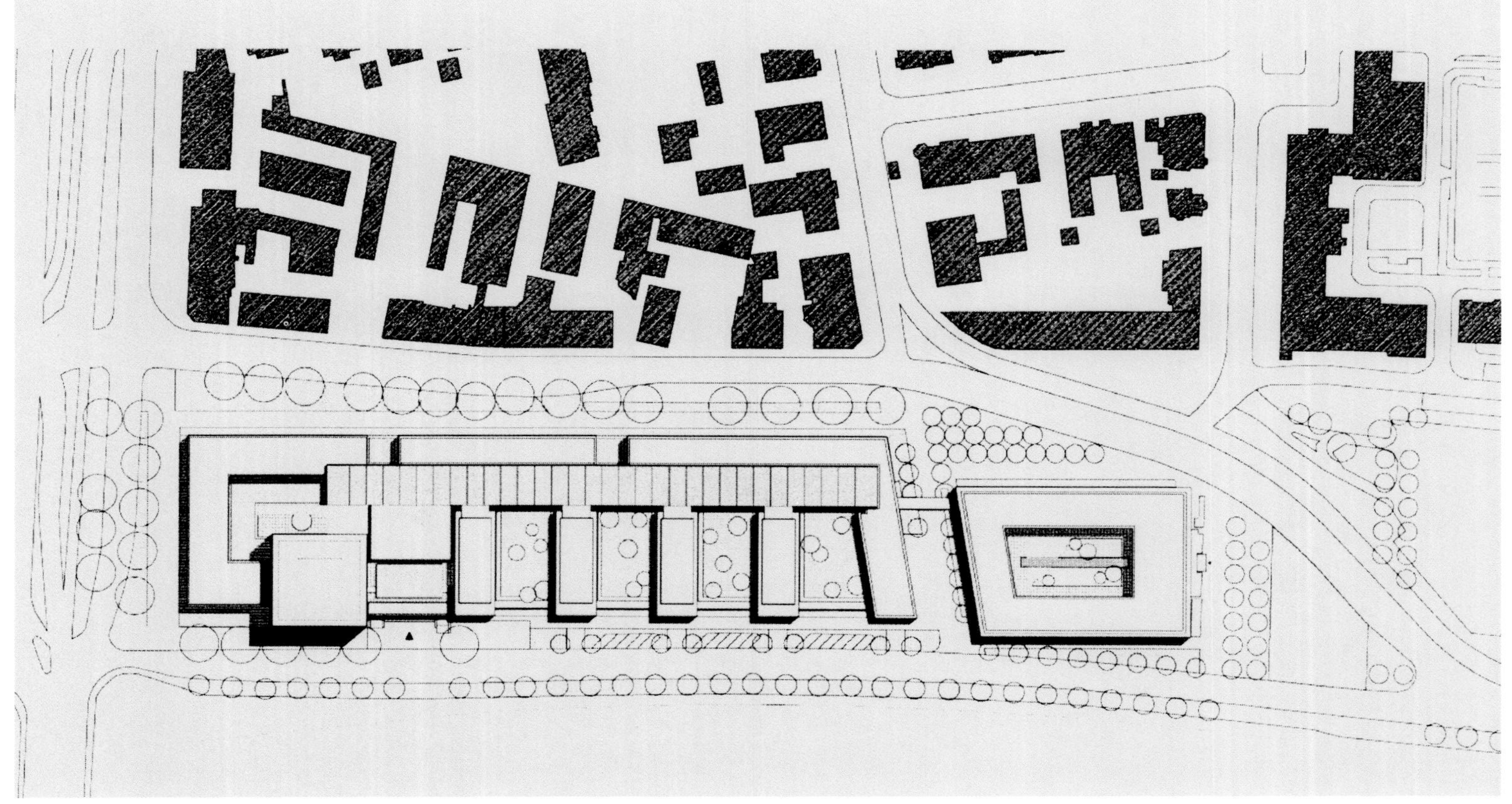

梳状结构延伸至高于一层的中央玻璃大厅，也就是主要通路。从这可通向楼内所有的楼梯和电梯。侧楼通过横跨大厅的钢架桥与主楼相连。大厅由大面积透明玻璃墙建造。从这里可欣赏到梅茵区的景色和法兰克福的地平线。

The comb-like structure ends in a glazed concourse on the first upper floor – the so-called 'Magistrale' (main thoroughfare) – which gives access to all the stairways and lifts. The wing levels are interconnected via steel bridges spanning the concourse. Large transparent glass walls close off the ends of this hall and afford views of the Main and, opposite, of Frankfurt's skyline.

种满植物的露台，把整个建筑展现给梅茵区，扩建低矮码头是为了使江边花园将来能够形成大规模的建筑区。

Planted patios open the building to the Main. The sunken quay is extended to form a further building block of the future Main riverside park.

HPP – HENTRICH-PETSCHNIGG & PARTNER KG, DÜSSELDORF

阿连兹－凯　美茵河畔法兰克福

ALLIANZ-KAI, FRANKFURT/MAIN

HPP 建筑设计事务所
杜塞尔多夫

外墙设计

FAÇADE DESIGN

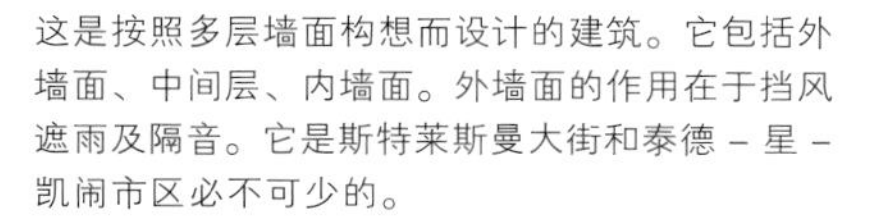

这是按照多层墙面构想而设计的建筑。它包括外墙面、中间层、内墙面。外墙面的作用在于挡风遮雨及隔音。它是斯特莱斯曼大街和泰德－星－凯闹市区必不可少的。

The design of the followed the idea of a multi-layer skin. It consists of the outer façade, the intermediate space (corridor) and the interior façade. The outer skin acts as a sun and rain shield and as a traffic noise barrier – a 'must' at the busy corner of Stresemannallee and Theodor-Stern-Kai.

每一层楼双墙面的外墙部分都有通风口，因此两墙之间的夹缝通道空气清新。整面墙的窗户可使办公人员呼吸到新鲜空气。双墙面夹缝安装了遮阳装置（百叶窗），对阳光的强烈照射起到足够的保护作用。室内为电脑工作人员安装的防眩晕屏，可以升高至约 0.7 米。

The outer skin of the double-layer façade has openings on every floor so that the cavity (and interior spaces) behind it can be aerated. Floor-to-ceiling window sashes enable office users to let in more fresh air. Sun-shading devices (slatted blinds) in between the two façade layers offer adequate protection against direct insolation. Inside the rooms, anti-dazzle screens may be raised to a height of approx. 0.7 metres for PC workstations.

建筑设计	HPP Hentrich-Petschnigg & Partner KG, Düsseldorf
施工单位	Allianz Lebensversicherungs- AG, vertreten durch Allianz Grundstücks GmbH
总体照明设计	Hochtief AG
工程编号	97012
设计伙伴	Duk-Kyu Ryang, Mag. Arch. Dipl.-Ing. KIA, BDA
监控参与者	Dipl.-Ing. Jochen D. Roll
项目计划参与者	Remigiusz Otrzonsek, Dipl.-Ing.
项目计划	Udo Wittner, Ing. (grad.)
项目目标监控	Stephen Minte, Dipl.-Ing.
摄影	H.G. Esch, Hennef/Sieg

工程参与者	ALU-SOMMER GmbH, Stoob (A)
铝玻璃外墙设计	Hofmann GmbH & Co. KG Natursteinwerke, Werbach-Gamburg
自然石	AFC Aluminium Fassaden Consulting, Wien (A)
Stahlbaukonstruktion Magistrale	Stahlbau Plauen GmbH, Plauen
Innere Verglasung/Magistrale，Aufzuge etc.	Glasbau Hahn, Frankfurt am Main
建筑物理	IFB Institut für Bauphysik, Mülheim/Ruhr

HPP 建筑设计事务所
杜塞尔多夫

LVM 保险公司塔楼
明斯特

LVM TOWER
MÜNSTER

维修与扩建

REFURBISHMENT AND EXTENSION

建筑成本	4 550 万欧元
计划和施工期	1994–1998 年
总体积	92 000m³
总面积	26 768m²
使用面积	9 500m²
占地面积	20 000m²

目前的扩建工程包括 1967 年的现代化建设及垂直部分还有一座 20 层新塔楼（现存的 A 座）用来代替原来的管理楼。

The present extension project included the modernization and vertical extension of the 1967 high-rise as well as a new twenty-storey tower (now called block A) to replace the management pavilion.

直到 1996 年，LVM 保险公司明斯特的办公大楼由以下部分构成：1967 建造的 10 层高连带经营仓库建筑（现存的 B 座）。1977 年扩建的 4 层和 6 层高十字交叉侧楼的 4 层建筑。1987 年由 HPP 公司设计的 U 形 6 层建筑，此项设计获得大奖。1994 年，HPP 公司在施陶芬贝格大街西侧建成又一座三角形 4 层侧楼。

Up until 1996, the building complex of the LVM insurance in Münster consisted of a ten-storey structure of 1967 (today called block B) with an annex pavilion housing the management, a 1977 extension with four and six-storey wings on cruciform and comb-like plans, and a U-shaped six-storey built in 1987 to the design by HPP which had won the competition. In 1994 HPP added another, triangular and four-storey, annex on the west side of Stauffenbergstrasse.

B 座扩建了 4 层，也就是在 13、14 层的行政办公室和会议室的顶部覆盖大面积单坡玻璃屋顶。以下的楼层及 A 座塔楼的 2 ~ 18 层设有标准办公区。设计师为塔楼设计的双底层采用玻璃结构。而领空的开放可以从考德环城路看到施陶芬贝格大街的主要干道。

The four-storeys added to building block B, i.e. the management offices and conference areas on the top 13th and 14th floors, were covered with a largely glazed single-pitch roof. The storeys below and the 2nd to 18th storeys of the block A tower contain standard office spaces. The architects designed the two bottom floors of the tower as glazed, but otherwise open 'air spaces' so that the main entrance on Stauffenbergstrasse can still be seen from Kolde Ring.

总共建成了 480 间新工作室。

A total of 480 new work stations were created.

LVM 保险公司塔楼
明斯特

LVM-HOCHHAUS MÜNSTER

外墙设计

FAÇADE DESIGN

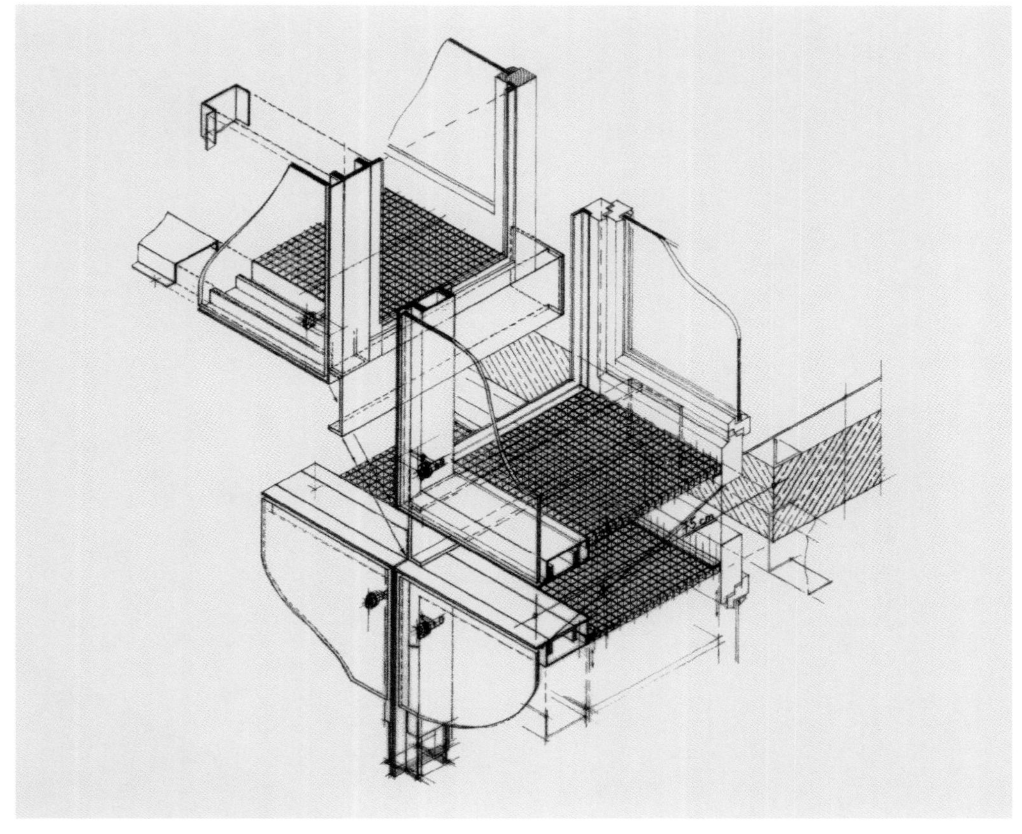

楼区的双层墙面不仅用于阻挡交通噪音，还可为办公室提供通风条件，利用高度灵活的百叶窗取得最佳遮阳效果。墙面夹道既可自由行走，又便于维修及清洁保养。

The double façade of the complex not only acts as a noise barrier against the traffic noise from Kolde Ring and Stauffenbergstrasse, but also provides aeration of the offices independent of the weather as well as optimal sun-shading by highly flexible horizontally slatted blinds. The façade cavity contains catwalks for maintenance and cleaning purposes.

建筑单位	HPP Hentrich-Petschnigg & Partner KG, Düsseldorf
施工单位	LVM Landwirtschaftlicher Versicherungsverein Münster aG
工程编号	94124
设计伙伴	Duk-Kyu Ryang, Mag. Arch. Dipl.-Ing. KIA, BDA
监控参与者	Walter Auer, Dipl.-Ing.
项目设计	Wolfgang Liebergesell, Dipl.-Ing.
	Rainer Friedrich, Dipl.-Ing.
项目目标监控	Martin Geimer, Dipl.-Ing.
摄影	H.G. Esch, Hennef/Sieg
	Manfred Hanisch, Mettman, S. 185, 186 oben
工程参与者	
基本结构	ARGE STRABAG / SCHÄFER, Münster
外墙技术	IB Memmert + Partner, Neuss
静力学	Ing. ARGE
	Nees & Degenhart
	Dipl.-Ing. Degenhart, Münster
热能建筑物理	Institut für Bauphysik
	Dipl.-Ing. Horst Grün, Mülheim/Ruhr
TGA	HL-Technik, Düsseldorf

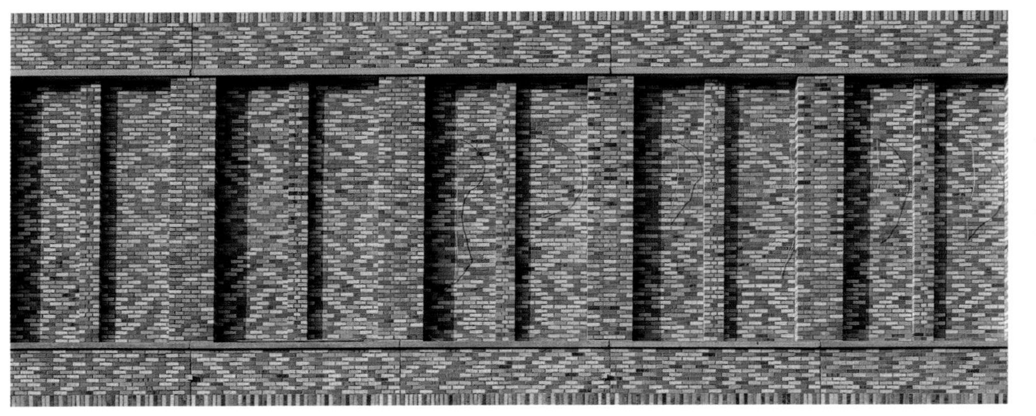

充满活力的墙面由 3 种硬砖构成，一些硬砖含有煤尘，在砖厂经过燃气高温烧制并在砖厂的托盘中根据设计类型预先挑选，以方便工地石匠的工作。

The lively façade consists of three types of clinker bricks, some of them containing coal dust and burnt in a gas-fired furnace. The bricks were pre-sorted on pallets at the brick field, following the architects' design pattern, to facilitate the masons' work on site.

克莱修斯兄弟建筑公司
柏林

艾尔斯特沃德尔广场
商业中心　柏林

COMMERCIAL CENTRE ON
ELSTERWERDAER PLATZ, BERLIN

外墙设计

FAÇADE DESIGN

建筑成本	4 500 万欧元
计划和施工期	1999–2003 年
总面积	94 233m²
总体积	483 925m³
使用面积	72 276m²
竣工验收	2003 年（一期）
占地面积	57 728m²

碧丝多夫的商业区是城市总体规划的一部分，包括购物中心、不同的专卖店、大型超市、小型零售店和餐饮场所。购物中心及邻近市场的广场，位于计划建成住宅区的起点位置，占据柏林东外围现代市区。

This commercial complex in Biesdorf is part of an urban masterplan. It includes a shopping mall, different specialized markets, a large supermarket, small retailers and catering outlets. The shopping mall and the market square (to be designed) form the starting point for planning a new residential neighbourhood, a modern urban quarter on the eastern periphery of Berlin.

克莱修斯兄弟建筑公司
柏林

艾尔·斯特沃德尔广场商业中心
柏林

COMMERCIAL CENTRE ON ELSTERWERDAER PLATZ, BERLIN

建筑设计

BUILDING DESIGN

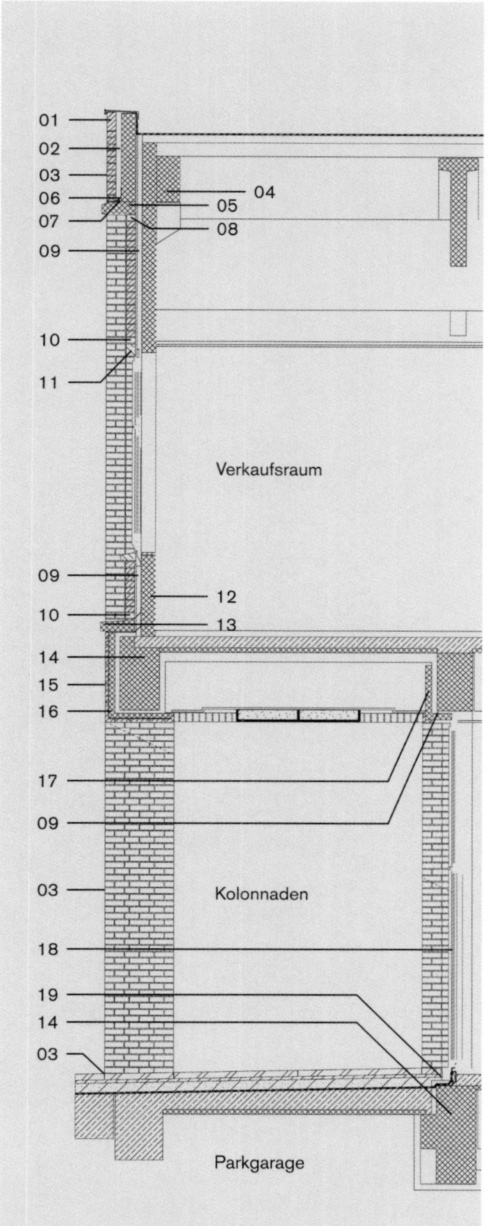

01 边缘层
verge course
02 空气层
air layer
03 外露砌砖，炼砖 NF
exposed brickwork, clinker NF
04 RC 预杠
RC prefab bar
05 预制合成屋顶低墙
prefab component roof parapet
06 防湿层
damp course (to German norm DIN 18195)
07 凹面制模
concave moulding
08 开敞式对接（用于通风）
open butt joint (for ventilation)
09 保温
insulation
10 敞开式对接（通风 / 排水）
open butt joint (for ventilation/drainage)
11 预制 RC 过梁（平滑）
prefab RC lintel, smooth
12 预制墙 18 厘米
prefab wall, 18 cm
13 预制组件（颜色、表面构造）
prefab elements, all of even colour/surface structure
14 RC 预制大梁
RC prefab girder
15 四分之一砖（粘着 3.5 厘米 +1厘米）
quarter bricks, glued, 3.5 + 1 cm
16 外层（8 厘米，每 4 毫米有一接合点）
facing layer (8 cm), joints every 4 mm
17 预制过梁构件
prefab lintel element
18 轻质金属组合框
light-weight-metal window element
19 排水道
drainage channel

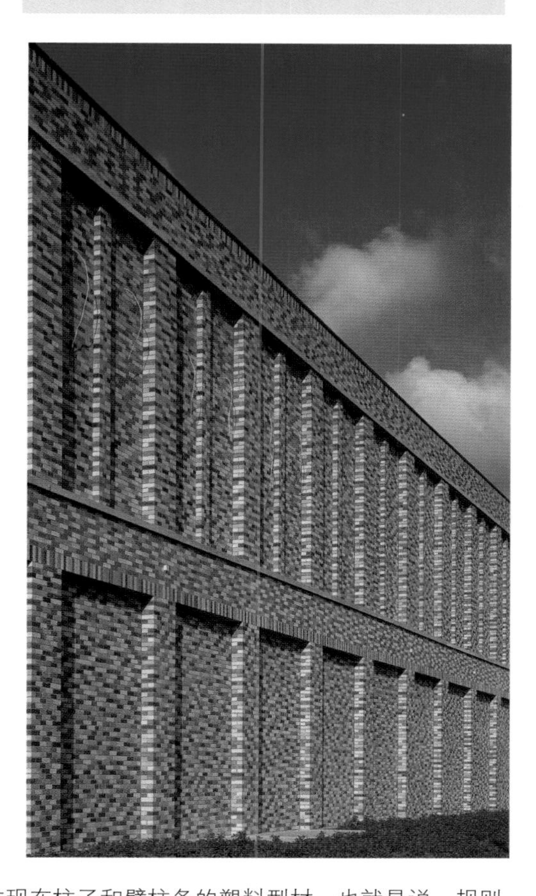

设计质量在于其清晰的结构和外形，主要体现在柱子和臂柱条的塑料型材，也就是说，规则的雕刻状栅格胜过在外墙窗户所建的不规则功能。硬砖的自然性与露天混凝土臂架、窗台和深色灰色金属轮廓相结合，体现出建筑的主要特色。

The quality of the design lies in its clear-cut form and structure, achieved by plastic modulation of pillars and pilaster strips. This regular sculptural grid ‘outplays’, so to speak, the functionally founded irregularity of façade openings. The natural quality of the clinker bricks, combined with exposed concrete ledges and window sills and the metal profiles coated a dark grey, determine the character of the building.

考虑到碧丝多夫商业中心的外墙，其设计创新在于把传统建筑工艺和现代生产技术紧密结合。砖是人类最早的建筑材料，技术创新增强了天然建材的特性。

With regard to the façade of the Biesdorf commercial centre, innovation represents the consistent combination of traditional building crafts and modern production methods. Brick is one of man’s oldest construction materials. Technical innovations serve to enhance the character of this natural material.

建筑设计	Prof. Josef P. Kleihues mit Norbert Hensel
员工	Götz Kern, Michael Alshut, Johannes Kressner, Rainer Wies, Susanne Weibrecht, Martina Wiesmann, Matthias Herrmann
施工单位	GEG Grundstücksentwicklungsgesellschaft H. H. Göttsch KG, Köln
摄影	Stefan Müller, Berlin
专业工程师	
塔吊安装	Ingenieurbüro Häussler, Illerkirchberg
房屋技术	K + S Haustechnik Planungsgesellschaft mbH, Meckenheim-Merl
大厦所有者	ARGE Handelszentrum Berlin Biesdorf (Gerdum und Breuer GmbH und Co. KG, Dechant Hoch- und Ingenieurbau GmbH und Co. KG, AKD-Bau GmbH), Berlin

科尔兄弟建筑事务所
杜伊斯堡 / 埃森
让·努维尔
格奥尔格·海克曼建筑师

科隆电视塔

COLOGNE TOWER

荒地转变为城市媒体公园

CONVERSION OF A WASTELAND INTO AN URBAN MEDIA PARK

塔楼建在 28.75 米 × 28.75 米见方的广场，它所在的方位反射出周边环境的特色，可看到全方位景色。43 层的塔楼高度为 148 米（算天台天线高 165.48 米）。

The orientation of the tower on a square plan of 28.75 x 28.75 m echoes the features of its immediate surroundings and follows a number of contextual sight-lines. The forty-three upper floors reach up to a height of 148 m (165.48 m counting the roof antenna).

建筑成本	1999–2001 年
总面积	36 000m²
总体积	125 000m³
使用面积	28 000m²
竣工验收	2001 年
占地面积	5 400m²

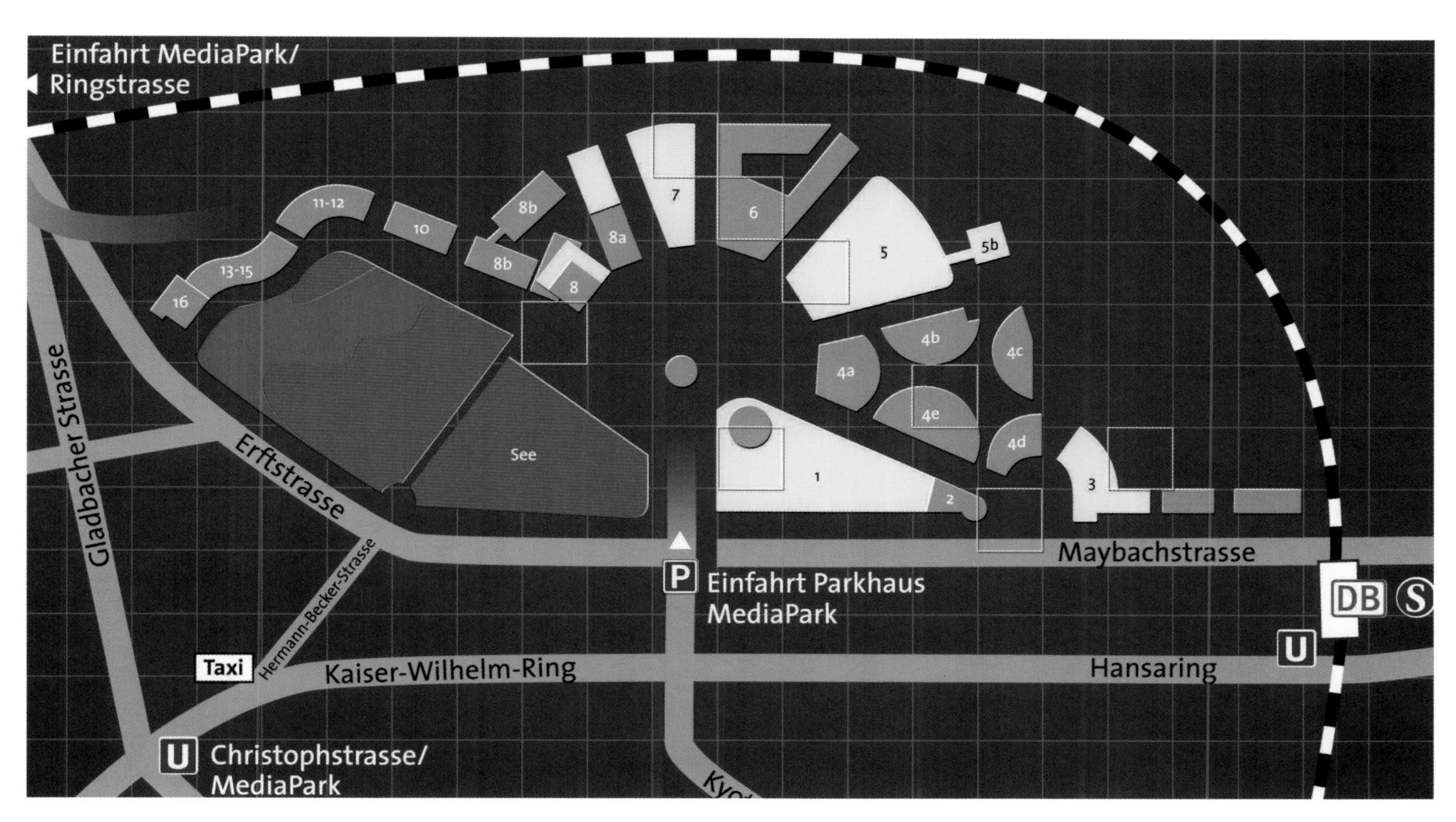

塔楼由三部分构成：楼基（2～7层）、楼体（8～29 层）和楼顶（30 ～ 42 层）。

The building has three parts: the base volume which cantilevers over the entrance from the 2nd to the 7th floor, the main tower section from the 8th to the 29th floor, and the top section from the 30th to the 42nd floor.

原为铁路货物仓库及集装箱堆场的杰里奥地区 20 世纪 80 年代末开始发展，主要用于建 IT 公司。从铁路荒地到现代媒体公园的转变产生一块新的市区，紧邻市中心。媒体公园分 8 个区域，以中央广场为中心扇形分布。1 区有 3 座建筑，一座 8 层，一座 7 层，还有科隆电视塔，是城市新发展的标志性建筑。

In the late 1980s, work started on developing the Gereon area, the site of a former railway goods depot and marshalling yard, mainly for IT companies. The conversion of the railway wasteland into a modern Media Park created a new urban quarter closely connected to the city centre. The Media Park has eight blocks arranged fan-like around a central square. Block 1 consists of three buildings, one with seven, one with eight storeys, and the Cologne Tower which forms the urban landmark of the new development.

在楼芯周围，塔楼提供每个楼层不同的设计。楼顶各层安装居住所需的各种设备。建有快餐店的大厅和会议室设在一楼，为所有用户开放。

Around its central core, the tower offers a broad range of different floor plans. At the top of the tower, the floors have been equipped with residential technical installations. The entrance hall with snack bars and conferences spaces on the first floor is open to all the tenants.

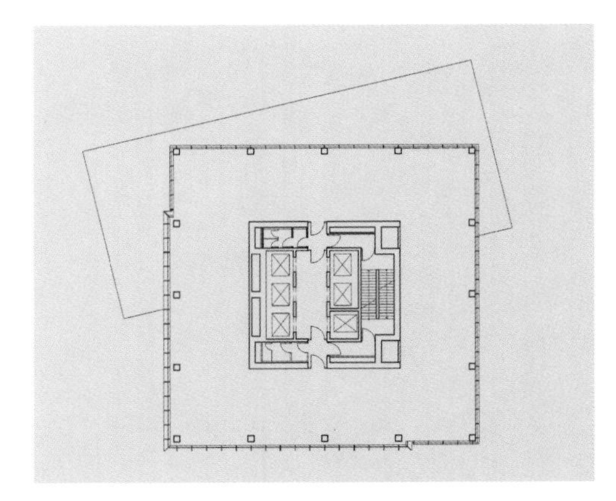

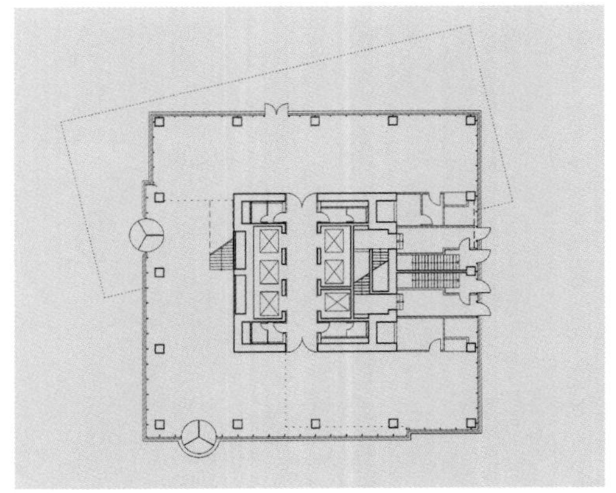

塔楼墙面由带有云彩图像和城市标语的屏幕式玻璃构成。城市的多彩形象从楼基延伸至12层。而单色云彩升至塔顶。墙体的艺术设计延伸至前方相邻的饭店和办公楼。双层外立面使楼内自然通风。

The façade is of glass screen-printed with images of clouds and urban motifs. The multi-coloured image of a city reaches from the base section up to the 12th floor, while the monochrome clouds rise up to the top of the tower. The façade printing was continued on the fronts of the adjacent hotel and office building. The two-leaf cavity façades (a 'primary' one with openable windows, and a 'secondary', outer one of glass) make it possible to ventilate the interior spaces naturally.

科尔兄弟建筑事务所
杜伊斯堡 / 埃森
让·努维尔
格奥尔格·海克曼建筑师

科隆电视塔

COLOGNE TOWER

外墙设计

FAÇADE DESIGN

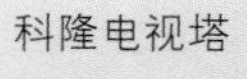

建筑设计	Projektgemeinschaft Kohl & Kohl Architekten, Duisburg/Essen Architectures Jean Nouvel Georg Heckmann
施工单位	Hypothekenbank in Essen AG
摄影	Christian Kohl
专业工程师	
规划房屋技术 HLS	Skiba Ingenieure, Herne
塔吊安装	BGS, Frankfurt am Main
外墙	IGS, Dr. Linder, München

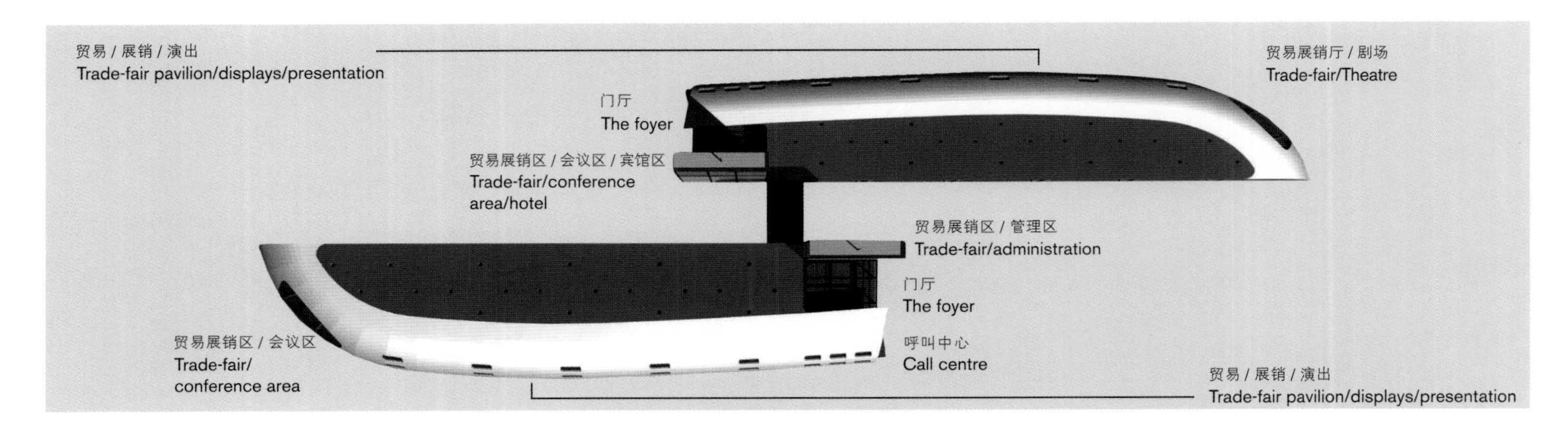

贸易展销会周围的大面积地区已成为生活小区和生态能源区。墙壁和楼顶覆满蔓生植物和草，还有部分太阳能集热器体现出这一点。对于塔楼来说，享用自然通风的两叶式空墙体是其优势之一。

The large areas of the trade-fair pavilions' enclosures were interpreted as a biotope and ecological energy resource. This means the façades and the roofs were partly covered with vines and grass, and partly with solar collectors. For the towers, experience from naturally ventilated two-leaf cavity façades were used to advantage.

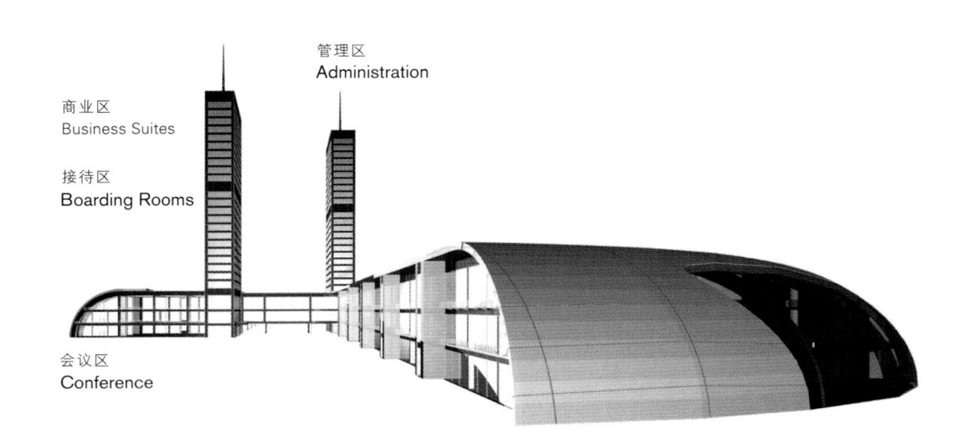

重建亚洲消费品在欧洲的分销渠道，除了已有的国际展销会外，还需要创新的贸易交易理念。新的贸易展销厅应该位于市中心的永久位置且交通方便。计划首先考虑到更有效地利用展区，为国外展销商提供与消费者有关的服务设施。

Restructuring the distribution of Asian consumer goods in Europe required an innovative trade-fair concept in addition to the existing international fairs. The new trade fair in turn required a permanent venue in a central location and at a logistic traffic junction. The project primarily aimed for a more efficient use of exhibition areas, and improved customer-relation facilities for foreign exhibitors.

亚洲展销商与欧洲贸易展销会来访人员之间的交流经过内外的设计，通过特殊的方式得以加强和促进。两个柱体由安装照明设备的过道相连，可引导来访者。墙面安有采光效果佳的景观窗。

The communication between Asian exhibitors and European trade-fair visitors is 'highlighted' and promoted in a special way by the interior and exterior design. The two pavilions are interconnected by a passage whose lighting system guides visitors along; and the façades include 'light-wells' and panorama windows.

科尔与福罗莫建筑师事务所
杜伊斯堡 / 埃森

通往亚洲的大门
杜伊斯堡

GATE TO ASIA
DUISBURG

照明与内部设计

LIGHTING AND INTERIOR DESIGN

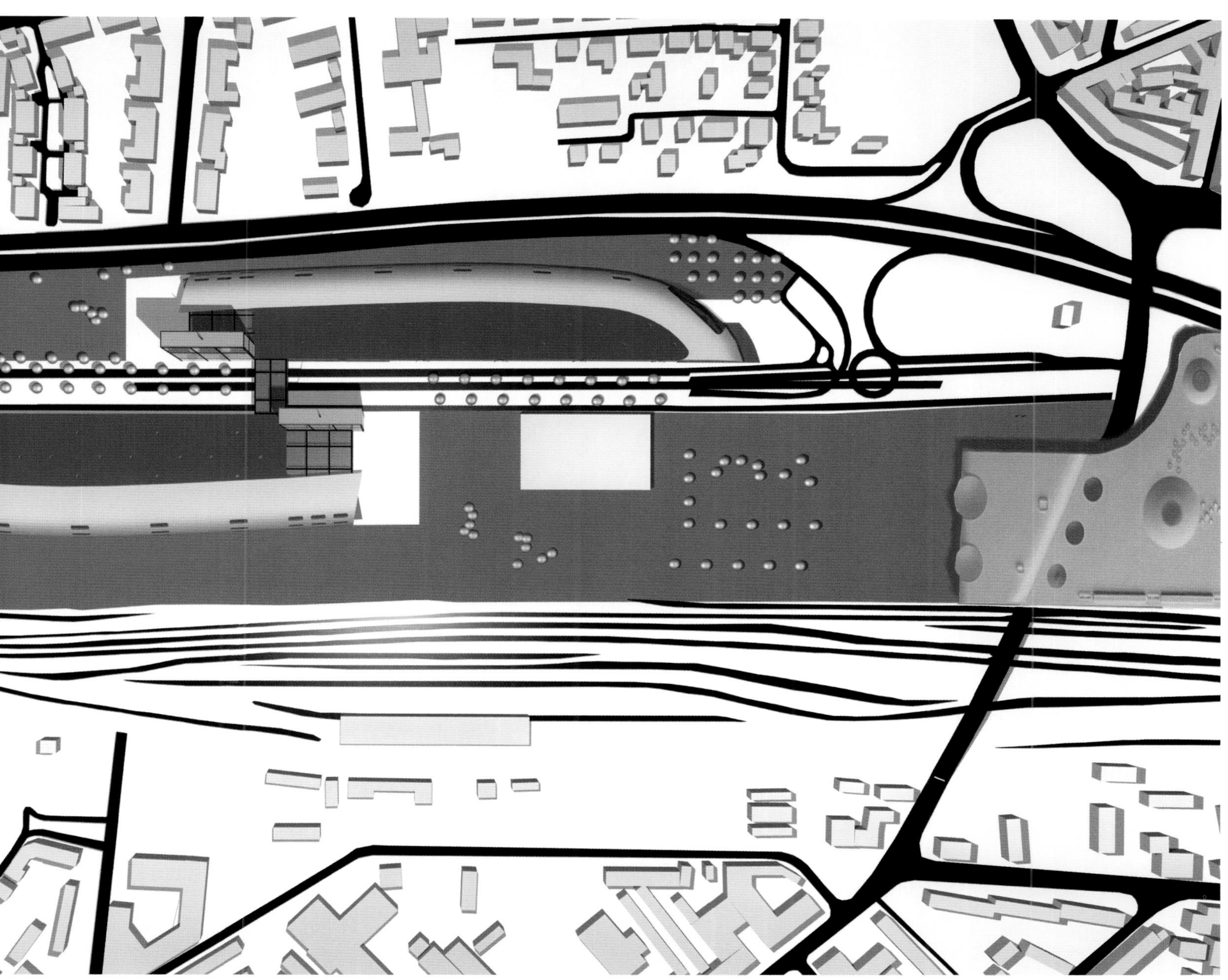

设计师考虑到贸易展销会所在地周围交通拥挤的特殊性，外墙面与楼顶面在接合处融为一体，这样可以保护它免受噪音与大气污染。

By integrating façade and roof surfaces at their 'joints', the architects took into account the special situation of the trade-fair grounds surrounded by heavy traffic in order to protect it from noise and air pollution.

建筑设计	Kohl & Fromme Architekten, Duisburg/Essen
施工单位	Gate to Asia Projektentwicklungs GmbH
透视图	Kohl & Fromme Architekten
发展规划	Spiekermann Ing.
继承权	Thau HG.
园艺设计 B 区	Kipar Landschaftsarchitekten

克拉姆和斯特里格尔
达姆施塔特

会展中心的灯饰
萨尔布吕肯

EXPOMEDIA LIGHT CUBE
SAARBRÜCKEN

城市设计理念

URBAN DESIGN IDEA

建筑成本	4 200 万欧元
施工期	1999–2000 年
总面积	3 184m²
使用面积	640m²
总体积	10 300m³
竣工验收	2000 年

两层杆式系统在不锈钢网与里层之间形成另一个界面。与发光二极管相匹配的专用玻璃管构成了该系统，并呈现出五光十色的光谱。

A two-level 'rod system' forms a second façade layer behind the stainless steel mesh and in front of the interior façade. The system consists of special glass tubes fitted with LED rods representing the rainbow colour spectrum.

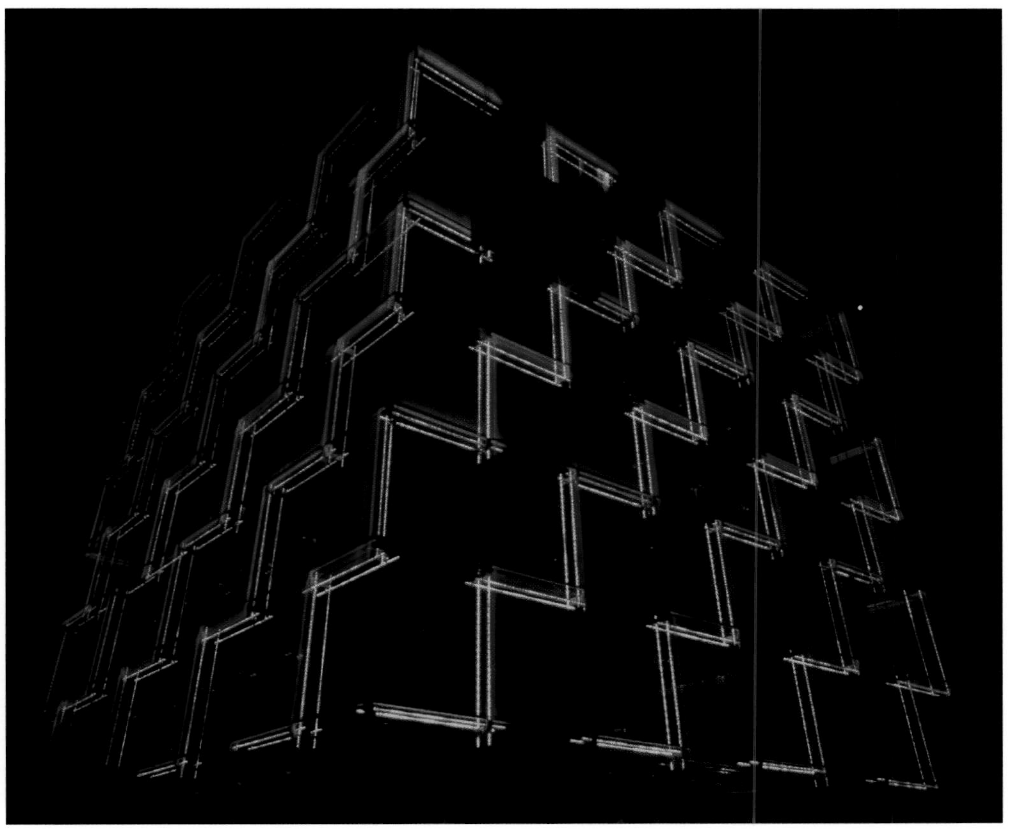

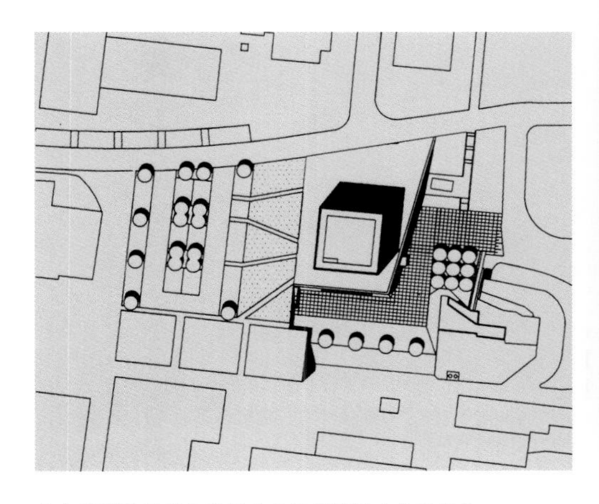

立方体形状及其日光闪变中的转换是立体光学的创新，这成为建筑形式的真实性和无形的光之间的媒介。这项工程被传统建筑学和媒体视为一次艺术实验。

Not the shape of the cube, but the transformation of the cube in the play of light – that is the innovation of the Light Cube which mediates between the reality of built form and the immateriality of light. The project is to be understood as an artistic experiment somewhere between conventional architecture and a media event.

建筑学仍然能够创造人类环境或者形成社会意识吗？这已是老生常谈。我们所考虑的地点正在发生变化，布尔巴克钢铁工厂位于市内建筑和工业建筑的边界，它正在发生潜在的变化。会展中心工程为该地区的发展树立了标志并继续影响该地区的发展。选择这个位置意味着占据了一个战略位置、一个交叉点以及今后工程的发展方向。

The age-old question is: can architecture still create human contexts or make social sense? The site concerned here is changing: the closed Burbach steel works, a borderland between urban structures and those still left over on the industrial site which is seen as a zone of potential transformation. The Expomedia project is to set an initial sign for, and is to continue to influence, the development of this place. Choosing this location meant occupying a strategic point – a CROSSING POINT – and point of orientation for subsequent projects.

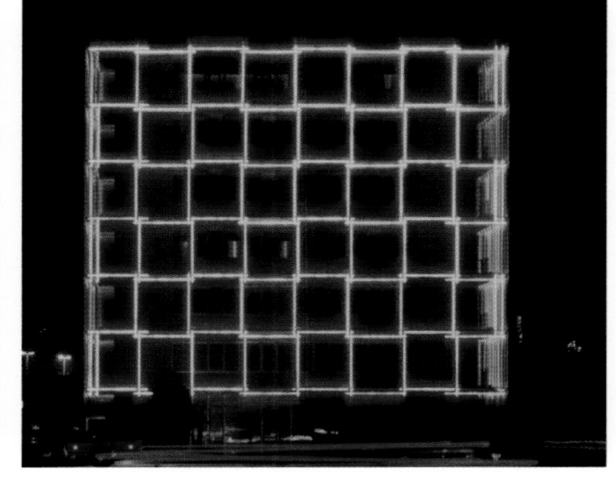
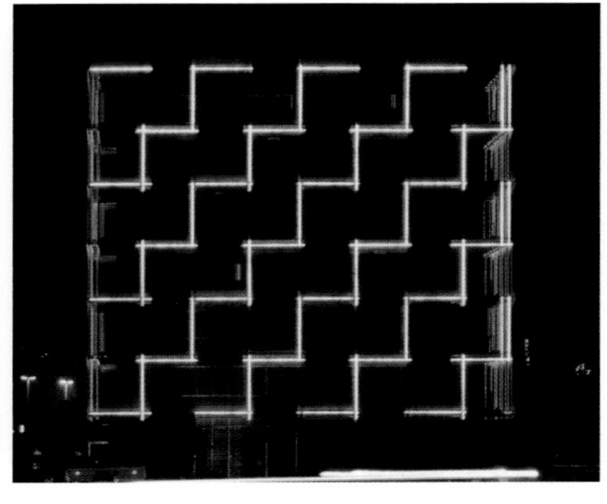

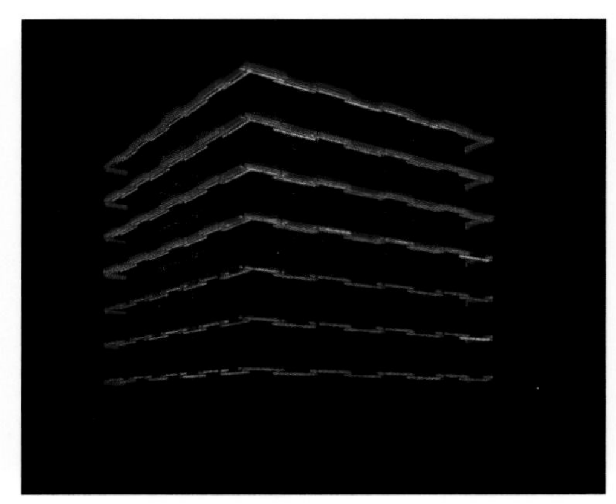
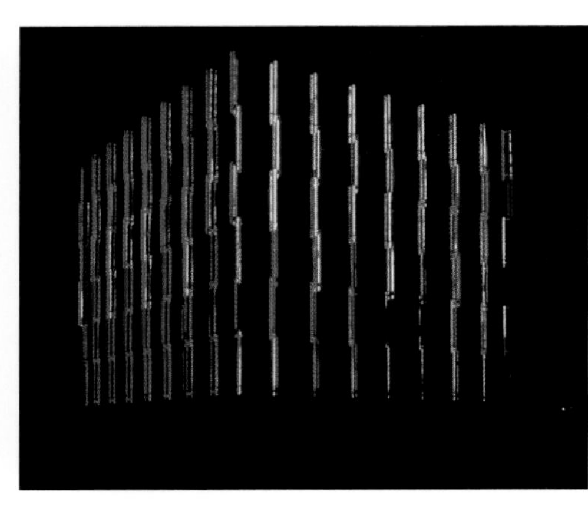
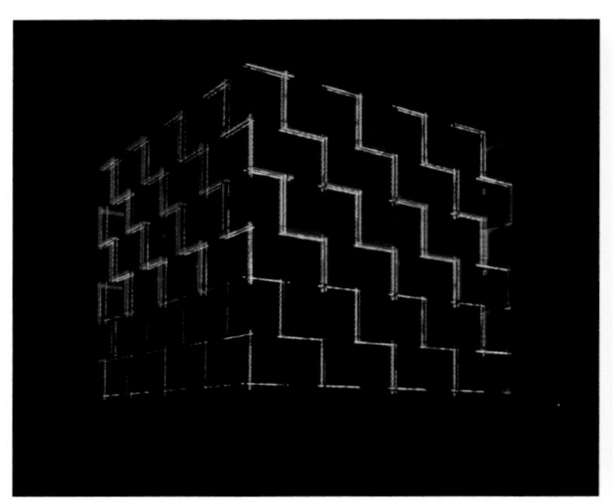

排放在两水平面的立体格状式管灯围成立方体。灯管可以被依次接通，或是连成一排，或是顺次连接。从抽象角度来说，他们象征着信息、运动和状态。介质板位于外立面，与具有背投光的外立面有所不同。随着现实的再现，它可能弥补或面临抽象的典型。

A three-dimensional grid of tube lights arranged on two levels surrounds the cube. The tubes can be switched on one by one, or all together in a row, or in sequence. On the abstract level, they symbolize information, movement and states. In contrast to a back-lit façade, the media board stands on the façade. It makes it possible to complement or confront the abstract images with reproductions of reality.

有力的形式语言中可见的等离子碎片是形成调节这些灯的图案的要素。等离子碎片表面上产生不规则轮廓，这些轮廓扩大和展示重复且相似的形状，类似于山区形态。

The visualization of plasma fractals in a dynamic formal language forms the basis for controlling these patterns of light. Plasma fractals generate seemingly irregular contours which, when magnified, reveal repeated similar shapes and therefore resemble the morphology of mountainous regions.

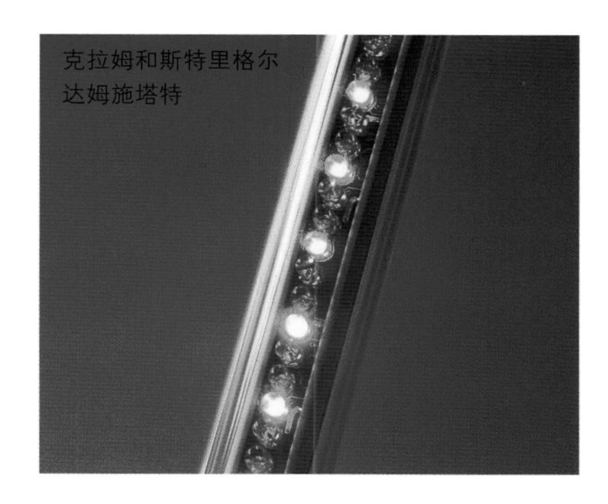

会展中心的灯饰
萨尔布吕肯

EXPOMEDIA LIGHT CUBE SAARBRÜCKEN

外观设计

FAÇADE DESIGN

由计算机控制的媒介墙体可以使各种活动和信息形象化，兼顾抽象和具体的艺术形式。

The computer-controlled media façade is able to visualize activities and information in abstract as well as concrete, figurative artistic form.

建筑单位 Kramm & Strigl, Darmstadt
业主 GIU Gesellschaft für Innovation und Unternehmensförderung mbH & Co, Saarbrücken
摄影 Prof. R. Kramm, Prof. Dieter Leistner, Madjd Asghari, Benjamin Kramm

技术工程师
支撑结构设计 WPW Ingenieure, Saarbrücken
建筑技术 KMW Ingenieurgesellschaft mbH, Saarbrücken
建筑物理 ITA, Wiesbaden
照明电路设计 Ingenieurbüro Zitnik, Frankfurt

公司灯光
VA- 经营 GKD - Gebr. Kufferath GmbH & Co. KG, Düren
经营外立面 K. M. Hardwork GmbH, Stuttgart
多媒体外立面 Paul Wagner et fils, Luxembourg
Richter Elektronik GmbH, Karben (LED Stäbe)
Mentzel und Krutmann, Wuppertal (Lon-works+Controller)

条形停车场结构和与之配套的办公区域沿着新开发的居住区形成新的“城市墙”。在东西的尽头，停车场墙壁或支架形成具有螺旋式进出口斜坡的圆柱形建筑，透过斜坡半透明的表面发出深蓝色的微光。

The long car park structure and the accompanying office blocks form a new 'city wall' along the entire length of the new residential development. At its western and eastern ends, this car park wall, or 'bracket', forms cylinders whose cores – spiralling access and exit ramps – shimmer ultramarine blue through the ramps' semi-transparent skins.

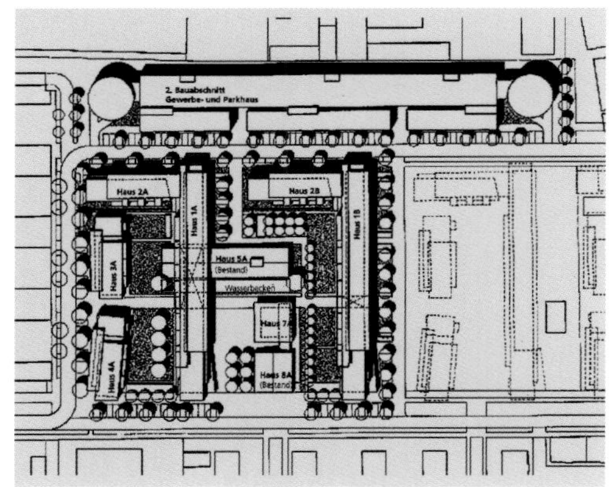

这是一项针对居住和小型工业的城市发展规划，这里以前曾是位于达姆施塔特外围的屠宰场。由于这个地区的地下水平线较高，传统的6层停车场成为设计的一部分。位于其前方的3个办公区在住宅区与高186米的停车场之间形成壁垒。新建的小路通过施工技术铺设，其中一个办公区完全由可循环材料建成。

This is an urban development project – residential and small industries – for the former slaughterhouse site on Darmstadt's northern periphery. Because of the high groundwater levels in this area, a conventional six-storey car park was part of the programme. The three office blocks in front of it form a 'shield' between the residential development and the 186-metre car park. New paths were trod in construction technology: one of the three office blocks was constructed entirely of recycled materials.

每一个办公区都能够通过相连的楼梯直接进入停车场，一楼能容纳几个大规模商店。公共区域以拱廊道路的城市设计理念为标志。

Every office level has direct access to the car park via joint staircases. The ground floor accommodates generously dimensioned shops. The public space is marked by the urban design theme of arcaded pavements.

行政楼与多层停车场
达姆施塔特

ADMINISTRATION BUILDING AND MULTI-STOREY CAR PARK, BÜRGERPARKVIERTEL (CIVIC PARK NEIGHBOURHOOD), DARMSTADT

能量概念

ENERGY CONCEPT

建筑成本	1 800 万欧元
施工期	1997–1998 年
总面积	26 666m^2
总体积	74 630m^3
获奖情况	1994 年一等奖
设计时间	1995 年
使用面积	4 756m^2

办公楼面朝南，并通过走廊内带有气孔的天花板进行排气通风。内部空间可通过强流控制机械装置在白天进行温度的调节，夜间能够同时冷却。在炎热的天气，夜晚到来之时，中央调节装置自动打开窗外的栏杆嵌板并运转走廊天花板上的抽气系统。

The office buildings open southwards and are cross-and-exhaust-ventilated via air cavity ceilings in the corridors. The interior spaces can be climate-controlled individually by day, and all together for night-time cooling, by means of a volume flow control mechanism. In hot weather, at night central controls automatically open the façade window parapet panels and operate the extract-air system in the corridor ceilings.

建筑单位	Kramm & Strigl, Darmstadt
业主	Bauverein AG, Darmstadt
摄影	Prof. R. Kramm, Andrea Stahl, Madjid Asghari
技术工程师	
支撑结构设计	Ingenieurbüro Kleinhofen + Schulenberg, Darmstadt
项目管理	Andrea Stahl
房屋技术	Ingenieurbüro Renner, Kehl
基本结构	Hochtief Construct AG, NL Rhein-Main, Frankfurt am Main
传送技术	ELT Electronic, Lift Thoma GmbH, Frankfurt am Main
顶棚施工	Dachland GmbH, Mainz-Hechtsheim
电气	Ch. Nuhn GmbH, Rüsselsheim/Main
空调	Holzapfel Lüftungsbau GmbH, Großwallstadt
玻璃隔离墙	Felmer Fensterbau, Darmstadt
立面	MBM Konstruktion GmbH, Möckmühl

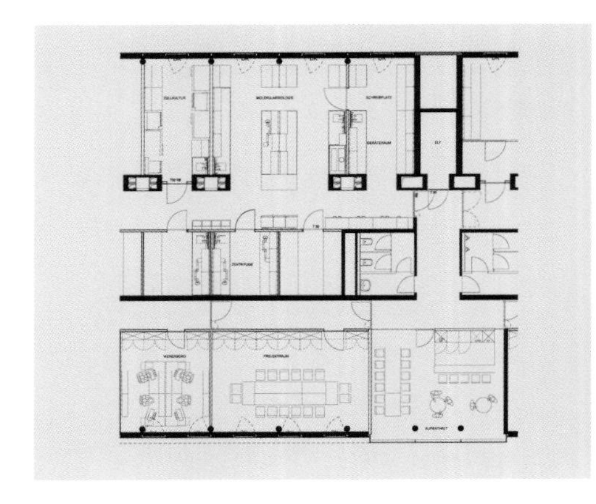

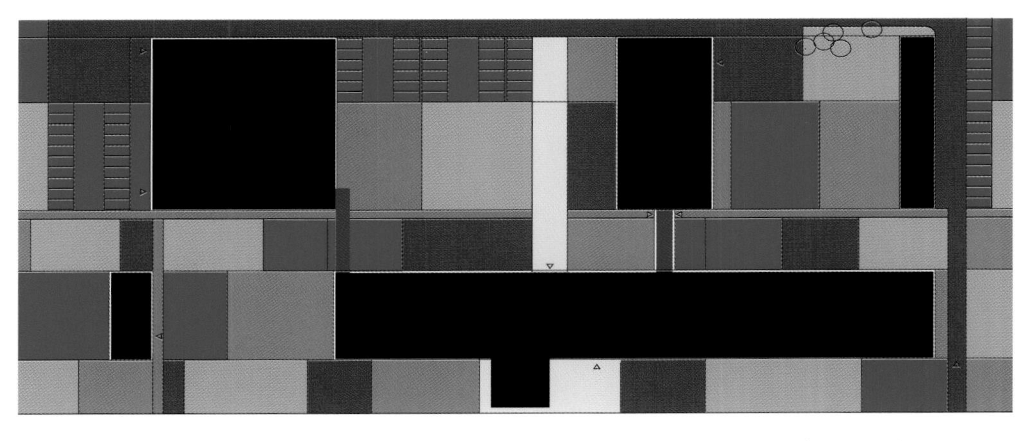

马克思普朗克分子生物研究所大楼由五部分组成：实验室、动物房、中央服务中心、宾馆和自行车车库。为了达到内与外、建筑与自然之间结合的目的，设计采用特殊而独立的透明思路。与基本的建筑理念和使用的材料形成统一。

The Max Planck Institute building consists of five sections: laboratories, animal house, central services, guest house and bicycle parking. With the aim of merging exterior and interior, building and nature, the design proposed transparent volumes which, while being distinct and independent, are unified by the basic architectural concepts and the materials used.

不同个体的开敞形结构设计也是出于对功能性的考虑。透明外墙与开敞式实验室的结合形成了很多“阁楼实验室”。这些都促进了优化功能和人类环境改造工作进程。

The open structure of the different units is also geared to functionality. The combination of generously glazed façades and open-plan laboratories has produced a number of 'loft labs'. These facilitate optimized functional and ergonomic work processes.

克莱森建筑师事务所
明斯特

马克思普朗克分子生物研究所
明斯特

MAX PLANCK INSTITUTE OF MOLECULAR BIOLOGICAL MEDICINE MÜNSTER

开放式结构

OPEN STRUCTURE

相通的方便交流区促进在这工作的科学家们进行信息交流。矮墙式的通道使大楼内的不同区域互相连接。

Clearly articulated communication areas promote informal exchanges between the scientists working here. Short walled passages interconnect the different building sections.

建筑成本	5 400 万欧元
施工期	2002–2005 年
总面积	18 450m^2
总体积	66 800m^3
竣工验收	2006 年
占地面积	19 000m^2
使用面积	3 900m^2
辅助使用面积	3 100m^2

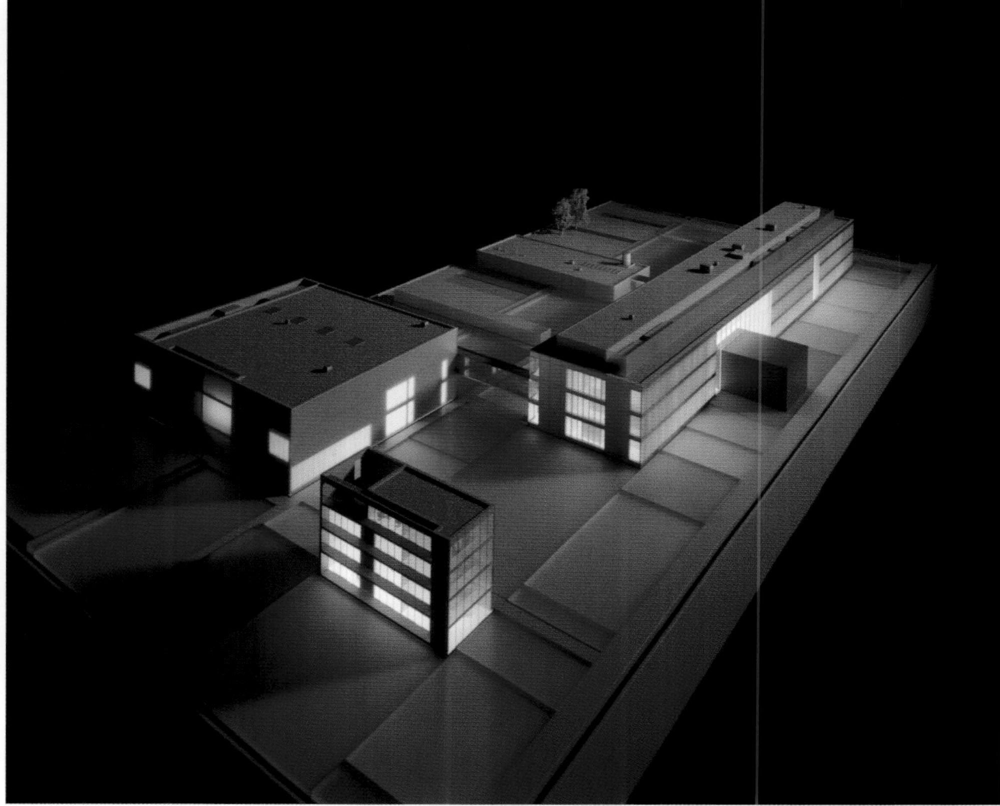

建筑单位	Kresing Architekten, Münster
业主	Max-Planck-Gesellschaft zur Förderung der Wissenschaften e.V., München
摄影	Christian Richters, Münster
可视化	Engel & Haehnel, Münster
技术工程师	
劳力规划	Dr. Heinekamp, Karlsfeld b. München / Berlin
HLS 房屋技术	IG Feldmeier mbH, Münster
ELT 房屋技术	Gerhard Loy + Partner GmbH, Hamburg
照明	Dinnebier Licht GmbH, Wuppertal
建筑物理	Peter Mutard Ingenieure mbH, Ottobrunn
静力学	Gantert & Wiemeler Ingenieure mbH, Münster

高清晰度的内部建造

BUILDING MASS AND INNER ARTICULATION

朝向院外的玻璃面由表面狭长的天棚所间隔。与接待处和光滑的玄武岩地面相适衬的大厅显得高贵典雅。在这工作的是金融方面的专家，使用面积不足 2 000 平方米的大楼非常适合他们办公。

The glass skin plane of the façades is interrupted only on the courtyard side by a narrow canopy projecting from the surface. The foyer, fitted with reception desk and polished basaltino flooring, is of modest elegance. The users are financial experts – quite fitting for a building that cost less than 2,000 €/m^2 of usable floor space.

建筑成本	1 170 万欧元
施工期	2000–2001 年
总面积	6 800m^2
总体积	22 700m^3
使用面积	6 018m^2
竣工验收	2001 年
占地面积	3 125m^2

具有动态效果的水平弧形窗与 1.15 米宽的垂直铝嵌板达到平衡效果，塔式建筑采用仰角设计，每层楼的开敞与封闭墙体各不相同。

The elevation design of the tower is based on the balance of dynamically curving horizontal strip windows and static elements in the form of 1.15 m wide vertical aluminium panels, i.e. of open and closed façade sections that vary from floor to floor.

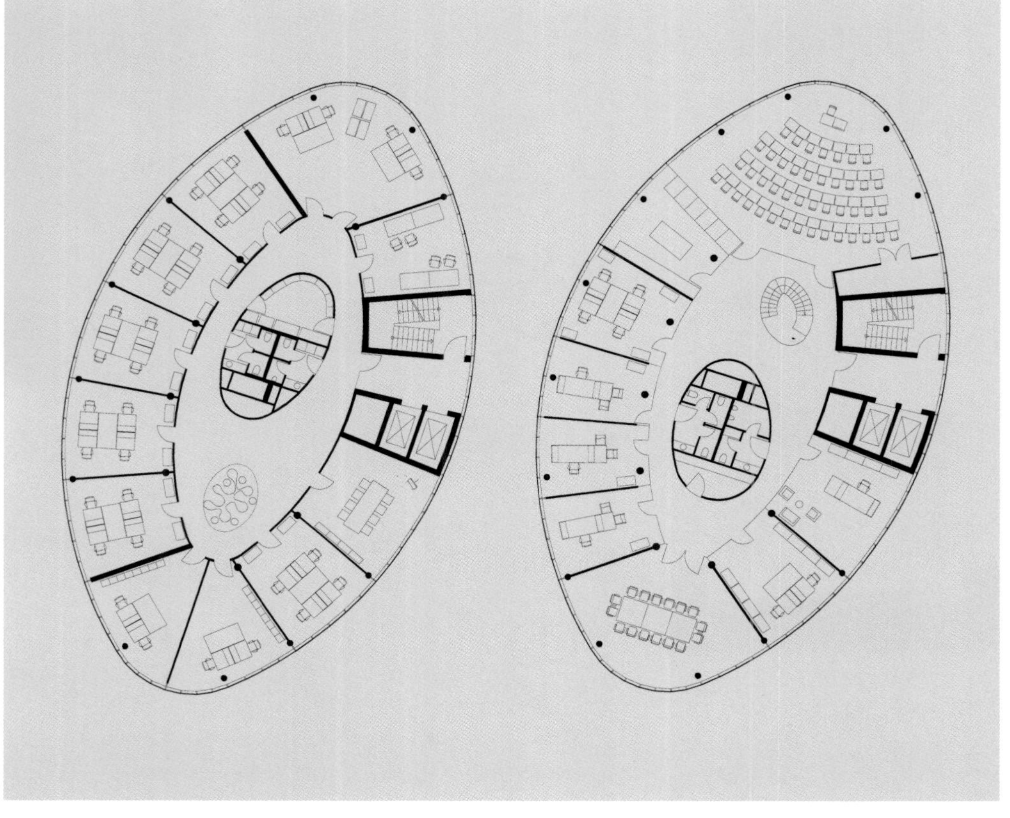

办公室外的长廊是大片的空地，这种设计构造与塔式建筑设计相符，包括中心部分或是多房间格局的构造，包括更衣室、卫生间和服务部门都能反映出大楼建筑形状。石灰质地的拱形外围安装了条形集束灯，使每两个到三个楼层就会产生变化。

The corridors in front of the offices are 'forecourts' whose plan configuration corresponds to that of the tower. They accommodate 'cores', or house-in-house structures, containing wardrobes, toilets and services and also reflect the shape of the building. Perimeter plaster coves with integrated strip lights form the top ends of these cores which change place every two to three floors.

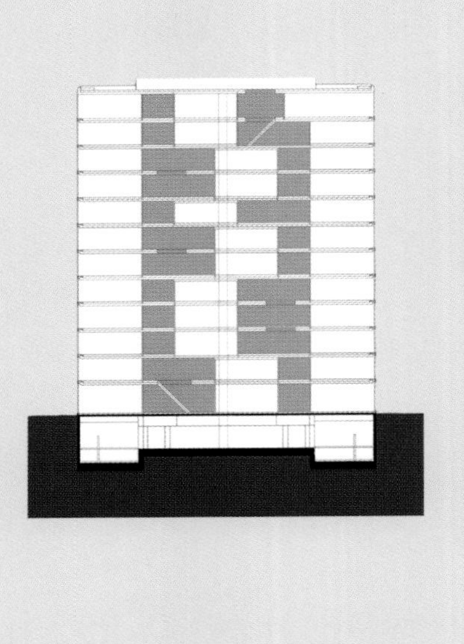

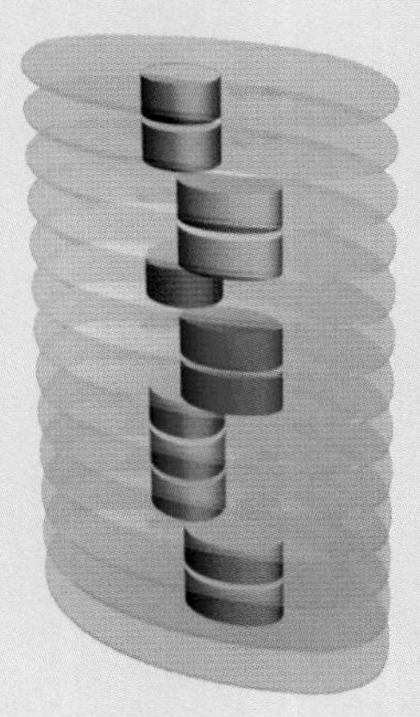

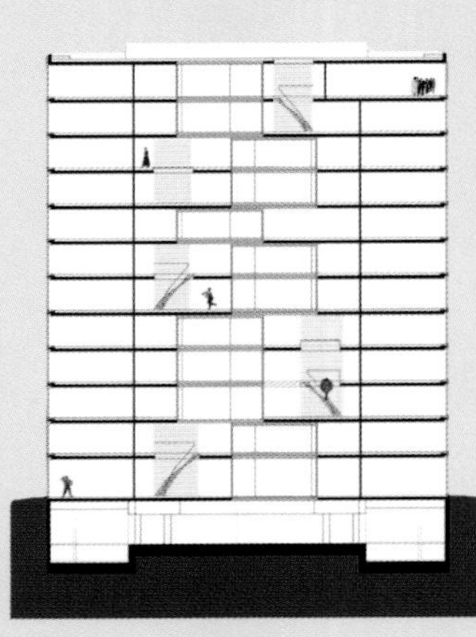

Haupt-
bahnhof
Luft
Wasser

克莱森建筑师事务所
明斯特

明斯特办公楼

OFFICE TOWER MÜNSTER

建筑与材质

CONSTRUCTION AND MATERIALS

这里没有昂贵的空调技术，没有自动控制的百叶窗及通风片，也没有双层外墙。通过铝制外层嵌板后的通风口，能够吸入自然通风。侧面的织布屏障能够抵挡强光的照射。

One will look in vain for expensive air-conditioning technology. There are no automatically controlled blinds or ventilation flaps, there is no double façade. It is possible to let in fresh air (controlled manually) via the ventilation openings behind the aluminium façade panels. Textile screens that move sideways provide protection from strong sunlight.

在大楼里有铝色涂层的纤维板后面，安装了桌面共享系统。大约只有 30% 的员工有固定的工作地点，其他 70% 的员工在其他地点开会或外出拜访客户。他们每周只有 1 天到办公室，然后根据个人所需选择规格不同的办公桌。

Behind the cladding of slit MDF panels with aluminium-colour coating, there is also an archive for the desk-sharing system used in the building. Only about thirty per cent of the employees have a fixed workplace, the other seventy per cent hold seminars elsewhere or go out to visit customers. They only come to the office one day per week and then choose a desk which they can adapt in size and height to their own needs.

大楼的外观总是能随着太阳的高度或是观看者的角度变化而变化。夜间，落地灯将室内照亮，灯光聚在办公桌周围并布满天花板，将大楼变成了灯塔。

The building constantly changes appearance, depending on solar altitudes or the perspective of the viewer. At night, the interior spaces are lit by floor lamps which spotlight the desks and flood the ceilings with light, changing the tower into a lighthouse.

建筑单位	Kresing Architekten, Münster
业主	BWN Bauträger, Münster West Invest, Düsseldorf
摄影	Christian Richters, Münster
技术工程师	
支撑结构设计	Ingenieurgesellschaft Degenhard, Münster
能源技术	Ingenieurgesellschaft Höpfner mbH, Köln/Münster
接管方	Oevermann GmbH + Co Kg, Münster
立面	Feldhaus GmbH, Emsdetten
立面系统	Schüco International KG, Bielefeld
电梯	Tepper, Münster

斯巴达银行
明斯特

SPARDA BANK MÜNSTER

办公楼与自然的相互联系

SYMBIOSIS OF OFFICE ARCHITECTURE AND NATURE

简单的框架能够给建筑勾勒出清晰的线条和轮廓。一层层按结构排列的部门构成有机整体，表现出积极进取的精神和对开发未来的憧憬。

A simple skeleton is responsible for the building's clear lines and profile. All the different parts and sections are structurally layered one on top of the other to form an effective whole conveying a sense of positive inspiration. Here, visions for the future are being developed.

建筑成本	3 930 万欧元
施工期	1998–1999 年
总面积	7 860m^2
总体积	28 500m^3
使用面积	11 400m^2
竣工验收	1999 年
占地面积	6 980m^2

延伸的长条形部分为公共区（部门办公室、研究室和会议室）。长长的斜坡将一楼与相邻的两层建筑相连接。斜坡通道光滑，建筑的大门对着日式花园。

The elongated bar-shaped section contains the public spaces (branch office, seminar and conference rooms). A long ramp mediates between this ground-floor and the adjoining two-level structure. The ramp passage is glazed and opens the building to the Japanese garden.

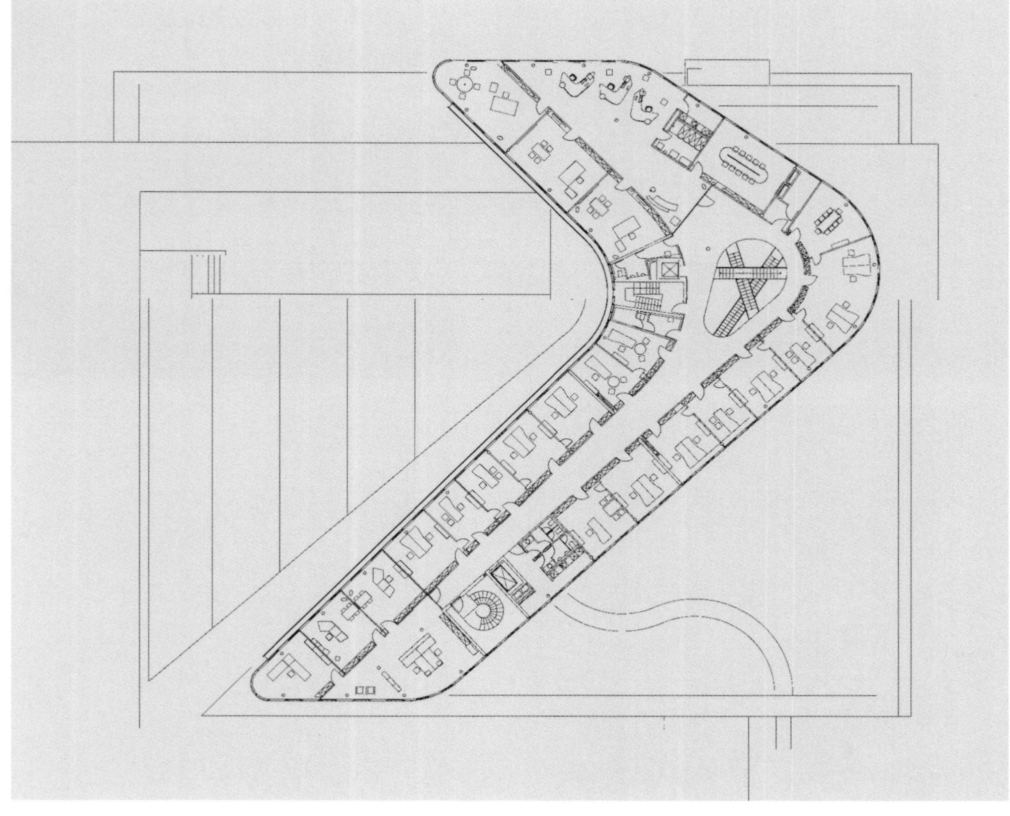

办公与自然形成了一种和谐融洽的关系，使整个田园风光与周围绿化区域保持协调。不同区域的划分和定位使城乡建筑结构互相联系，具有同等价值，并使城市区域欣欣向荣。联系着城乡建筑并使城乡建筑发挥同等作用，形成一个新的市区。

Business and nature thus form a harmonious symbiosis, keeping the balance between an intact rural landscape and the landscaped ambient site areas. The zoning of the site and the siting of the different sections of the complex interconnect the urban and rural structures, treating them as being of equal value and revitalizing an urban area.

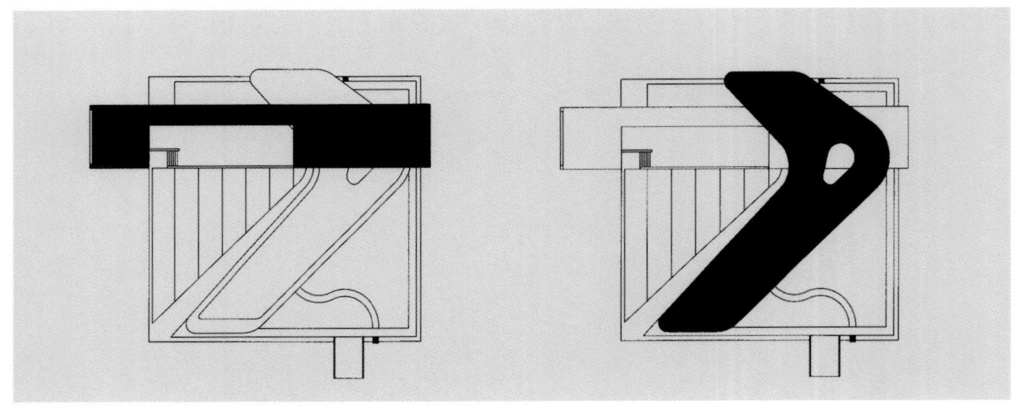

主楼的不等边曲线像是张开的双臂，这样设计象征着银行随时准备向顾客提供服务和善于接受新方法的态度，是与客户主动交流和相互沟通的标志。

The unequal-leg angled curve of the main building appears like arms opening and symbolizes the bank's readiness to offer its services and openness to new ways of doing so, a sign of proactive communication and interaction with its customers.

克莱森建筑师事务所
明斯特

斯巴达银行
明斯特

SPARDA BANK MÜNSTER

建筑设计与室内流通

ARCHITECTURAL DESIGN AND INTERIOR CIRCULATION

庞大的长条形建筑和安装玻璃的办公室的分界处是玻璃圆顶覆盖的错位式楼梯，一直伸到3楼。楼梯台阶由具有安全性的糙面玻璃制成，栏杆由具有安全性能的复合玻璃制成，扶手则由不锈钢所制。透明的、镶有金丝边图案的楼梯构造，使阳光从玻璃圆顶一直照到底层。

At the interface between the massive bar-shaped building and the glazed office wings, offset staircases covered by a glass dome lead up to the third upper floor. The steps are of matte safety glass, the parapets of laminate safety glass and the handrails of stainless steel. This filigree and transparent stairway structure lets the daylight from the glass dome through to the ground floor.

布局清晰，大面积的空间设计体现了委托公司的特征，并有利于产生清晰的思路和效果。开放灵活的内部布局产生了不同的视角效果，具有重要意义的细微之处使室内生机勃勃。

The clearly arranged and generously dimensioned interior was designed to represent the client's corporate identity and to promote clear thinking and actions. The open and flexible interior arrangement offers different perspectives and is enlivened by many details that are given new significance.

建筑单位	Kresing Architekten, Münster
业主	Sparda-Bank Münster (Westf.) eG, Münster
摄影	Christian Richters, Münster
技术工程师	
支撑结构设计	Gantert & Wiemeler Ingenieure mbH, Münster
建筑技术	IG Feldmeier mbH, Münster
玻璃 / 铝立面	Feldhaus GmbH, Emsdetten
水泥立面	Imbau Industrielles Bauen GmbH, Rinteln

KSP 恩格尔・齐默尔曼建筑设计事务所
美茵河畔法兰克福

朗根导航控制中心总部

DFS – GERMAN AIR-TRAFFIC CONTROL HEADQUARTERS IN LANGEN

低能源办公楼

LOW-ENERGY OFFICE BUILDING

建筑成本	7 200 万欧元
施工期	2000–2002 年
总面积	57 800m²
总体积	230 000m³
能源面积	44 500m²
竣工验收	2002 年
使用面积	23 600m²
获奖	“Energieland” 黑森，2002 年

建筑师为德国航空运输协会建造了一座高质量的工作场所和最大灵活性办公楼。结构的突出特点是建筑设计、功能有效性和通过宽敞的庭院充分利用日光。地热能源探测系统的安装迎合了低能源办公楼的需求。

The architects erected an office building with high-quality workplaces and maximum flexibility for the German air-traffic control association. The structure is outstanding for its architectural design and functional efficiency as well as for its maximum use of daylight through spacious patios. Equipped, as it is with a geothermal energy probing system – on a scale unique in Germany – it meets the demand for a low-energy office building.

尽管投资费用较高，这种节能的加热和冷却系统要比传统的系统先进，这是由于长时间的运转能够降低能源成本：每年的地热能源设备所达到的总利润为初始投资的16%。

Despite higher investment costs, this energy-saving heating and cooling system is superior to conventional systems because it saves energy costs in the long run: the annual profit for the geothermal energy plant amounts to 16% of the initial investment.

朗根导航控制中心总部

DFS – GERMAN AIR-TRAFFIC CONTROL HEADQUARTERS IN LANGEN

KSP 恩格尔・齐默尔曼建筑设计事务所
美茵河畔法兰克福

地热能源探测系统

GEOTHERMAL ENERGY PROBE SYSTEM

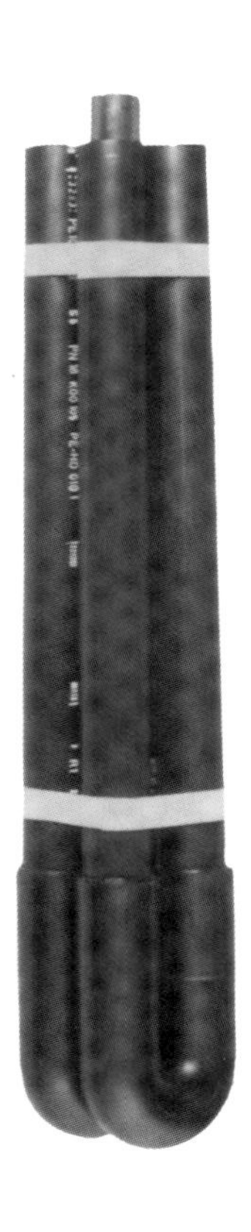

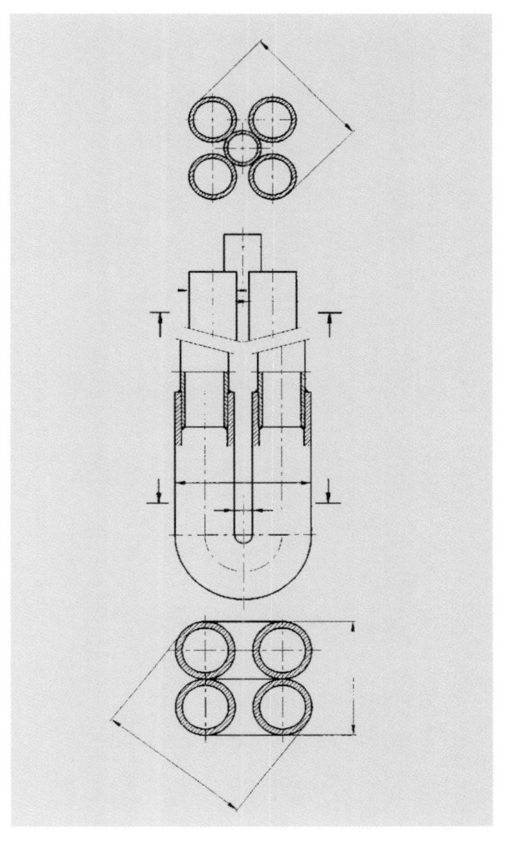

该地热能源探测系统是欧洲最大的探测系统之一，是 DFS 建筑全部能源的重要组成部分。系统由 154 根能够达到地下 70 米的地热能源子探测器（聚乙烯 U 形管）组成。他们可以吸收并传导能源进行加热和冷却，所需功率分别为 380 千瓦和 340 千瓦。冬季，系统能够从可再生能源中提取超过 50% 以上的用于加热和冷却的可再生能源。

The geothermal energy probe system – one of the largest in Europe – is the central element of the overall energy concept for the DFS building. It supplies the basic energy load for heating and cooling the building. The system consists of 154 geothermal energy probes (U-shaped tube loops of polyethylene) reaching a depth of 70 m altogether. They draw and transmit a wattage of 380 kw for heating and 340 kw for cooling. In winter, the system supplies over 50% of the consumed heating and cooling power from a renewable source of energy.

建筑单位	KSP Engel und Zimmermann, Frankfurt am Main
业主	DFS – Deutsche Flugsicherung GmbH, Langen
景观规划	Fenner · Steinhauser · Weisser, Düsseldorf
摄影	Jean-Luc Valentin, Frankfurt am Main
技术工程师	
能源技术	Lemon Consult GmbH, Zürich, im Untervertrag von Amstein + Walthert AG, Zürich
地热检测	UbeG GbR, Wetzlar-Nauborn
支撑结构设计	Krebs und Kiefer, Darmstadt

The basic design idea was to convert the former brewery complex into a mixed-use urban centre in an industrial monument. This led to the creation of a 'neighbourhood in a neighbourhood' on Hanauer Landstrasse in Frankfurt's eastend that takes into account its new users, their working methods and requirements. With its architectural variety, it offers scope for office suites and lofts; retail spaces and workshop floors; a 'factory of fiction' and artists' studios; catering outlets; a jazz club; photo and casting agencies; a hairdresser's, a discotheque and a dressmaking studio – in an open ensemble that attracts people in the way a whole urban neighbourhood does elsewhere.

基本的设计思路是将过去的酿酒厂改建为工业历史中的综合性市中心。这使得法兰克福东部的 Hanauer Landstrasse 街道形成“邻里之间的关系”。建筑具有多样性，它提供了办公房间和阁楼、零售空间和车间、“虚构”工厂和画家工作室、餐饮场所、爵士乐俱乐部、摄影和铸型代理处、一个理发店、一个迪斯科舞厅和裁缝店，这些全部都对外开放，因此，吸引人们来此光临，促进了城市及其他地区的邻里关系。

兰德斯及其合伙人建筑设计事务所
美茵河畔法兰克福

酿酒厂
美茵河畔法兰克福

'UNION' BREWERY COMPLEX
FRANKFURT/MAIN

生活空间的邀请

INVITATION TO LIVING

Baukosten	25 Mio. Euro
Planungs- und Bauzeit	1996–2000
Bruttogeschossfläche (BGF)	34.000 m^2
Bruttorauminhalt (BRI)	147.000 m^3
Hauptnutzfläche	26.400 m^2
Fertigstellung	2000
Grundstücksfläche	12.500 m^2
Bebaute Grundstücksfläche	8.100 m^2

建筑成本	2 500 万欧元
施工期	1996–2000 年
总面积	34 000m^2
总体积	147 000m^3
使用面积	26 400m^2
竣工验收	2000 年
占地面积	125 000m^2
可建筑地表面积	8 100m^2

对于来访者而言，这个联合体本身就像是一座城市的缩影。广场内生长着成熟的栗树，这里是整个建筑的中心，四周被不同的设施包围着。过去酿酒厂建筑的细微不同之处就是被人们刻意地保护和扩建。

To visitors, the Union complex presents itself like a city in miniature. The piazza with its mature chestnut trees (to the right) is the heart of the ensemble, surrounded by different facilities. The small-scale heterogeneity of the older brewery buildings was deliberately preserved and extended.

尽管每一个独立的建筑都有各自的特点，除了现在形成的公共区域和整个建筑之外，单个建筑之间并不具有相互转换性，最后形成的不是一个已建成的住房，也不是前卫的建筑产物，而是一个充满生气的具有良好职能的城市。

Although every detached building has its own character, the individual structures were not the priority of the conversion, but the ensemble and the (public) space it now forms. The result is not a built manifesto, not an architectural avant-garde creation, but a piece of a vibrant and well-functioning city.

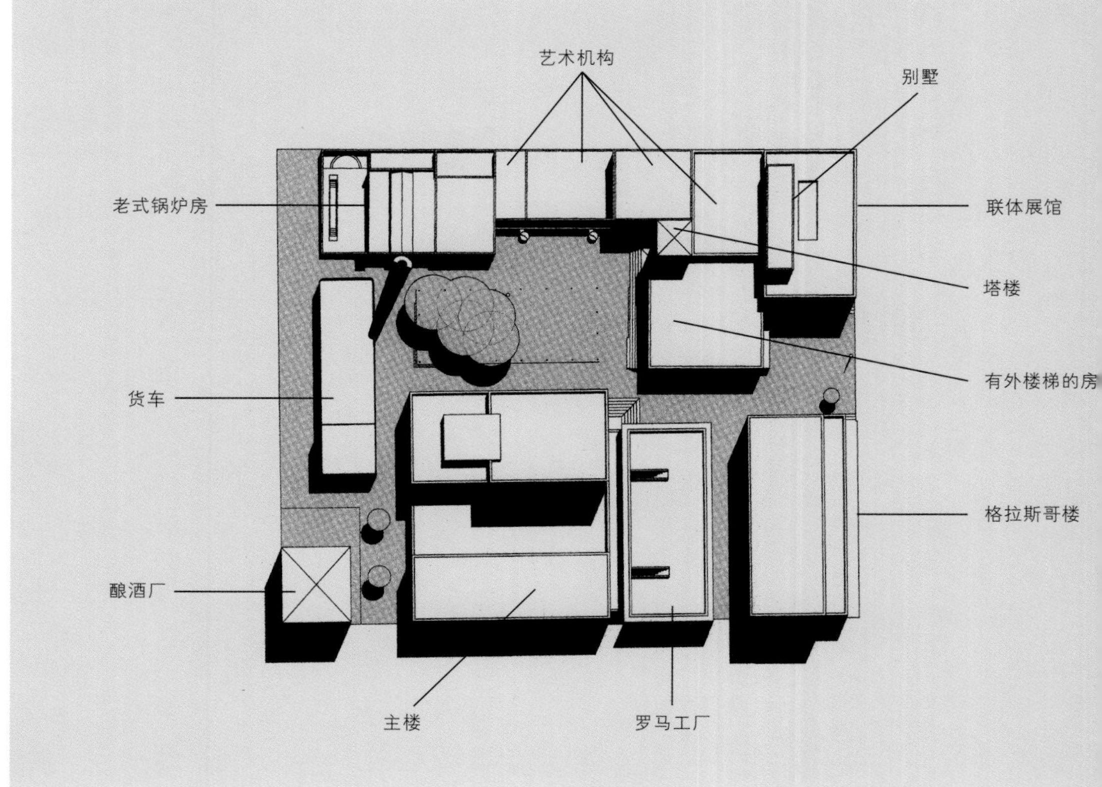

有意地调整市场概念可以适应灵活性和波动性，同时也能适应与流动的员工、生产商及其合作者做生意的新形成。

The marketing concept was deliberately geared to flexibility and fluctuation, to new forms of doing business with changing staff, production teams and partners.

LANDES & PARTNER ARCHITEKTEN
FRANKFURT AM MAIN

酿酒厂
美茵河畔法兰克福

'UNION' BREWERY COMPLEX
FRANKFURT/MAIN

兰德斯及其合伙人建筑设计事务所
美茵河畔法兰克福

多样性的结合

'UNION IN DIVERSITY'

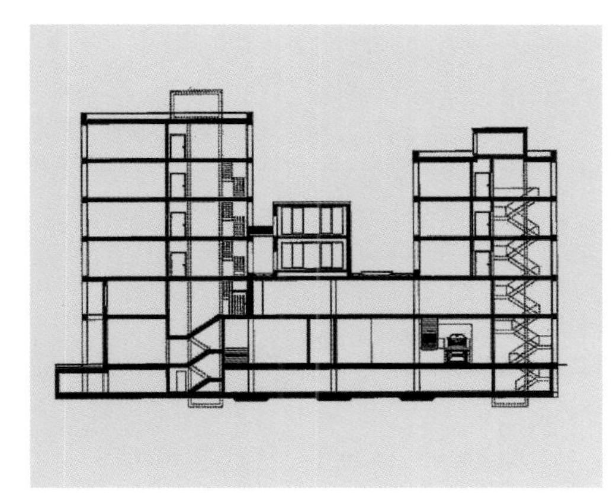

让用户容易掌握使用的建筑新设计创造了新颖的阁楼，如开阔的矩形空间可以被单独划分或按规格设置。外露的混凝土砖墙，裸露的管子，涂有透明漆的散热器和石木样板使室内凉爽，这种凉爽受每一个富于创造的工人欢迎，并在联合体大楼开业前就被全部租用。

The user-friendly architectural redesign created lofts in the original sense, i.e. large rectangular open spaces that can be individually partitioned and fitted. Exposed concrete or brick walls, unclad pipes, transparent-lacquer-coated radiators, and stone-wood screed radiate the cool charm every creative worker likes and resulted in the Union complex being fully rented out even before its opening.

无论你在哪儿，都可看到每个角落或在每个角落都能被看到：从供人们坐下休息的台阶、从大窗户的前面或后面和锅炉房内。锅炉房建于1908年，建筑的主体高高耸起并突出了建筑工业生产的历史。

From everywhere one can see and be seen: from the steps that invite people to sit down, from in front of or behind the large windows and inside the former boiler house, built in 1908, whose tall stack sets a vertical dominant element and underlines the industrial past of the complex.

建筑单位 Landes & Partner Architekten, Frankfurt am Main
业主 Benjamin Goldmann Nachlass, Frankfurt am Main
摄影 Ivan Nemec, Berlin

技术工程师
建筑工程管理 B & S Consulting GmbH, Frankfurt am Main
支撑结构设计 Thürauf & Partner, Frankfurt am Main
灯光设计 Daniel Zerlang-Rösch, Frankfurt am Main

参与企业
电气 / 灯光安装 Nohl Darmstadt GmbH & Co. KG, Darmstadt
砖面 Klinkerwerke Wittmund GmbH, Wittmund
空调 Kälte Steudter, Frankfurt am Main
金属安装 Schölch GmbH Stahl- und Metallbau, Hardheim
Kührener Metallverarbeitung GmbH, Kühren
主体建筑 Philipp Holzmann AG, Neu-Isenburg
保险技术 Limburger Sicherheitstechnik
Hillebrand & Lotz GmbH, Limburg a.d. Lahn

克利斯多夫·麦克勒教授建筑设计事务所
美茵河畔法兰克福

办公楼
艾斯克米尔大街
美茵河畔法兰克福

OFFICE BUILDING
ESCHERSHEIMER LANDSTRASSE 6
FRANKFURT/MAIN

市内环境

URBAN SITUATION

建筑成本	1 200 万欧元
施工期	2000–2002 年
总面积	6 854m^2
总体积	23 992m^3
规划草案	1999 年
占地面积	2 434m^2
建筑用途	出租
建筑面积	963m^2

双倍高度的门廊迎接着来访者。

A double-height entrance hall welcomes visitors.

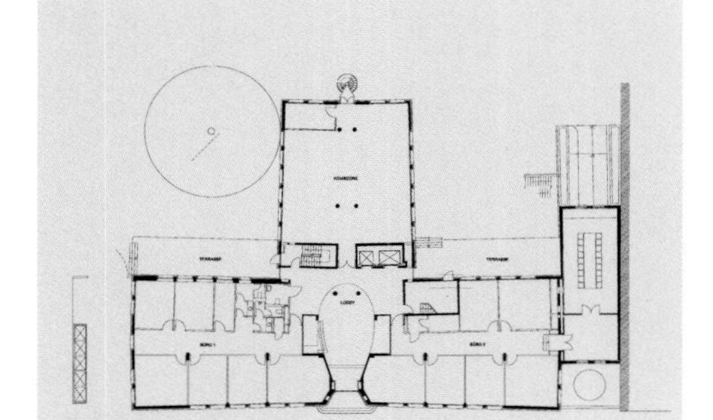

这个 7 层的建筑是个 T 形行政办公楼，全部被利用而且人们经常光顾。它与这条街上的其他建筑一样，正面有相同的邻街道路，类似于独立的别墅。因为相邻近的建筑设计有不同的外立面，并且顺序“后退”才这样说。

This seven-storey structure is a fully occupied and highly frequented T-shaped administration building. It has the same frontage lines as other buildings on this street, yet is experienced as an independent building, similar to a free-standing villa, because the section adjoining the neighbouring building has a different façade design and 'steps back in line', so to speak.

对称格局的外立面，按照传统方式被分为上、中、下三部分。面向大楼的正中间，可以轻轻地推门而入。这种设计手段将一楼的楼梯前留出空间。

The symmetrical main façade is divided in the classical way into base, middle and a striking top. Towards the middle, the building is slightly turned inward to emphasize the entrance. This design step made room for a small flight of front stairs up to the raised ground floor.

凹进的部分是一个会议室，其窗户为木制窗框，高 5.6 米。

The recessed building section contains a conference room with wood-framed windows 5.6 m high.

克利斯多夫・麦克勒教授建筑设计事务所
美茵河畔法兰克福

办公楼
艾斯克米尔大街
美茵河畔法兰克福

OFFICE BUILDING
ESCHERSHEIMER LANDSTRASSE 6
FRANKFURT/MAIN

外观设计 / 能源概念

FAÇADE DESIGN / ENERGY CONCEPT

所有能够俯瞰到大街的窗户都是双层的。另外，装在这些玻璃前面的玻璃嵌板将办公室与四条繁忙小巷上的交通喧嚣隔离开来。

All the windows overlooking the street are double windows. In addition, glass panels mounted in front of them shield the offices from the traffic noise of the busy four-lane thoroughfare.

大楼安装了中央通风系统，通过隔热的等热制冷系统能够获得大部分所需的冷却效果，即使空间流量最热，也能达到卫生效果。同时，建筑师还考虑了耗能或是节能方面问题，形成一个在夏季也能使室内保持凉爽的混凝土区。

The building is equipped with a central ventilation system. In order to achieve the 'hygienical' minimum air circulation, a large portion of the necessary cooling effect is achieved by way of the adiabatic cooling process. At the same time, the architects also considered energy consumption, or rather energy saving aspects and activated the concrete core for continual interior cooling during in the summer.

建筑单位 Prof. Christoph Mäckler Architekten, Frankfurt am Main
业主 Groß und Partner, Grundstücksentwicklungsgesellschaft mbH & Ca. Solitär KG, Frankfurt am Main

技术工程师
房屋技术 IBK Ingenieurbüro Klöffel, Bruchköbel
静力学 Ing.-Büro Kannemacher & Dr. Sturm, Frankfurt am Main
木质窗户和百叶窗 Schreinerei Löb, Sinn-Fleißbach
立面 Marmorwerk Grünzig GmbH, Aachen
照明 Belux, Wohlen (Schweiz)
摄影 Christoph Lison, Frankfurt am Main

克利斯多夫·麦克勒教授建筑设计事务所
美茵河畔法兰克福

工厂改造的办公室
哈瑙

DMC2
CONVERSION OF A FACTORY HALL
INTO OFFICES, HANAU

设计

DESIGN

建筑成本	1 250 万欧元
施工期	2000–2001 年
使用面积 I	2 041m²
使用面积 II	3 522m²
设计	1999 年
占地面积	5 000m²
用途	出租

一个建于1910年的老式的火炉生产大楼要被改成现代的办公室。为了能够保持工厂原有的结构、吸引力，其宽敞的空间结构也给人们留下深刻印象，建筑师根据“房中房”的理念进行设计，取得一些明显的突破。此外，老工厂外墙的铜绿仍然保持完好。现在它为新建立的独立办公室结构形成了一个挡雨膜。

An old stove manufacturing hall of 1910 was to be converted into modern offices. In order to preserve the structure and appeal of the old factory, including its impressive spaciousness, the architects based their design on the idea of a ‘house in a house’. Except for just a few obvious ‘penetrations’, the façade of the old factory hall with its ‘patina’ was preserved unchanged. It now forms a weather-protective membrane for the new free-standing office structure erected inside.

老式火炉生产大厅的外墙为这个“房中房”起到保温膜的作用。屋内温度从来都不会低于11℃。这种能源理念节省了一笔可观的取暖费用。

The outer walls of the old stove manufacturing hall function as a temperature insulation membrane for this 'house in a house'. The interior hall temperature never sinks below 11°C. This energy concept led to a considerable reduction in heating costs.

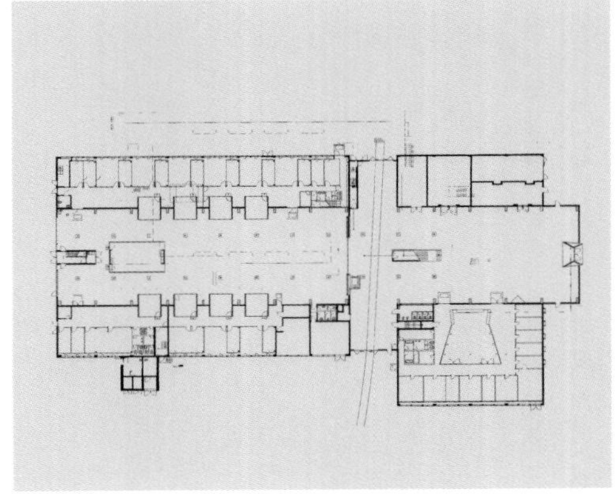

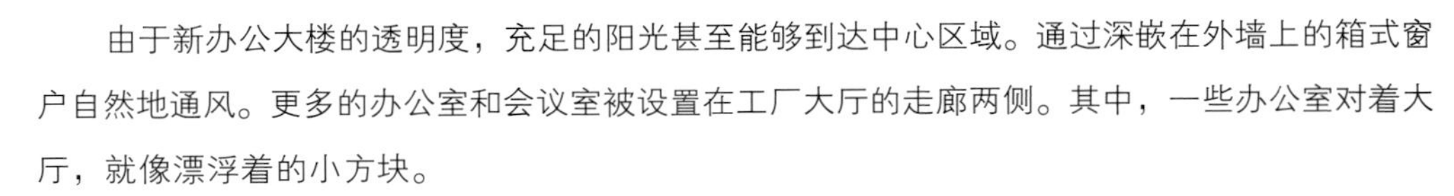

由于新办公大楼的透明度，充足的阳光甚至能够到达中心区域。通过深嵌在外墙上的箱式窗户自然地通风。更多的办公室和会议室被设置在工厂大厅的走廊两侧。其中，一些办公室对着大厅，就像漂浮着的小方块。

Due to the transparency of the new office building, ample daylight reaches even its central spaces. They are naturally ventilated through deep box-type windows perforating the outer walls. More offices and conference rooms are placed on two levels in the 'side aisles' of the factory hall, a few of them projecting into the large hall like floating cubes.

工厂改造的办公室
哈瑙

DMC²
CONVERSION OF A FACTORY HALL INTO OFFICES, HANAU

克利斯多夫·麦克勒教授建筑设计事务所
美茵河畔法兰克福

通风 / 结构

VENTILATION / STRUCTURE

巨大的悬臂遮篷保护着入口，从这里可以直接进入楼内。由于新结构并没有完全占据整个工厂，进入休息大厅的拜访者能够感受到旧工厂的巨大规模。透明的办公建筑，作为一个独立结构加插于楼内。通过工厂内的楼梯和人行天桥，可以进入同高的格孔开敞式办公室。

The building is accessed through a portal protected by a large cantilever canopy. Inside, visitors entering the foyer are able to experience the large dimensions of the old factory hall to the full because the new structure does not fill it completely. The filigree, transparent office building was inserted as a free-standing structure into the hall. The level containing cellular and open-plan offices is accessed via stairs anchored in the factory floor and a 'skywalk'.

建筑师 Prof. Christoph Mäckler Architekten, Frankfurt am Main
业主 dmc² division, verkauft an OMG, verkauft an Umicore Konzern

技术工程师
房屋技术 Schmidt Reuter Partner, Köln/Offenbach
照明 Belux, Wohlen (Schweiz)
FSB Franz Schneider Brakel GmbH, Brakel
隔板系统 Bless Art, Dürnten (Schweiz)
摄影 Christoph Lison, Frankfurt am Main

这座新的建筑吸收了先前建筑的建筑理念：虽然和它相邻的建筑同等高度，并不紧凑。除了在地基连接的部分有三个独立的塔楼和邻近的老建筑基本相似。

The new building takes up the architectural concept of the old: though of the same height as the neighbouring existing building, it does not form a compact block, but consists of three separate towers rising from a joint base structure, again in analogy to the base course of its neighbour.

对于提到的相邻的乌尔斯泰因议院大楼，设计了三个在乌尔斯泰因大街上独立的塔楼和一个能鸟瞰整个泰尔托运河的大厅式建筑。在进行平面设计的时候，这个建筑的功能还没有被确定，因此，设计师采用了“混合”建筑型设计。

Respecting the neighbouring listed Ullstein House building, the design comprises three separate towers on Ullsteinstrasse and a hall-type building overlooking the Teltow Canal. At the time of planning, the function(s) of the complex had not yet been fixed, the designers therefore opted for a 'hybrid' building type.

纳尔巴奇兄弟建筑设计事务所
柏林

乌尔斯泰因房产交易中心
柏林

ULLSTEIN HOUSE BUSINESS CENTRE
BERLIN

似混合物的办公区

OFFICE BLOCK AS A HYBRID

建筑成本	7 500 万欧元
施工期	1991–1995 年
总面积	69 260m^2
总体积	242 400m^3
占地面积	33 460m^2
使用面积	34 000m^2
使用面积	6 800m^2

塔楼之间的安全梯，Gosz Hall 和大厅式建筑。

Emergency staircases between towers, Gosz Hall and hall-type building.

从乌尔斯泰因大街所看到的全景。

View from Ullsteinstrasse.

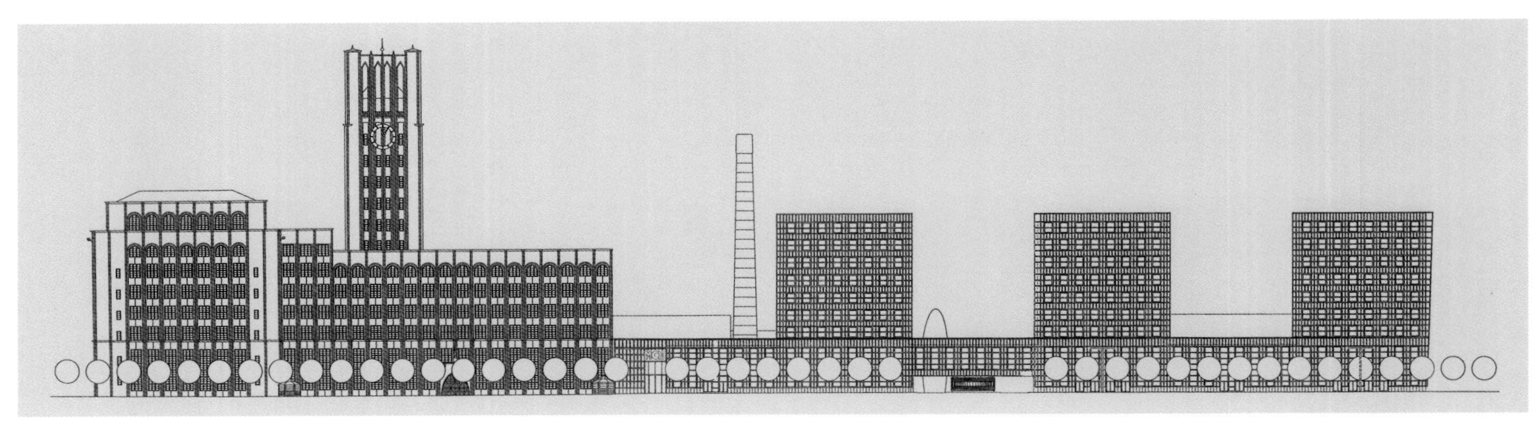

二楼平面图。

1st-floor plan.

一楼平面图。

Ground-floor plan.

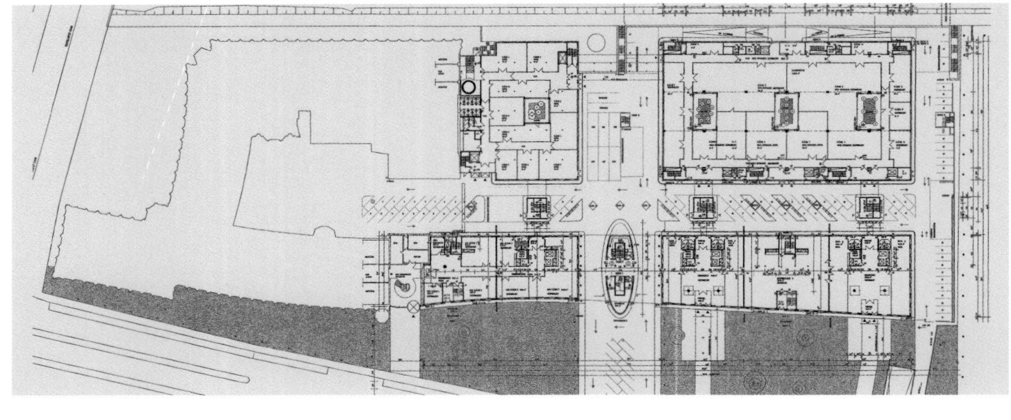

整体建筑不仅包括交叉使用区，还有开敞式设计的办公室结构，办公室未装修时有 3.5 米高，安装技术设施并吊棚之后的净高为2.75~3米。对此而言，遵守劳动法律规则是很有必要的。这个“综合建筑”具有很大的灵活性，提供了从小到大的可以租用的办公单元。

The brief included not only mixed uses, but also open-plan office structures with unfinished rooms at least 3.5 m high with a clearance height of 2.75 to 3 m after suspended ceilings had been fitted behind which service technology was installed. For this, it was absolutely indispensable to comply meticulously with labour law regulations. The ‘hybrid’ offers great flexibility for letting everything from small to large rental office units.

NALBACH + NALBACH ARCHITEKTEN
BERLIN

乌尔斯泰因房产交易中心
柏林

ULLSTEIN HOUSE BUSINESS CENTRE BERLIN

纳尔巴奇兄弟建筑设计事务所
柏林

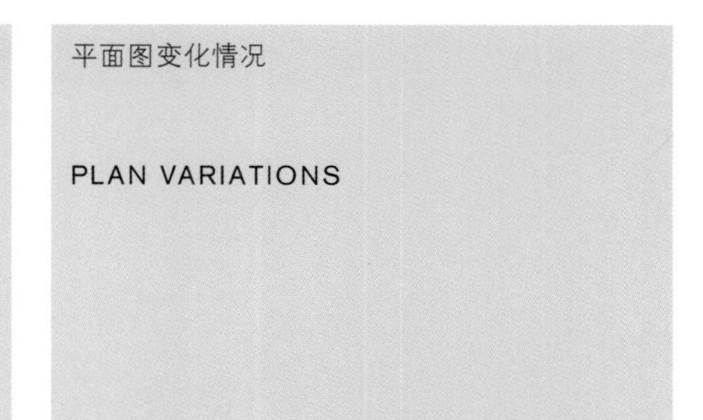

平面图变化情况

PLAN VARIATIONS

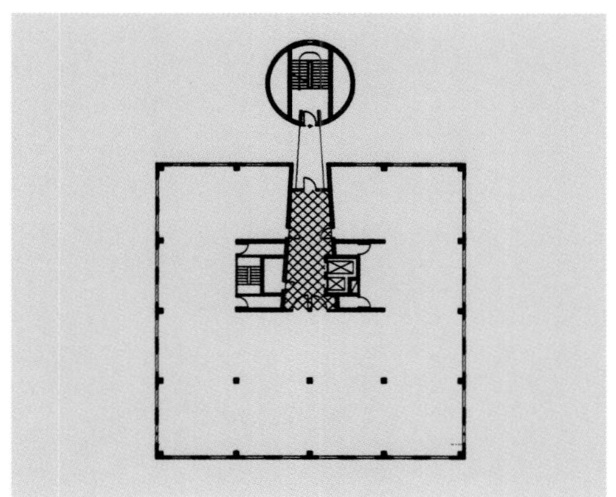

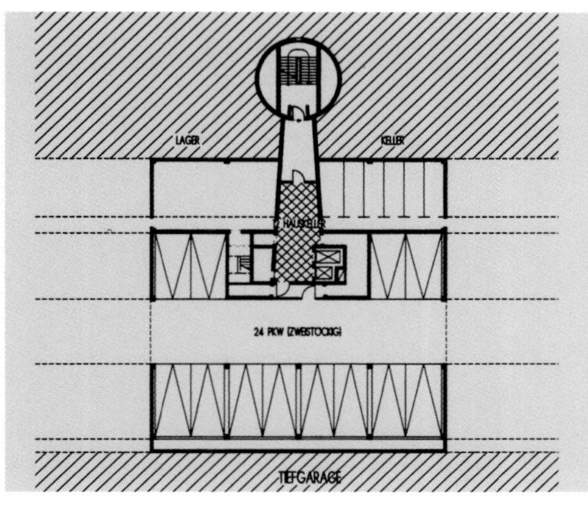

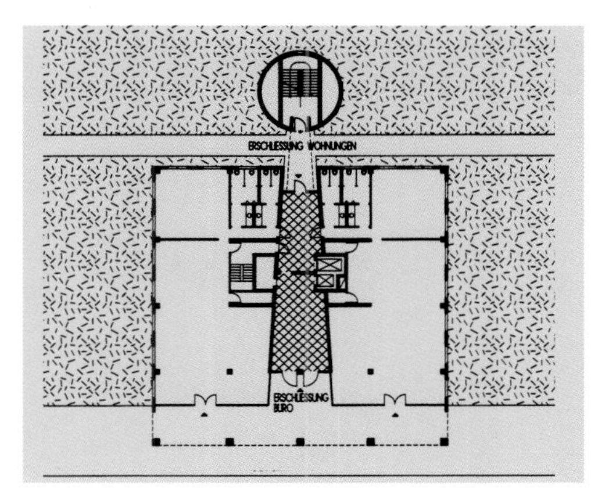

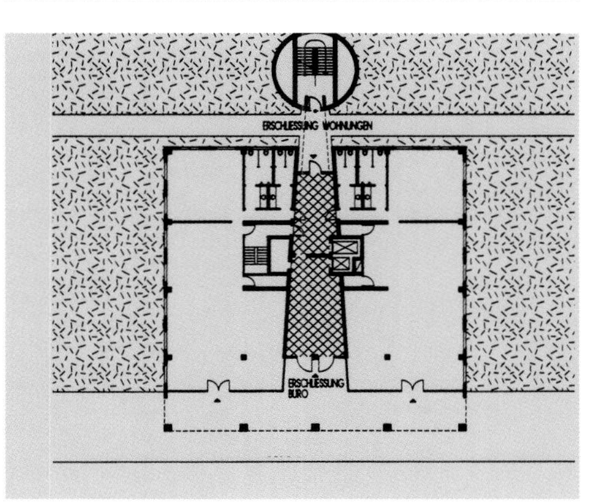

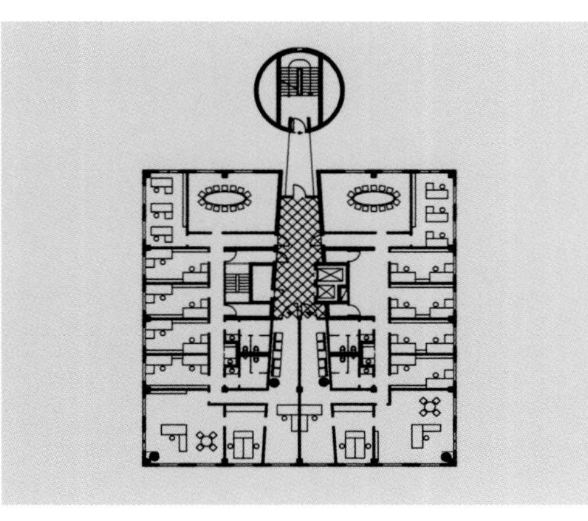

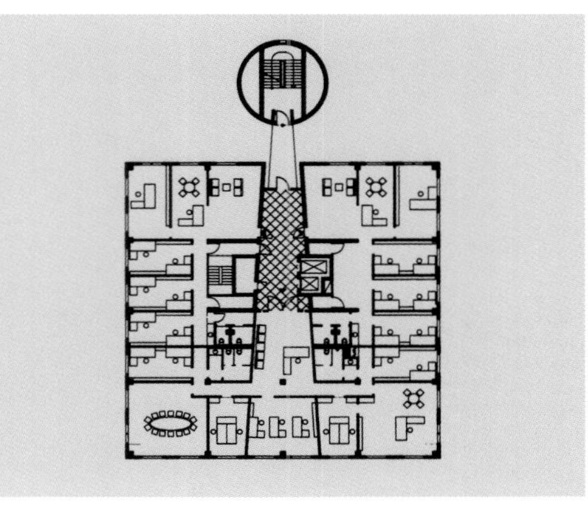

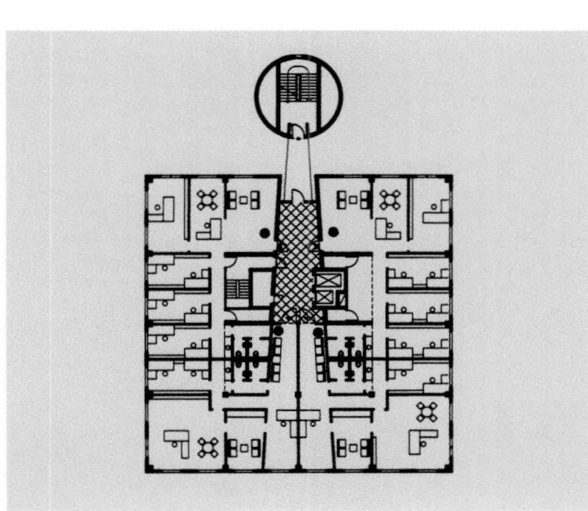

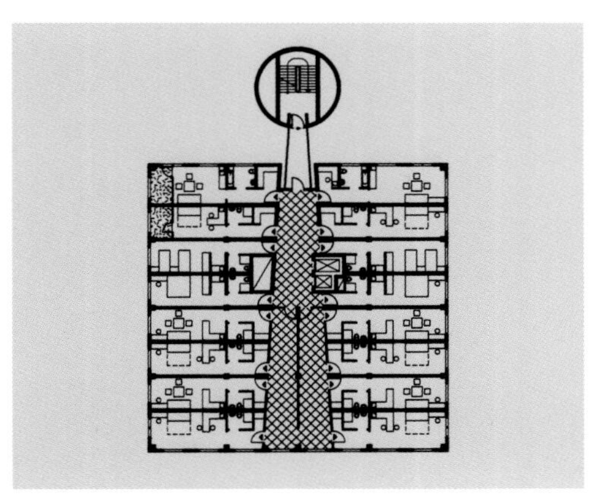

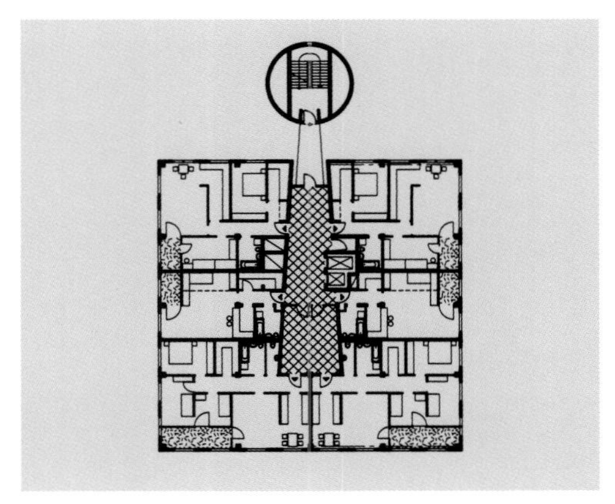

建筑平面图的变化（从左上角顺时针）：构架—地下停车场—业务用区—办公用区；开敞式平面布置—办公用区；两个单元—办公用区；三个单元—办公用区；四个单元—带有16个客房的旅馆—住宅用区；6个公寓住宅。

Floor plan variations (clockwise from top left): Carcass – underground parking – business use – office use: open-plan – office use: two units – office use: three units – office use: four units – hotel use with 16 guest rooms – residential use: six apartments.

除了这个功能型大楼的投资问题，建筑的设计利用混合用途的解决办法使其简化，安全方面的问题尤显重要，甚至决定了设计因素。外部的安全楼梯通道在设计时缩短了每个单元的逃生路线。这使得设计者忽略了辅助逃生路线，给用户带来不便，尤其是小型区的用户不得不利用他人地方通行。

Apart from the problems of funding a functional building – in this case substantially simplified by means of a hybrid-purpose solution – safety aspects represented important, even determining design factors. Exterior emergency staircases were designed to shorten escape routes for every unit. This enabled the designers to omit secondary escape routes which, especially in the case of small-scale interior partitioning, would inevitably have led to users having to cross the areas occupied by other tenants.

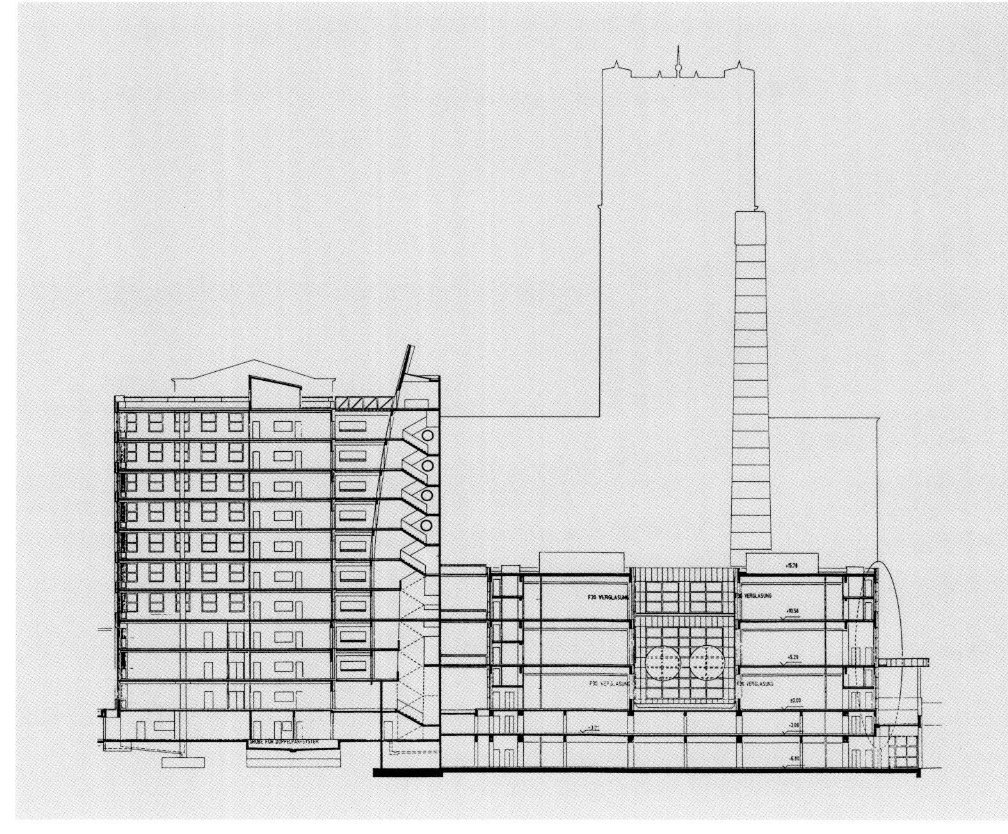

每座塔式大楼都可由一个雅致的门廊进入，每个楼层都能划分成 4 个租用单元。塔形建筑由共同的地基相连，使整幢建筑在未来有可能成为一个大的承租网点。

Every tower block is accessed via an elegant entrance hall and can be divided into up to four rental units per floor level. The towers are interconnected by their joint base level; this makes it possible to let the entire building to one large corporate tenant at some future date.

厅式建筑的内部高度为 5.25 米，是塔式大楼楼面高度的 1.5 倍。这种构造从厅式建筑的 2 楼到塔楼的 3 楼形成了径直且水平的过渡区，用作时尚中心的环形区域。厅式建筑的每层都有 500 平方米的 6 个租用单元，外设 3 个庭院。

The interior height of 5.25 m in the hall-type building is exactly 1.5 times the floor level heights of the tower buildings. This produced direct, level transition spaces from the 2nd floor of the hall-type building to the 3rd tower floors, which are used as part of the fashion centre's 'circuit'. The hall-type building contains a maximum of six rental units of 500 m^2 each per level surrounding three courtyards.

乌尔斯泰因房产交易中心
柏林

ULLSTEIN HOUSE BUSINESS CENTRE
BERLIN

纳尔巴奇兄弟建筑设计事务所
柏林

建筑与材料

CONSTRUCTION AND MATERIALS

整个建筑的全景，1998。

承重结构：
扩建：现浇钢筋混凝土构造
大厅式建筑：分别为装配式钢筋混凝土构造和组合钢架构造
外墙：混凝土制壁板，铝制窗框或铝制梁柱结构

Overall view of the completed complex, 1998

Load-bearing Structure:
Extension: cast-in-situ RC structure
Hall-type building: prefab RC and composite steel structure respectively
Façades: curtain wall of concrete blocks, aluminium-framed windows, or aluminium post-and-beam structure.

在三维空间方面，最有趣的一点就是新老建筑之间的连接处。它是一个带有中央电梯的3层的"玻璃密封圈"，一个可以到达新老建筑不同楼层的观光电梯和现存的烟囱。

In terms of three-dimensional space, the most interesting point is the connection between the old and the new buildings – a three-storey 'glass joint' which accommodates the central stairway, a panorama lift with stops at the different levels of both old and new buildings, and the existing chimney stack.

建筑单位 Nalbach + Nalbach Architekten, Berlin
业主 Becker & Kries Grundstücks GmbH & Co., Berlin
重要客户 Mode-Center-Berlin Management-Verwaltungs GmbH, Berlin
主营业建筑 Strabag AG Direktion Hochbau Ost, Berlin
摄影 Reinhard Görner, Berlin

技术工程师
静力学 Ingenieurbüro Fink, Berlin
供暖 / 空调 PIN Planende Ingenieure, Berlin
讨论会 system ingenieur planung, Berlin
电气 Ingenieurbüro Wolfgang Bormann, Berlin
外部设备 Bernd Slopianka, Berlin

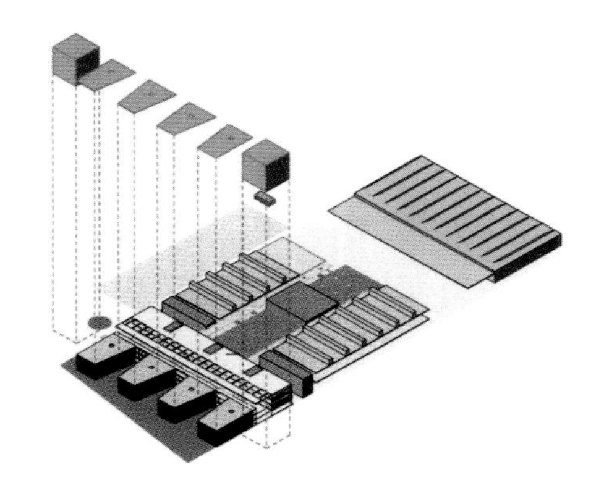

办公大楼大门朝东。沿着湖边有 4 个侧楼。西面是有关工程样本、检验、质量管理和下部构造的大厅。办公室和车间由玻璃架桥相互连接，并能在第二个建筑阶段进行扩建。

The office block completes and closes the complex to the east. Wings formed like the 'teeth' of the GETRAG logo are arranged on a lakeside. To the west are the halls for sample engineering, testing, quality controls and infrastructure. The office and workshop buildings are interconnected by glazed bridges and can be extended in a second construction stage.

诺伊格鲍尔与罗斯
斯图加特

格特拉克研发中心
安特格拉彭堡

GETRAG INNOVATION CENTRE UNTERGRUPPENBACH

统一与审美观

IDENTIFICATION AND AESTHETICS

施工期	1999–2001 年
总面积	16 000m^2
总体积	69 000m^3
竣工验收	2001 年
使用面积	27 000m^2
占地面积	50 000m^2

汽车配件制造商格特拉克的创意中心作为公司的研发部和行政总部，位于安特格拉彭堡地区附近的城郊。该建筑具有感官上的吸引力，并集功能性、适应性和经济效益于一体。与其相匹配的建材和有清晰的建筑线条代表着格特拉克优质产品。无论是公司的员工还是客户都能容易产生主人翁意识。

The Innovation Centre of the automobile spare parts manufacturer GETRAG's serves as the company's research and development section and administrative headquarters. It is situated in the countryside near Untergruppenbach. The building unites functionality, flexibility and economic efficiency with aesthetic appeal. Just a few well-matched materials and clear lines are representative of GETRAG quality products. Both staff members and customers have a sense of 'ownership' of the company.

NEUGEBAUER + RÖSCH
STUTTGART

格特拉克研发中心
安特格拉彭堡

GETRAG INNOVATION CENTRE
UNTERGRUPPENBACH

诺伊格鲍尔与罗斯
斯图加特

交流与功能

COMMUNICATION AND FUNCTION

我们称之为市场大厅的是个吊顶大厅，里面充满阳光。这里是整个大楼的中心。大厅和走廊为员工和客户的相互交流提供便利设施。

A high-ceilinged light-flooded hall, the so-called market hall, forms the centre of the complex. It is the entrance hall and lobby offering attractive facilities for staff and customers communicating with each other.

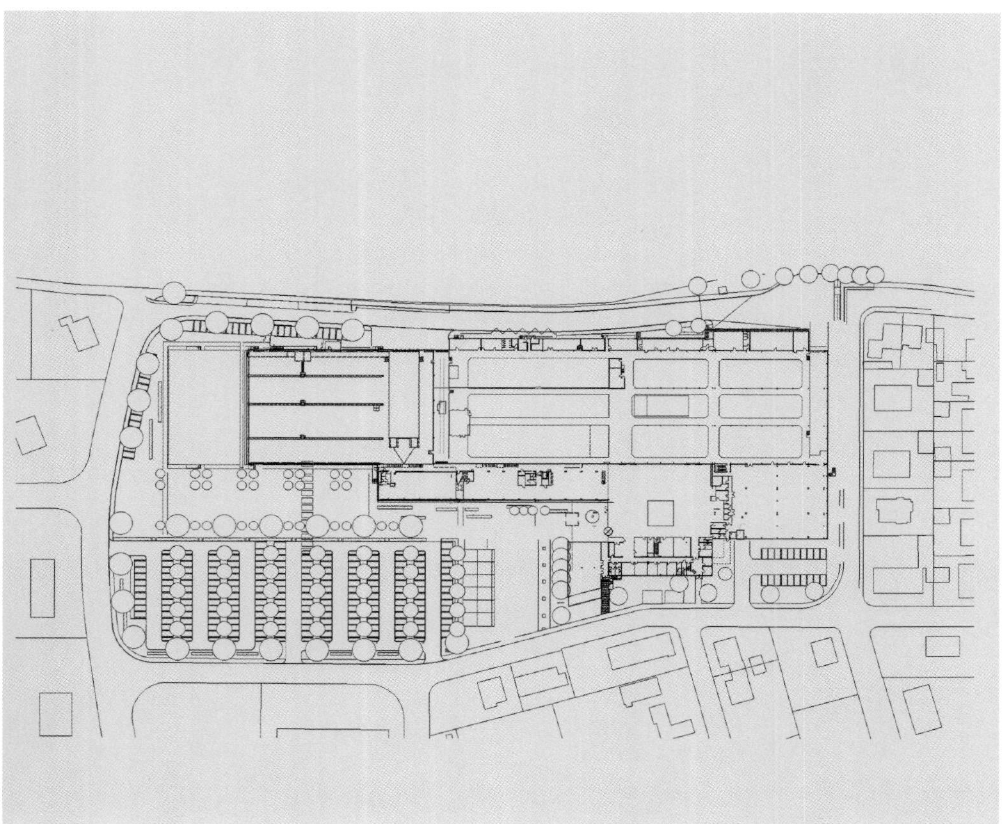

不同的功能区由开放式楼梯和通道相互连接着。4 个楼层所有办公室向周围的环境敞开，并可以通过楼梯和镶玻璃的升降梯进入。

The different functional areas are interconnected by open stairways and bridges. All the offices accommodated on four levels open to the surroundings and are accessed via stairs and a glazed lift.

建筑单位	Neugebauer + Rösch Architekten, Stuttgart
业主	GETRAG, Untergruppenbach
摄影	Roland Halbe und Zoey Braun, Stuttgart
景区建筑	Michael Heintze, Konstanz
技术工程师	
静力学	Boll und Partner, Stuttgart
建筑技术	Interplan, Gerlingen

清晰可见的办公室设施使得策划部和生产部得到视觉上的沟通。因此，建筑形成的开放性促进了生产并加强了有关部门之间的交流。通过这种方式产生的协作优势使公司获得整体效益。

The transparency of the offices facilitates direct visual contact between the planning and the production teams. Thus, the new openness of the architecture promotes production and strengthens communication between those involved. The synergies created in this way benefit the company as a whole.

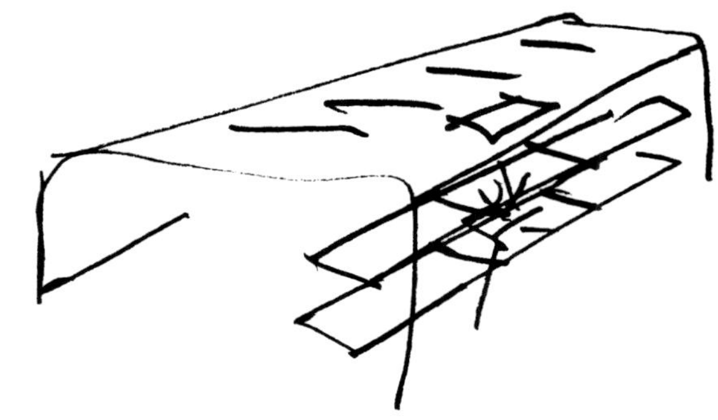

通过对生产程序和内部工作过程的分析，设计师产生灵感，比如埃马克公司完全客户化的工厂。在萨拉赫市工厂施工阶段中，这些灵感得以实施。包括会议中心在内的其他部分将在以后建成，其目的在于促进生产进程。宽敞的新办公区供各生产部门的行政与策划部使用。

By analyzing the manufacturing procedures and internal work processes, the designers developed the 'ideal', i.e. perfectly customized, factory for EMAG. Findings were implemented in several construction stages of the factory in Salach. Further building sections, including a congress centre, are to be erected in the future. The factory premises were restructured to improve production processes. Spacious new office suites were built for the administration and planning teams assigned to the different production teams.

以大厅为轴心，使讨论问题的公共场所范围扩大。大厅通向新的小组办公室和生产车间（专用场所，比如员工餐厅和会议室）。带过道的公共场所是连接的主要部分。透明的会议室和生产部办公室被设置在一楼。

The forum (i.e. the area for special functions, staff restaurant and conference rooms) was extended by a concourse axis leading to the new team offices and to the production hall. The forum with its access balconies is the connecting element. Transparent conference rooms and production offices are located on the ground floor.

诺伊格鲍尔与罗斯
斯图加特

格平根附近的埃马克集团
“理想工厂”

EMAG 'IDEAL FACTORY'
SALACH NEAR GOEPPINGEN

综合性与透明度

INTEGRATION AND TRANSPARENCY

朝南的室内空间用一种新发明的反光墙面抵挡强光，这是一个被叫做“聪明遮阳玻璃”的遮阳金属窗，能够抵挡阳光的直射而不影响户外的景色。从外表看起来，窗户是关闭的；当“聪明遮阳玻璃”不工作时，人们可以看到内部景物。

The south-facing interior spaces are protected from bright sunlight by an innovative anti-glare façade with integrated sun-shading – a metal grille called 's_enn' which keeps direct sunlight out, without blocking the view from inside. Seen from outside, the façade then appears closed; when 's_enn' is not activated, one can see inside.

施工期	2003–2004 年
总面积	8 000m^2
总体积	58 000m^3
竣工验收	2004 年
使用面积	6 000m^2
占地面积	34 000m^2

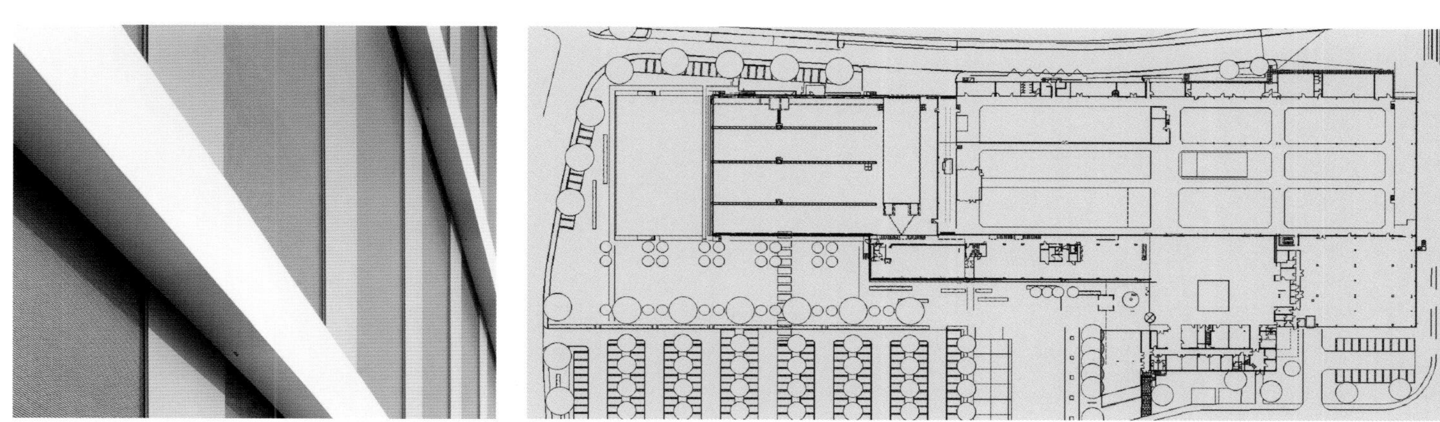

业主	Emag, Salach
摄影	Zoey Braun, Stuttgart
景区规划	Möhrle und Partner, Stuttgart
静力学	Boll und Partner, Stuttgart
房屋技术	Interplan, Gerlingen

在设计时，位于卡塞尔的大众汽车制造厂垃圾收集中心优化了制造产生的污物处理。不同的工作车间被集中在一个楼内，设计师利用该处的斜面为汽车坡路提供有利条件。这使得熔渣包层的地基能够支撑整个玻璃大楼。通过采用太阳能供热和最小化使用人工光源，规格化的玻璃块的使用，减少了能量消耗 / 成本。这要经过将人造光源的最小化和太阳能的使用。

The new refuse collection centre of the Volkswagen factory in Kassel was designed to optimize the logistics of the manufacturer's waste disposal. The different operating units were grouped in one building and the architects used the slope of the site to advantage for a vehicular ramp. This leads into a clinker-clad base which supports a fully glazed building volume. The use of structured glass blocks reduces energy consumption/costs by minimizing artifical lighting and by using solar heat gains.

欧普士建筑设计事务所
达姆斯塔特

卡塞尔大众汽车工厂垃圾收集中心

REFUSE COLLECTION CENTRE OF
VOLKSWAGEN FACTORY KASSEL

规划

PROGRAMME

测量尺寸	69m × 49m × （7.5~10.5）(Hanglage)
总面积	3 350m^2
总体积	24 080m^3
建筑成本	179 万欧元
施工期	2001–2002 年
竣工验收	2002 年

水平板条安装在外墙的里层，即使没有统一的玻璃墙或是在通风不好的情况下，它也能发挥作用。

Horizontal slats installed in the underside of the façade projection made it possible to do without awkward ventilation units in an otherwise unified glass façade.

欧普士建筑设计事务所
达姆斯塔特

卡塞尔大众汽车工厂垃圾收集中心

REFUSE COLLECTION CENTRE OF VOLKSWAGEN FACTORY KASSEL

作为建筑设计的功能型建筑

FUNCTIONAL BUILDINGS AS ARCHITECTS PROJECTS

院子周围和单层房顶下的不同功能区的组合，形成了清新宽敞的环境，一直延伸到很大的生产车间。室内的加工过程在视觉和听觉上都与外界隔离。建筑材料硬砖和密封玻璃窗与生产车间所使用的材料相符。

Grouping all the different functional spaces around a patio and under a single roof, created a clear and spacious environment which asserts itself next to the large production halls. The internal processes are screened from the outside, both visually and acoustically. The construction materials, clinker bricks and gasket glazing, correspond with those of the factory halls.

在德国，这种所谓的功能型建筑有90%都不是由建筑师建造的。令人担忧的是它不断对我们国家感知认识产生消极的影响。卡塞尔大众汽车工厂的垃圾收集站也没用建筑师。只靠工厂管理者的机遇和勇气，到"最后一刻"，作品才能检查通过建筑行业的评估。工业、垃圾处理公司、铁路、公共交通、城市社区的功能型建筑都必须由建筑师重新设计，以使不断发展的环境具有或保持其吸引力。

Ninety percent of all so-called functional buildings in Germany are built without architects. An alarming situation, which has an increasingly negative effect on the visual perception of our country. Also the waste handling station of the Volkswagen plant in Kassel would have been built without architect. Only by chance and the courage of the plant management, opus were able to architecturally review the already planned facility 'at the last minute'. Functional buildings in the industry, in utility and disposal companies, at the railroad and public transportation, in cities and communities must be designed by architects again, so our developed environment becomes and remains as attractive as possible.

建筑单位	opus Architekten, Darmstadt
业主	Volkswagen AG, Wolfsburg
摄影	orendt Studios, Niedenstein/Kirchberg
技术工程师	
支撑结构设计	Ingenieurbüro S+P GmbH Schlier und Partner, Darmstadt
支撑结构设计	Ingenieurbüro Altmann GmbH, Löhlbach
草图	Arcadis, Büros Darmstadt und Braunschweig
企业	
总公司	Hermanns HTI Bau GmbH, Kassel
专业建筑安装，送货	Pilkington Bauglasindustrie GmbH, Schmelz
玻璃加工，装配	Glasbau Rieser, Nürtingen

尽管表层大部分是玻璃窗，热激发水泥板和通风系统共同为室内环境提供了宜人的空气调节和舒适的温度。左：办公室走廊。

The combination of thermo-active floor slabs and ventilation system provides the interior with environmentally friendly air-conditioning and pleasant temperatures, despite the relatively large proportion of glazing in the façades. Left: office gallery.

目前，欧洲最大的苹果麦金塔电脑批发商堪康姆 IT 系统股份公司在位于奥格斯堡和乌尔姆之间的乡下建造新的总部和后勤中心。新公司的椭圆形建筑反映了标准的动态经济学，从远处望去就像个分开的建筑在地面上漂浮，而后勤中心利用并融合了所处位置的地形学。

At the time of writing, the Cancom IT Systeme AG, Europe's largest distributor of Apple Macintosh computers, was building its new headquarters and logistics centre in the country between Augsburg and Ulm. The standing and economic dynamism of the young company is reflected by an oval building which appears to float above ground as a detached building visible from afar, while the logistics centre takes advantage of and is integrated into the topography of the site.

行政办公大楼主楼。

Main administration

OTT 建筑设计事务所
奥格斯堡

堪康姆股份公司总部及中央区商店

CANCOM AG, SCHEPPACH
CORPORATE HEADQUARTERS
AND CENTRAL STORE

工程简介

COMMISSION BRIEF

建筑成本	550 万欧元
施工期	2000–2001 年
总面积	5 200m²
总体积	25 500m³
使用面积	4 690m²
竣工验收	2001 年
用途	中心管理，物流中心，赌场，会议室

使用预制构件能够协调紧缩预算和进度。不规则椭圆几何学和行政楼的褐色不锈钢倾斜外表具有独特的挑战性。

By using prefab elements it has been possible to keep within the extremely tight budget and schedule. The geometry of the irregular oval and inclined façade of browned stainless steel of the administration building represented a particular challenge.

OTT 建筑设计事务所
奥格斯堡

堪康姆股份公司总部及中央区商店

CANCOM AG, SCHEPPACH
CORPORATE HEADQUARTERS
AND CENTRAL STORE

车间性质与能量效率

WORKPLACE QUALITY
AND ENERGY EFFICIENCY

充满阳光的中庭四周是开敞式办公室。不同的楼层由宽敞的走廊互相连接，从而促进了室内的沟通。

The open-plan offices surround a central light-flooded atrium. The different levels are interconnected by spacious galleries to facilitate inhouse communication.

员工餐厅；左页：中厅内景观。

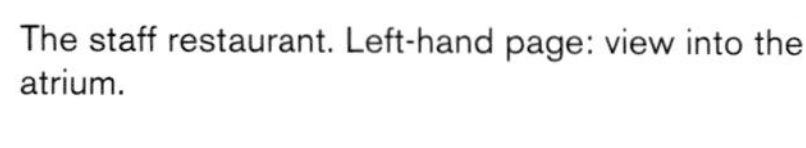

The staff restaurant. Left-hand page: view into the atrium.

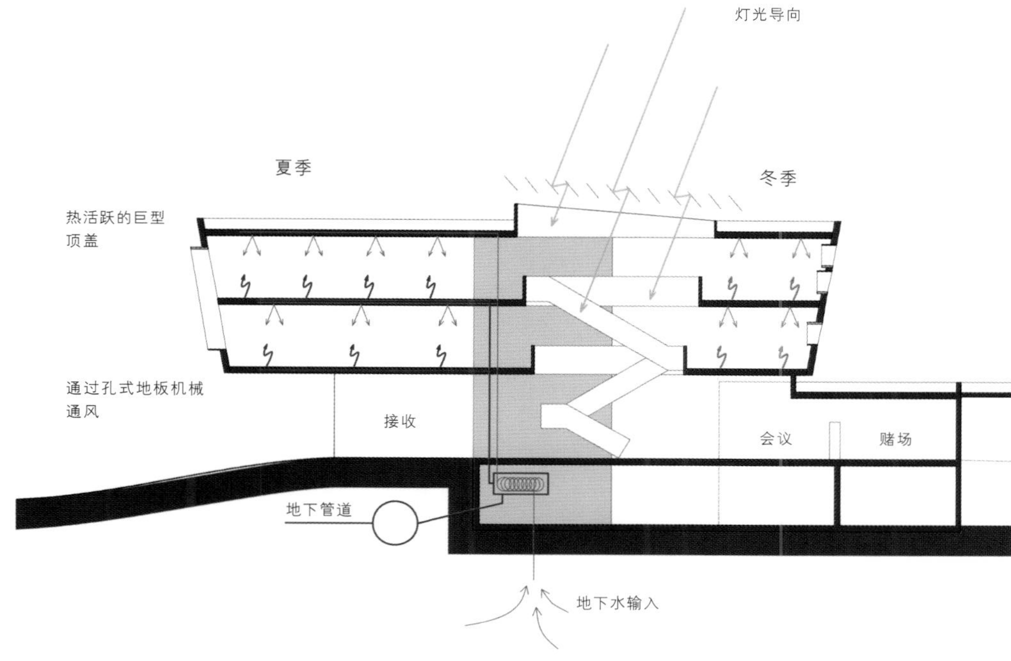

与建筑的动态几何学相比，表层材料显得尤为简单干净；室内采用白色瓷砖墙；外墙被覆以混凝土和不锈钢。与客户产品的设计性质类似，建筑师的设计尤为重要。尽管有相当高的功能和技术要求，该建筑仍然表达了动力学的原始设计理念，在户外看起来就像是个耀眼的宝石。

In contrast to the 'movemented' geometry of the building, the surface materials are deliberately simple and clear; glass, concrete and white walls inside; concrete and stainless steel cladding on the outer façades. Analogous to the design quality of the client's products, the architectural design was of special importance. Despite the extremely high functional and technical standards required, the building still expresses the initial design idea of a dynamic and significant 'solitaire' in an open area.

建筑单位	Ott Architekten, Augsburg
业主	Cancom IT Systeme AG
项目管理	Richard Laeverenz
共事	Daniela Sacher
摄影	Eckart Matthäus, Augsburg, S. 246, 247 unten Christian Richters, Münster, S. 246 oben, 248/249
专业设计	Elektrotechnik Ingenieurbüro Daschner, Augsburg

对于室内，设计时首先考虑的是营造出愉快的工作氛围和促进内部人员的相互交流。这种处理手段的主要目的是培育公司的员工、客户与公司之间主人翁意识和独特意识。

As to the interior, the design priorities were to create attractive workspaces and improve in-house communication. The main goal of the management is to foster a sense of ownership of and identification with the company – among staff and customers alike.

OTT 建筑设计事务所
奥格斯堡

罗马自动卷帘有限公司
罗斯托克

ROMA ROLLLÄDEN UND TORE LTD.
FACTORY, ROSTOCK

设计方法

DESIGN APPROACH

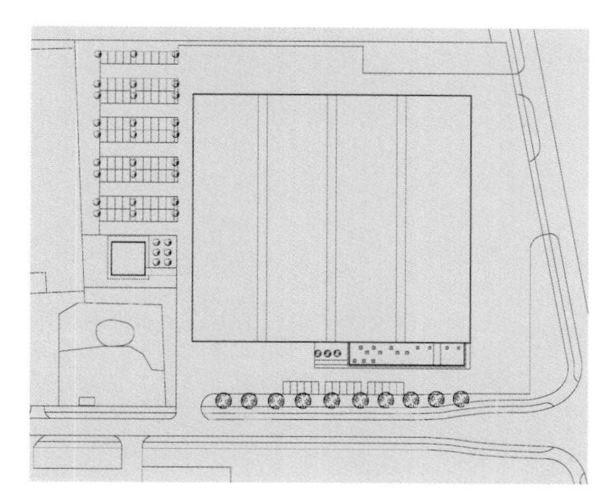

建筑成本	380 万欧元
施工期	2004 年 1–6 月
总面积	6 000m²
总体积	48 000m³
使用面积	5 520m²
竣工验收	2004 年
用途	生产工厂，管理

位置图

Site plan

窗户细节

Window detail

半透明的不锈钢网状壁板形成的幕墙，将新旧建筑连接起来形成一个整体。在黑色钢架的外表打上孔。随着光线的变化，网状表层或是发出深红的微光，或是像银锭一样闪闪发光。

A curtain wall of translucent stainless steel mesh connects and unifies old and new buildings. This skin is punctured by openings in black steel frames. Depending on light incidence, the mesh façade shimmers dark red or glitters like a silver ingot.

德国最大的百叶窗和电动卷门制造商计划将位于罗斯托克的分厂房占地面积扩建到 12 000 多平方米，是原占地面积的2倍。客户不但重视生产设备效率的最大化，而且还重视建筑外部设计。现有的结构和面积可以改变，以便使工厂形成截然不同的外观。因此，陈旧且极其平庸的建筑和车间的规模是对特殊设计的挑战。

Germany's largest manufacturer of roller shutters and roll-up doors planned to double the floor space of its branch factory in Rostock to over 12,000 m^2. The client not only set great store by maximum efficiency of production facilities, but also by a 'signifying' architectural design of the exterior. Existing structures and proportions were therefore altered so as to give the factory a completely different appearance. The old, rather mediocre buildings and the dimensions of the plant therefore represented a special design challenge.

左：入口区域

Left: entrance area

OTT 建筑设计事务所
奥格斯堡

罗马自动卷帘有限公司
罗斯托克

ROMA ROLLLÄDEN UND TORE LTD.
FACTORY, ROSTOCK

预制构件建筑方法

PREFABRICATED
CONSTRUCTION METHOD

咖啡厅

The cafeteria

高高的薄钢板横墙是主入口的标志并通向2倍高的接待大厅，一面不规则的多孔墙将与它一楼敞开式的办公室隔离开，通过楼梯可以进入到楼上的华美的自助餐厅。楼梯台阶为悬臂式木制构造，扶手为钢网护栏。由于建筑过程中全部使用预制构件，大楼会在5个月内建成。虽然预算紧张，内部设计却十分精致。

A tall slender steel-sheet cross-wall marks the main entrance and leads into the double-height reception lobby, separated from the open-plan ground-floor offices by an irregularly perforated wall. The upper floor with the colourful cafeteria is accessed via a stairway with cantilever wooden steps and steel-mesh balustrades. With its consistently use of prefabricated elements throughout, the design made it possible to erect the building in less than five months and to achieve a sophisticated interior despite a tight budget.

左：入口大厅

Left: the entrance lobby

建筑单位	Ott Architekten, Augsburg
业主	Roma Rolladen + Tore GmbH, Burgau
项目管理	Annabelle Hermann
员工	Stephan Wehrig, Stefanie Schmidt
摄影	Christian Richters, Münster
技术工程师	
支撑结构	Ingenieurbüro Reisch, Augsburg / Donauwörth
提议 / 分发	Ingenieurbüro Bleyer, Augsburg
电子技术	Elektrotechnik Ingenieurbüro Daschner, Augsburg
企业 咨询和公司	Fa. Haver & Boecker, Oelde

贯穿中庭的架桥将大楼两边的 8 个楼层的侧楼相互连接起来。

The bridges through the atrium interconnect the two eight-storey wings.

潘茨卡·平克建筑设计事务所
杜塞尔多夫

卡尔阿诺得办公楼
杜塞尔多夫

OFFICE BUILDING,
KARL-ARNOLD-PLATZ 1
DÜSSELDORF

工艺建筑

TECHNOLOGICAL ARCHITECTURE

建筑成本	2 000 万欧元
施工期	2001–2002 年
总面积	18 200m^2
总体积	75 200m^3
建筑标准	35.10m × 39.15m × 29.54m
占地面积	2 439m^2
净地表面积	16 300m^2
竣工验收	2002 年
获奖情况	BDA Auszeichnung “Guter Bauten”，2003

等距

Isometry

建筑第二外层

Second building skin

办公区及商业区

Offices and commercial spaces

中厅内人行天桥和地下室

Skywalks in the atrium, and basement floors

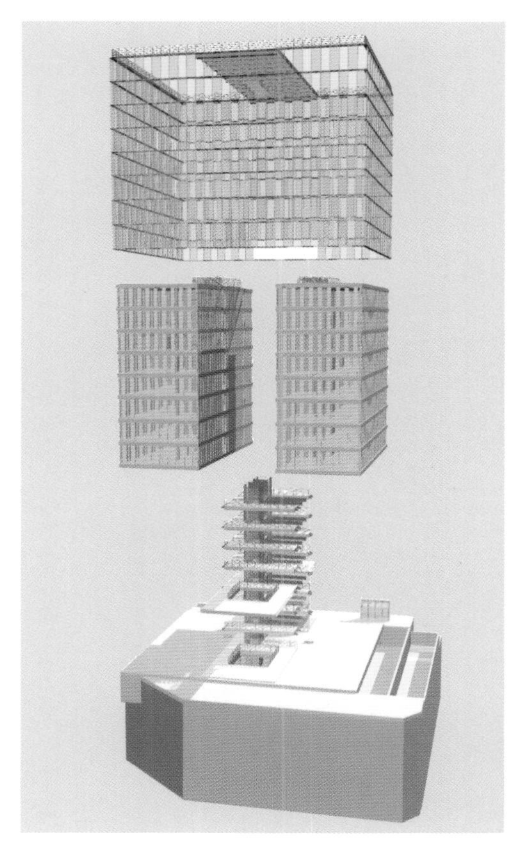

自然与工艺、木材与玻璃构成了办公室与商业大楼的设计全体。透过高度不同的玻璃嵌板的棋盘式外墙能够看到大楼的内部。透明外墙将内部空间和中庭围了起来，中庭阳光充足并种有大面积植物。室内露台将两侧不同功能的建筑与“空中花园”的美景交融于一体。

Nature and technology, wood and glass informed the design of this office and commercial building. A chess-board-type façade of split-level-mounted glass panels affords views into the depth of the building.The transparent façade encloses the interior spaces and the light-flooded, extensively planted atrium. Indoor terraces combine the functional architecture of the different wings with the beauty of ‘hanging gardens’.

潘茨卡·平克建筑设计事务所
杜塞尔多夫

卡尔阿诺得办公楼
杜塞尔多夫

OFFICE BUILDING, KARL-ARNOLD-PLATZ 1 DÜSSELDORF

外立面结构

FAÇADE STRUCTURE

自然通风、日光的有效利用和低能源消耗的发展，节省能源的观念得到证实。

The energy concept is convincing with natural ventilation, efficient use of daylight and low power consumption.

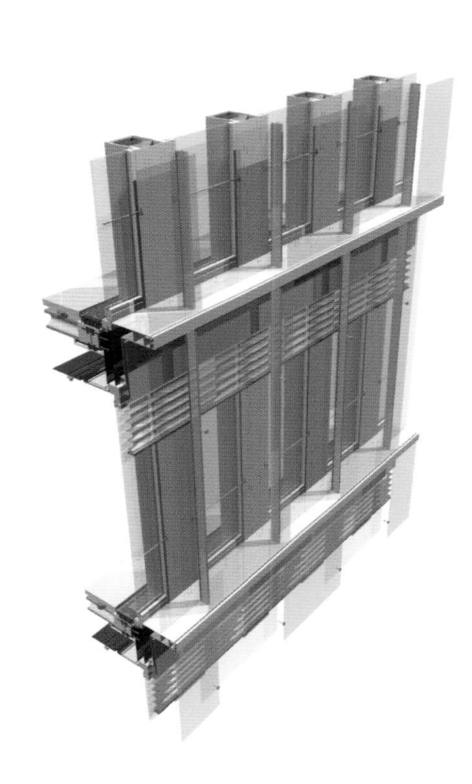

中庭外墙等距：双层的外墙平面使建筑充满动态效果且错落有致。

Isometry of atrium façade: the two façade planes make the building appear dynamic and intricate.

大楼两侧的外墙上有合成的木制框架的法国旋转窗。窗前的第二层外墙面由嵌入承重钢轨的等高的玻璃嵌板组成。每个第二层玻璃嵌板都向内移 15 厘米，从而使整个外立面成为精细的、凹凸不平的格状玻璃表面。木制构件上的挡风膜也能够使新鲜空气进入室内。

The two wings have façades with integrated wood-framed revolving French windows. The second façade layer in front of these windows consist of storey-high glass panels set in bearing rails. Every second glass panel is set back by 15 cm so that the entire façade shows a finely gridded split-level glass surface – a weather-protective membrane for the wooden façade elements that also lets fresh air into the interior spaces.

潘茨卡·平克建筑设计事务所
杜塞尔多夫

卡尔阿诺得办公楼
杜塞尔多夫

OFFICE BUILDING,
KARL-ARNOLD-PLATZ 1
DÜSSELDORF

景观美化

LANDSCAPING

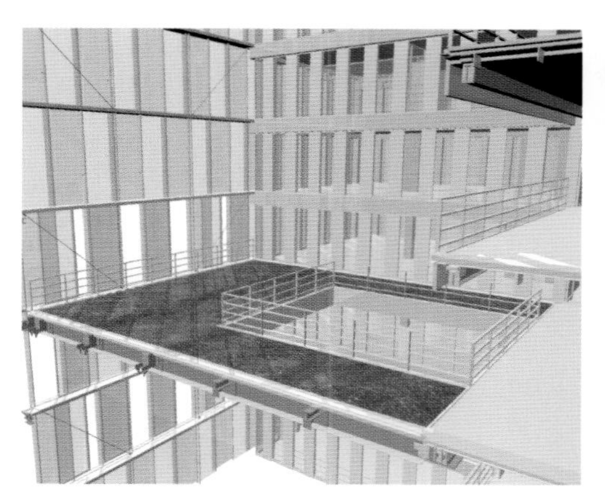

空中花园的细长结构只能靠钢板来实现。

The slender structures of the hanging gardens could be executed only in steel.

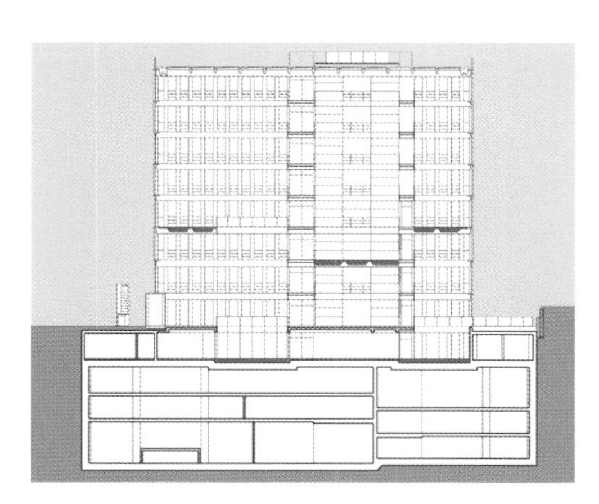

所有楼层内的可出租空间都能被灵活地划分。在写此书的同时，1 楼和 2 楼被用做样品陈列室，3 楼到 7 楼为办公室。

The rentable spaces on all the floors can be flexibly partitioned. At the time of writing, the ground and first upper floor are used as showrooms and the second to seventh upper floors as offices.

空中花园为中庭的氛围添加了新意与乐趣。3 楼和 4 楼的大面积植被面对大楼内的所有工作人员开放并吸引他们来到木制平台休闲娱乐。这里种植的大量松树在相当大程度上不仅改善了室内气候，而且创造了一种地中海风情。

Hanging gardens add originality and interest to the atrium's atmosphere. Large planted terraces on the second and third upper levels are open to all the users of the building and invite them for recreation on their wooden decks. The many pines planted here not only improve the interior climate to a considerable extent, but also create Mediterranean flair.

建筑单位 Petzinka Pink Architekten, Düsseldorf
业主 Bernd Voswinkel GmbH & Co., Düsseldorf
项目管理 Carpus + Partner, Aachen
景区建筑 Prof. Dr. Jörg Dettmar, Bochum
Planergruppe Oberhausen GmbH, Oberhausen

技术工程师
支撑结构设计 Petzinka Pink Tichelmann, Darmstadt
立面技术 IGF, Mülheim a.d. Ruhr
建筑物理 Trümper-Overath, Bergisch Gladbach
房屋技术 Ebert Ingenieure, Düsseldorf
周边保护 Berg. Universität Wuppertal
Prof. Dr.-Ing. Wolfram Klingsch
Entrauchungsstudie I.F.I., Aachen
检测 Dipl.-Ing. Gerd-Joachim Töpfer, Düsseldorf
地面检测 Reducta GmbH, Düsseldorf
摄影 Hermann Fahlenbrach, Neuss

潘茨卡·平克建筑设计事务所
杜塞尔多夫

北莱茵—威斯特法伦州政府代理机构
柏林

REPRESENTATION OF THE STATE OF
NORTH-RHINE WESTPHALIA
AT THE SEAT OF GOVERNMENT
BERLIN

合作建造

'CO-OPERATIVE BUILDING'

建筑成本	2 970 万欧元
施工期	2000–2002 年
总面积	12 500m²
总体积	48 135m³
竣工验收	2002 年
建筑标准	57.10m × 38.20m × 16.25m
占地面积	5 550m²
净地表面积	9 800m²
获奖情况	1999 年一等奖
其他奖项	Deutscher Holzbaupreis，2003

这里采用了“物质相溶”和持续性的创意来建造重要建筑。

The idea of 'dissolving materiality' and of sustainability were applied here to create significant architecture.

新颖的设计诠释了生态建筑的原则，也就是未来持续性发展和创新发展以获得高水准的美学效果。木材和钢铁的混合构造原型将两种材料的结构优势相结合，创造出新的具有生态、经济和美学效果的建筑并能使资源得到有效利用。

The innovative design translated the principles of ecological architecture, i.e. sustainability and pioneering developments for the future, into high aesthetic quality. This prototype of a hybrid structure of timber and steel combines the structural advantages of both materials to create a new architectural quality that is ecological, economical and aesthetic and uses resources efficiently.

潘茨卡・平克建筑设计事务所
杜塞尔多夫

北莱茵—威斯特法伦州政府代理机构
柏林

REPRESENTATION OF THE STATE OF
NORTH-RHINE WESTPHALIA
AT THE SEAT OF GOVERNMENT
BERLIN

外立面

FAÇADE

外立面的等距诠释了自然通风的设计。

The isomety of the façade explains the natural ventilation design.

建筑的玻璃围栏不仅保护了木质抛物线结构，而且防护喷层能对户外的气候环境作出反映并与之相适应。双层外墙外层上的通风口和排气口总是开着的。它们和空中花园的调制设备是外立面和能源观念的一部分。

The glass enclosure of the building not only protects the wooden parabolic structure, but functions like a cocoon in responding and adapting to outdoor climatic conditions. The air intake and exhaust openings in the outer 'membrane' of the double façade are permanently open. Together with the controllable ones in the 'hanging gardens' they are a part of the façade and energy concept.

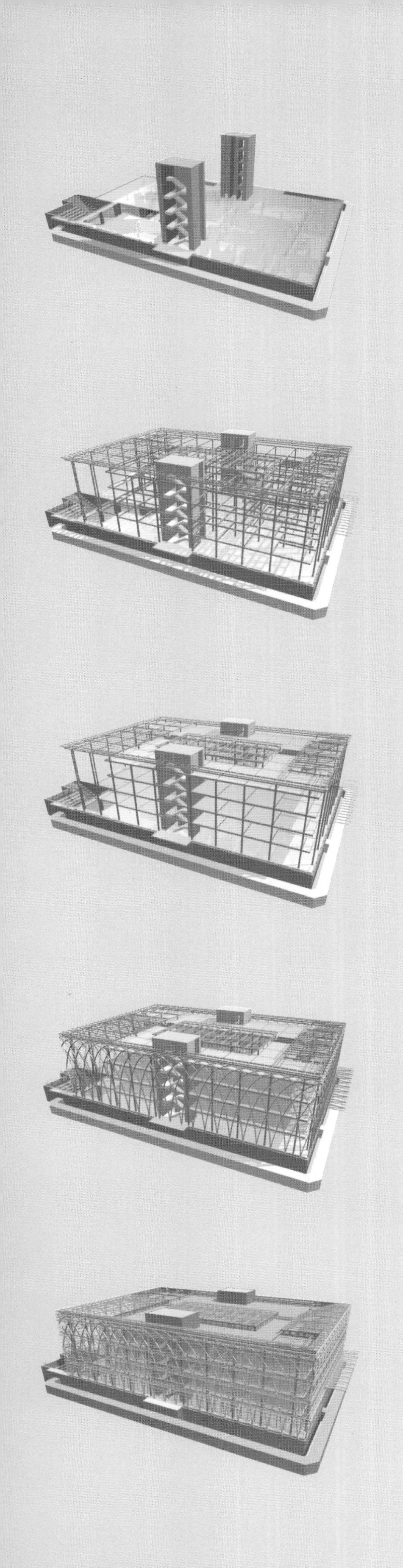

有地下管道的地下室：175 米的地热输送管使吸入的新鲜空气在流入通风系统和中庭之前就提前处理。

Basement with ground duct: the 175-metre geothermal heat transmitter duct preconditions incoming fresh air before it is fed into the ventilation system and the atrium.

屋顶结构：建筑设计要求高度透明且空间封闭的二维框架。

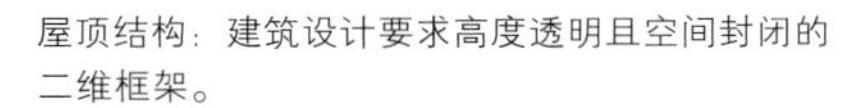

Roof structure: the architectural design required a highly transparent, space-enclosing two-dimensional framework.

注重防火设施的天花板显示出建筑的特殊性：最初，对同样规模的行政楼安装了用于木构架建筑的木制天花板元件。

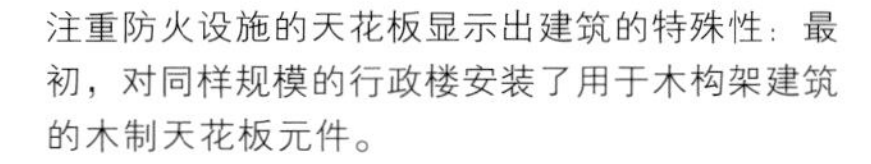

The ceilings represent a speciality of the building with regard to fire-precautionary measures: for the first time in an administration building of this size, light-weight wooden ceiling elements, adapted from timber-frame construction, were installed.

为了让建筑能承受垂直载荷，建筑师采用一种符合静力学的动态二级框架：一个由弯曲薄板形成的抛物菱形结构。

For the building to withstand vertical loads, the architects created a statically highly indeterminate secondary framework: a parabolic rhombus structure of bent laminated boards.

建筑的垂直性和水平的多功能需求，促使建筑师设计了无柱的轻型木钢结构。它不仅能灵活使用，而且经济美观。木材是再生资源中最高效的建筑材料。

The demand for a vertical and horizontal multi-functionality of the building prompted the architects to develop a column-free primary light-weight structure of timber and steel. It is not only flexible in use, economical and aesthetically pleasing, wood is the construction material with the highest efficiency in terms of sustainable resources.

潘茨卡・平克建筑设计事务所
杜塞尔多夫

北莱茵—威斯特法伦州政府代理机构
柏林

REPRESENTATION OF THE STATE OF NORTH-RHINE WESTPHALIA AT THE SEAT OF GOVERNMENT BERLIN

组织

ORGANIZATION

中央大厅向来宾展现了建筑的美感、技术和生态观念。

The aesthetic, technical and ecological dimension of the building is revealed to visitors in the central hall.

建筑单位	Petzinka Pink Architekten, Düsseldorf
业主	Ministerium für Städtebau und Wohnen, Kultur und Sport, NRW, Düsseldorf, vertreten durch den: BLB Bau- und Liegenschaftsbetrieb NRW NL Düsseldorf
规划者	Petzinka Pink GmbH & Co, KG
对象监控	Assmann Beraten + Planen, Dortmund
空闲设备	Prof. Dr. Jörg Dettmar, Bochum Planergruppe Oberhausen GmbH, Oberhausen
技术工程师	
支撑结构设计	Petzinka Pink Tichelmann, Darmstadt
立面技术	IGF Zimmermann, Mülheim a.d. Ruhr
能源管理	DS-Plan, Stuttgart
技术设备	INTEG GmbH, Berlin
技术设备，电工	UEP, Berlin
空调安装	ag licht GbR, Köln
热学建筑物理	DS-Plan, Mülheim a.d. Ruhr
门窗	Petzinka Pink Tichelmann, Darmstadt
边缘保护	BPK, Wuppertal
安全防火	I.F.I. – Institut für Industrieaerodynamik, Aachen
摄影	Taufik Kenan, Berlin

早在大约1900年，摩汀制造公司就为福特公司著名的“Tin Lizzy”车型生产散热器。目前，全世界有近乎1万名工人生产渐进式发动机冷却系统。在瓦克多夫，摩汀制造公司目前专门为宝马汽车的第三代和 X 系列车型生产散热器。它的合同数额到 2012 年将超过 5 000 万欧元，新工厂的建筑投资费用总额为 2 300 万欧元。

As far back as around 1900, the Modine company produced the radiators for Henry Ford's legendary 'Tin Lizzy'. Today, almost 10,000 workers around the world produce progressive motor-cooling systems. In Wackersdorf, Modine Montage Ltd. currently exclusively produces radiators for all new third-series and X-series BMWs. Its contract volume is in excess of 500 million euros until 2012, the investment costs of building a new plant amount to an estimated 23 million euros.

J・雷查特教授建筑设计事务所
埃森

摩汀汽车公司
瓦克多夫

MODINE AUTOMOTIVE WACKERSDORF

设计因素与建筑观念

DESIGN FACTORS AND ARCHITECTURAL CONCEPT

建筑成本	1 050 万欧元
施工期	2002–2003 年
总体积	98 238m³
使用面积	10 959m²
竣工验收	2003 年

从最初的蓝图到完成的过程中，建筑设计者和生产加工工程师共同合作，两者的特殊需求让合作车间在早期就保证建筑能最大化地支持线性生产。结构、外表、服务和内部装修的有效交互作用能使建筑保持持久性、适应性和沟通性。

The special requirement of both architectural planners and production process engineers co-operating from the first sketch through to completion made joint workshops necessary from early on to ensure that the architecture optimally supports the linear production process. The sustainability, flexibility and communicativeness of the building are now the result of the efficient interaction of structure, skin, services and interior finishing work.

沿着高吊棚走廊内的空间构架，透过立式玻璃窗，阳光能照射到生产车间的每一个角落。为保证使用时充分发挥灵活性，尽量减少从地面通向屋顶的支柱，并与三个垂直加劲装置将建筑加固。

All areas of the production hall receive daylight through bands of upright glazing along the space-frame section of the high-ceilinged aisles. In order to ensure maximum flexibility in use, the structure was stiffened by means of a minimum ground-floor bracing via the roof structures and only three vertical stiffening systems.

摩汀汽车公司
瓦克多夫

MODINE AUTOMOTIVE WACKERSDORF

结构

STRUCTURE

J・雷查特教授建筑设计事务所
埃森

生产车间由两个宽36米、高10.3米的走廊和两个18米宽的下层走廊组成。在纵向，有一个9米宽的支撑办公室和教研室的墩座墙。自由跨度的梁式桁架被悬挂在高屋顶的钢制桁架上并由下层走廊内的铁柱支撑着。这种结构实现了生产材料沿着高屋顶走廊的线性无柱流程。

The production hall comprises two ‘aisles’, each 36 m wide and 10.3 m high, and two lower ‘aisles’, each 18 m wide. To one longitudinal side, there is a podium, 9 m wide, which supports the offices and staff rooms. The free-span podium girder beams are suspended from the steel trusses of the high-hall roof structure, and supported by steel columns in the area of the lower aisles. This structure allows a linear column-free flow of production materials along the high-ceilinged aisles.

项目参与者
建筑单位 Prof. J. Reichardt Architekten BDA, Essen
业主 MODINE Wackersdorf GmbH, Wackersdorf
建筑规划 Prof. J. Reichardt Architekten BDA, Essen
项目管理 BePro Udo Wieczorek, Alteglofsheim
支撑结构 Baum und Weiher
Ingenieurbüro für Bauwesen, Bergisch-Gladbach
房屋技术 Planungsgesellschaft Karnasch mbH, Essen
门窗 Ingenieurbüro Ulrich Schwarz, Velbert
周边保护 SV. Zahn, Sachverständigenbüro für Brandschutz, Mönchengladbach
摄影 Klaus Ravenstein, Essen

参与企业
接管方 Wiemer & Trachte AG, Dortmund/Crimmitschau
钢结构 Stahlbau Söll, Hof/Bayern
铝—玻璃立面 Metallbau Korsche, Weiden/Oberpfalz
天棚密度 Koch Dachtechnik GmbH, Meerane
立面 Tahedl Dach + Wand GmbH, Lappersdorf
系统隔离墙 SOMETA Bauelemente GmbH, Mannheim
解锁工作 Stahlbau & Bauschlosserei Ernst Walther, Wilsdruff
平墙 Holz Bösl, Ursensollen

彼得面包房的设计和细节设计是要与客户合作开拓新的领域和探索新的途径。在设计初期，用3D城市规划、结构、服务和处理模拟程序设计空间排列、结构、建筑外表、服务、处理技术和高质量工作空间。

The designing and detailing of the Peter Bakery was to open up new territories and explore innovative approaches – in co-operation with the client. At the beginning of the design stage, 3D urban layout, structural, services and process simulation programmes were used to devise the best possible combination of spatial arrangement, structure, building skin, services, process technology and high-quality workspaces.

J · 雷查特教授建筑设计事务所
埃森

彼得面包房
埃森

PETER BAKERY, ESSEN

面包房的典型设计

MODEL PROJECT OF THE BAKERY TRADE

建筑成本	180 万欧元
施工期	1996–1998 年
总面积	16 295m²
总体积	2 350m³
使用面积	1 820m²
占地面积	5 079m²

烘烤大厅是无柱的。延伸的屋顶形成前置驱动的天棚，并为通过卷帘门的运货车挡风遮雨。多种材料的结合被证明是对建造不同功能结构最有利的，被认为是可循环的复式钢骨构架。个建筑没有使用聚氯乙烯地板材料，也没有夹层混合物或是硝化纤维涂漆。

The bakery hall is a column-free space. The roof extends to form a front-drive canopy and weather-protection for loading delivery vans with bakery products via roll-up gates with air locks behind them. A combination of different materials proved to be the most advantageous for constructing the different functional structures, conceived as recyclable modular skeleton framework. No PVC flooring materials, no sandwich compounds or nitro-cellulose lacquers were used in the entire building.

在彼得面包房，每年的采暖耗能已经减少到62%，制冷耗能减少到39%。烤箱附近的工作区温度不超过22～27℃。大约100千克面粉消耗80千瓦小时的电量（其中70%电量用于面包房生产）也减少到60千瓦小时。

At the Peter Bakery, the annual consumption of heating energy was reduced by up to 62 per cent, and the consumption of cooling energy by up to 39 per cent. Temperatures in the working areas at the ovens do not exceed 22 to 27° C. The energy value of approximately 80 kW/h per 100 kg flour (70 per cent of which is used for bread) is thus reduced to about 60 kW/h per 100 kg flour.

LUFTTEMPERATUR

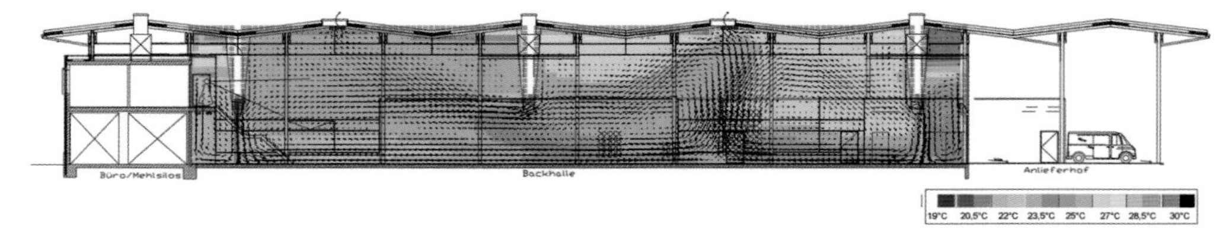

STRAHLUNGSTEMPERATUR

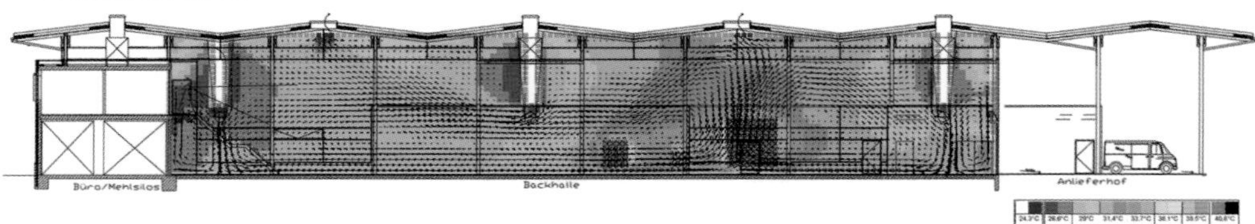

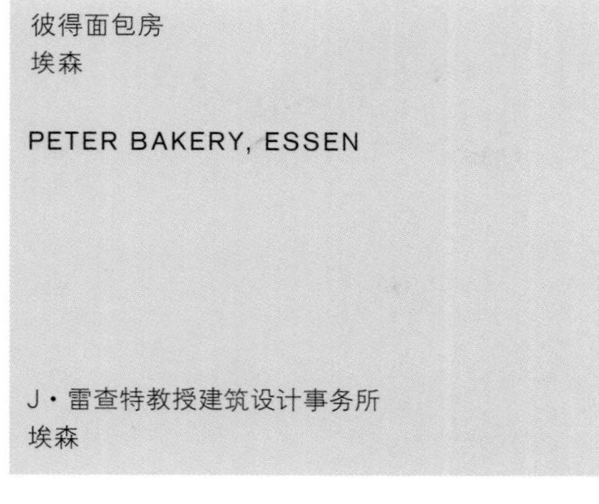

彼得面包房
埃森

PETER BAKERY, ESSEN

J・雷查特教授建筑设计事务所
埃森

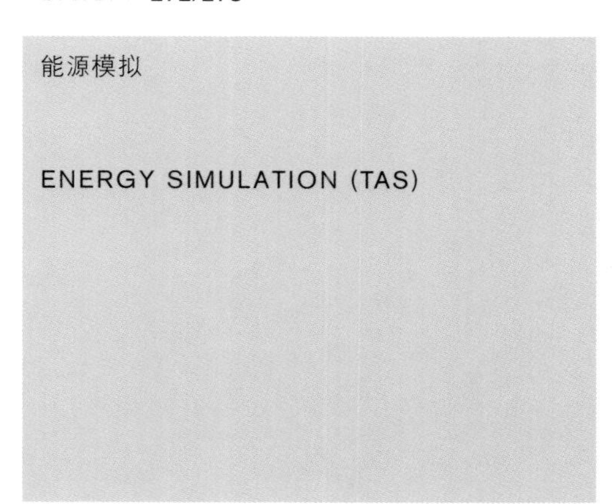

能源模拟

ENERGY SIMULATION (TAS)

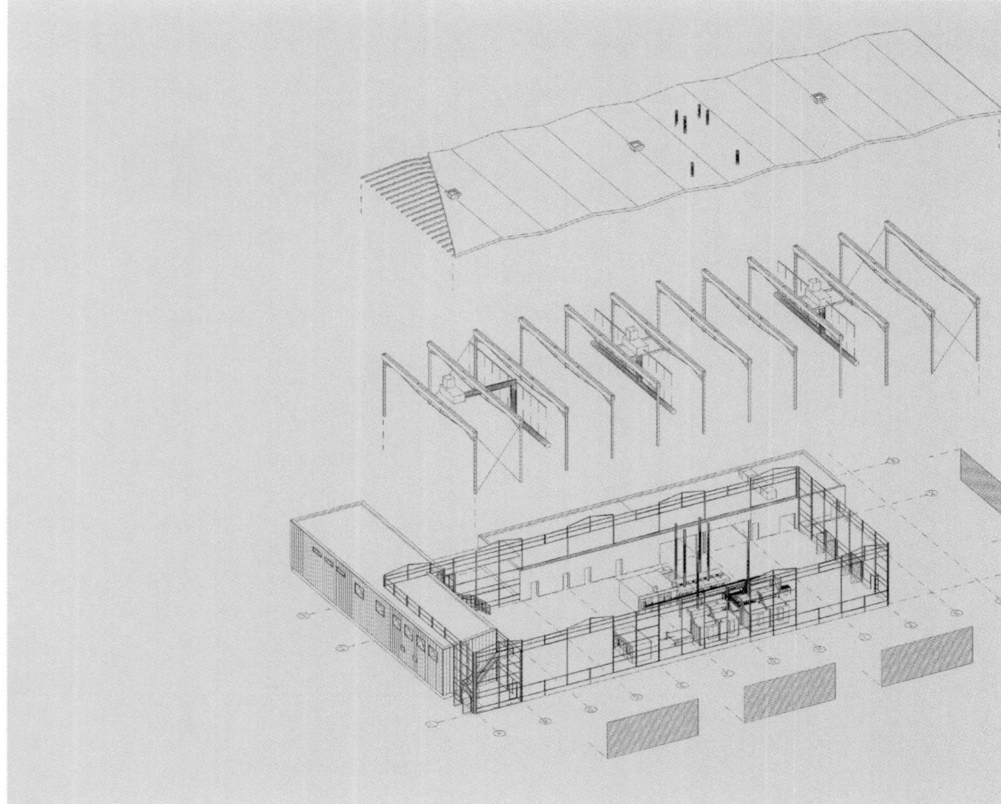

人们使用 TAS 建筑模拟程序来估算建筑的抗低温性和抗热性。对温度曲线、模拟通风和建筑物的全部检测的部分结果是地基层的正常隔热被忽略，由于烤箱的安装负荷升到1 000千瓦所产生的保温问题，可能会使室内温度升高 2 ~ 3℃。

The structural simulation programme TAS was used to calculate the cold and heat resistance of the building. A partial result of the integral reviewing of temperature curves, simulated ventilation and construction was that the usual thermal insulation of the foundation slab was left out, as it would have increased interior temperatures by about 2 to 3° C – due to the problem of 'heat retention' from the ovens' installed loads of up to 1,000 kW.

项目参与者	
建筑单位	Prof. J. Reichardt Architekten BDA, Essen
业主	Christa Peter, Essen
所在地	Gewerbegebiet M1, Essen
概念，方案	Prof. J. Reichardt Architekten BDA, Essen
许可规划，执行规划	
艺术总管	MA: A. Schöpe, S. Czech
工程管理	agiplan, Mülheim/Ruhr
	MA: B. Fürst
摄影	Klaus Ravenstein, Essen
支撑结构设计	Baum und Weiher, Bergisch-Gladbach
房屋技术	Planungsgesellschaft Karnasch mbH, Essen
TAS 能源模拟	G. Hoffmann, Frechen

玻璃屋顶的前庭、露台和“层叠式办公室”创造性地给人留下科学迷宫的印象。在这里，人们不仅可以发挥自己的想像力，而且还能“摆弄”电子设备或用它进行实验。办公室和车间、试验室以及艺术试验设备结合在一起。光束隧道、环境检测室、声音试验室和骑马模拟器都能为试验新的想法和测试原型提供可能的设备。

Glass-roofed atria, terraces and 'cascading offices' together give the impression of a scientific maze in which people not only think creatively, but also experiment with and 'fiddle at' electronic devices. Offices mingle with workshops, laboratories and state-of-the-art test rigs. Light beam tunnel, environmental test chamber, sound laboratory and ride simulators offer every conceivable facility for trying out new ideas and testing prototypes.

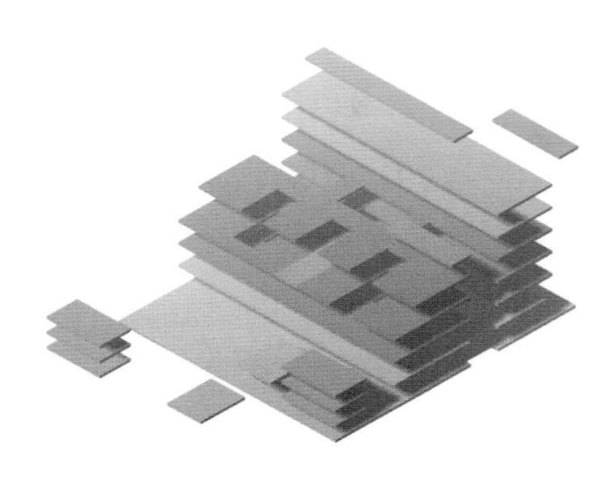

RKW 建筑设计事务所及城市规划所
杜塞尔多夫

奥迪电子公司与汽车电子系统开发中心
因戈尔施塔特

AUDI ELECTRONICS CENTRE, INGOLSTADT
DEVELOPMENT CENTRE FOR AUTOMOBILE ELECTRONIC SYSTEMS

可延伸单元在交流上的灵活性

FLEXIBILITY THROUGH COMMUNICATION
MODULAR – EXTENDIBLE

建筑成本	约 5 000 万欧元
施工期	2001–2003 年
总体积	200 000m³
使用面积	约 36 000m²
主要地表面积	约 31 000m²
竣工验收	2003 年

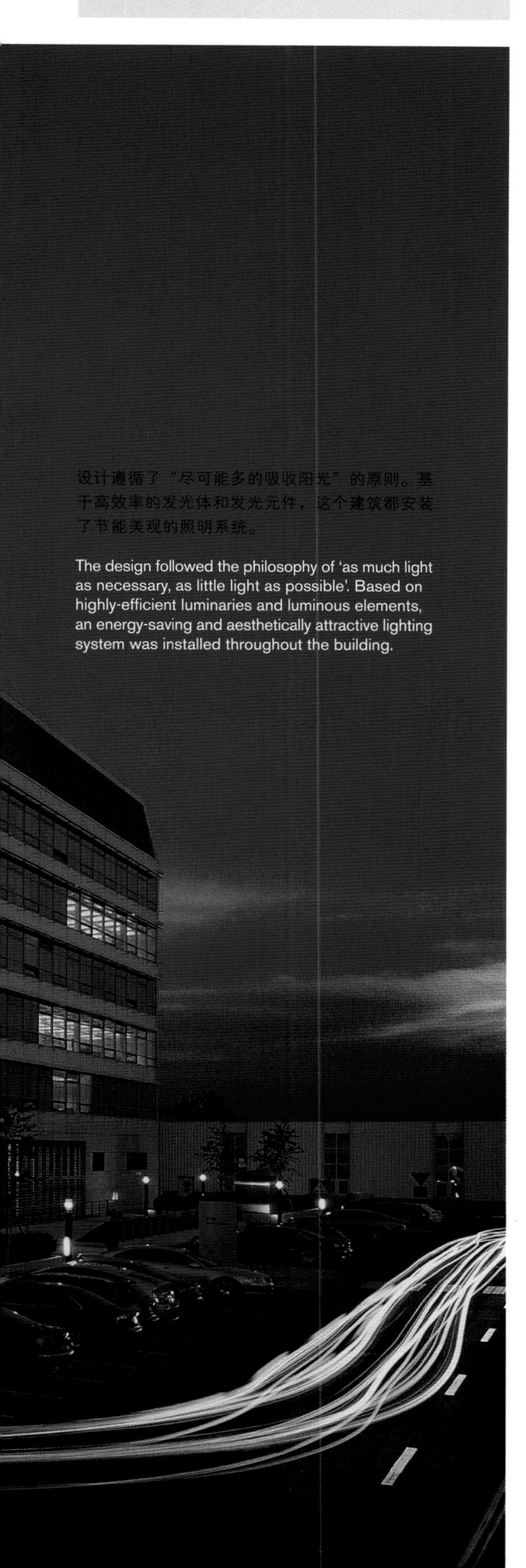

设计遵循了“尽可能多的吸收阳光”的原则。基于高效率的发光体和发光元件，这个建筑都安装了节能美观的照明系统。

The design followed the philosophy of 'as much light as necessary, as little light as possible'. Based on highly-efficient luminaries and luminous elements, an energy-saving and aesthetically attractive lighting system was installed throughout the building.

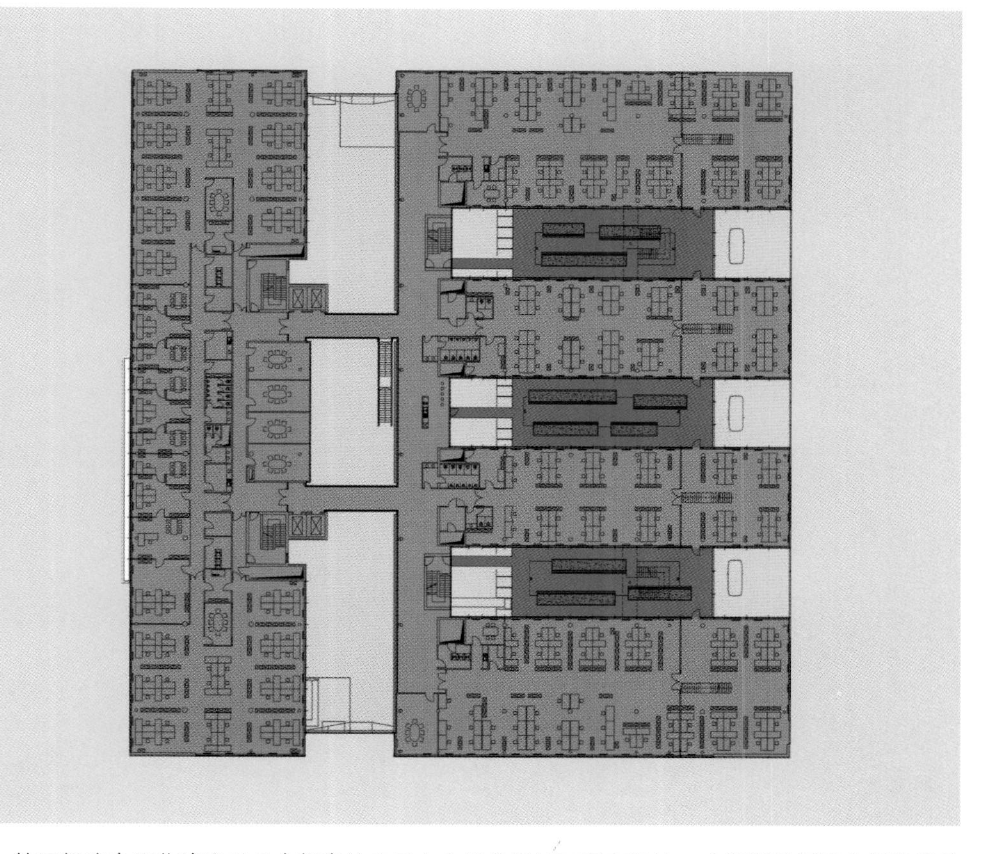

德国经济合理化建议委员会将奥迪电子中心当做建筑工具来设计。功能型地板平台能够满足建筑快速变化的灵活性的需求，从而使不同的使用者能够尽可能的不用预先确定就将他们再分成单元化办公室、联合办公室或是敞开式办公室。建筑以满足个体空间的多样化使用为主，低于目前可以建筑平均水平的 20%。

RKW planned the Audi Electronics Centre as if it was an architectural tool. The demand for maximum flexibility in the fast-changing use of the building was met by functional floor platforms as little predetermined as possible to allow different users to subdivide them into cellular, combination or open-plan offices. The building and its parts live on the multiple use of individual spaces and therefore stay below the current usual cost levels for comparable buildings by about twenty per cent.

RKW 建筑设计事务所及城市规划所
杜塞尔多夫

奥迪电子公司与汽车电子系统开发中心
因戈尔施塔特

AUDI ELECTRONICS CENTRE, INGOLSTADT
DEVELOPMENT CENTRE FOR AUTOMOBILE ELECTRONIC SYSTEMS

建筑能源观念

BUILDING ENERGY CONCEPT

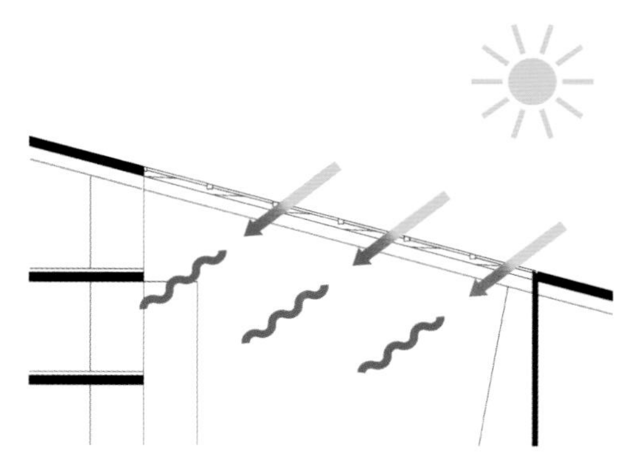

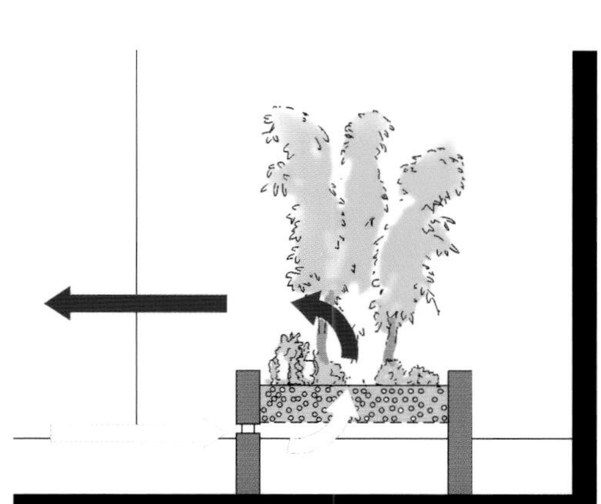

通过发热系统和智能气候控制器以及设备管理系统能够有效地使用热能，大厅起排气作用，达到自然通风的效果。所有这些意味着建筑只需要外界供应最小能量。虽然有很多的系统试验台和试验区集约使用，建筑却分别只需消耗 5 000 千瓦加热和 4 900 千瓦冷却。

Heating energy is used efficiently via a heat recovery system in connection with intelligent climate controls and a facility management system, and supported by natural ventilation via a hall which functions as an exhaust stack. All this means that the building needs a minimum of energy from external suppliers. Despite the many test rigs and intensive use of experimental areas, the building consumes only 5,000 kW for heating, and 4,900 kW for cooling.

建筑单位 RKW Rhode Kellermann Wawrowsky Architektur + Städtebau, Düsseldorf
业主 AUDI AG, Ingolstadt
摄影 Bernd Nörig, Heimbach

技术工程师
静力学与房屋技术 DBN Dröge Baade Nagaraj, Salzgitter
接管方 M. + W. Zander, Nürnberg
空调安装 Dreist + Partner, Düsseldorf

RKW 建筑设计事务所及城市规划所
杜塞尔多夫

医学博士楼
杜塞尔多夫

HOUSE OF THE MEDICAL DOCTORS
DÜSSELDORF

多样性的统一

UNITY IN DIVERSITY

建筑成本	9 440 万欧元
施工期	1999–2003 年
总体积	265 470m³
总面积	56 517m²
使用面积	18 709m²
占地面积	15 500m²

划分为很多部分的玻璃中庭是建筑的主题，它将建筑内的不同部分互相连接起来。它由一个主厅也就是一个宽敞的会场和服务与功能一体化的机构所组成。

The theme found expression in the multi-part glazed atrium concourse which interconnects all the different parts of the building. It comprises a main hall which is a spacious meeting place and houses joint services and functions.

北莱茵医疗协会和机构的客户联盟希望将他们的办公室安排在同一个大楼内。这就是设计的出发点。将他们的活动集中在同一区域不仅为用户机构提供“短距离”，而且还能为联合功能提供大面积空间。“多样性的统一”保持了每个机构个体的统一性，这决定了城市规划和建筑设计。

The clients – consortium of different North-Rhine medical associations and institutions – wished to concentrate their offices in one building. This formed the starting point of the design. Concentrating their activities in one place offered the user organizations not only ‘short distances’, but also generously dimensioned spaces for joint functions. The demand for the individual identity of each institution to be preserved was expressed in the formula ‘unity in diversity’ which determined both urban planning and architectural design.

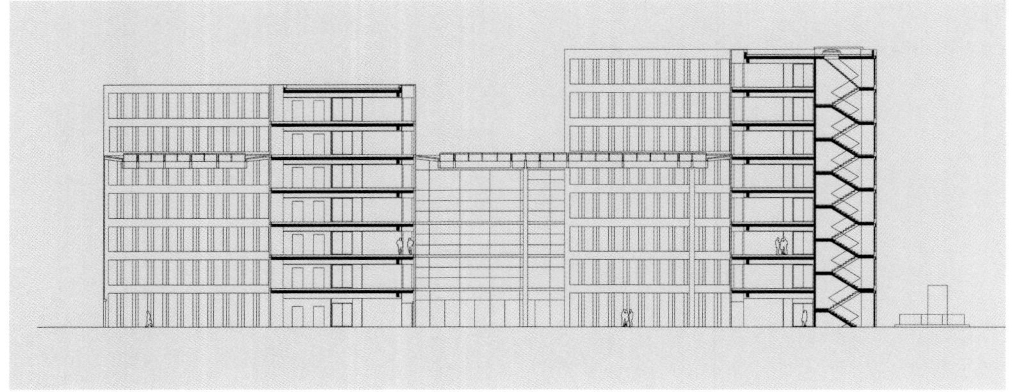

清晰、简洁的几何图形使建筑环境井然有序，而它的连接方式、结构及空隙能够与城市周围环境充分融合。

The clear, simple geometry of the structure brings order into its environment, while its marked articulation and composition of masses and voids also enter into dialogue with the urban surroundings.

线条简洁清晰的立体结构和连接方式的多样化（不同高度的 8 幢侧楼共同形成了 4 幢 L 形的建筑）展示了建筑的统一性。通过公共大厅的屋顶，侧楼之间富于变化的空间相互影响更加明显，而且能使人们立刻产生最佳的感性认识。

Unity is expressed in the building's compact clear-edged cubic form, and diversity in its articulation as eight wings of different heights joined to create four L-shaped volumes. The varied spatial interplay between these volumes is made more conspicuous through the concourse hall roof and is responsible for the high quality of sensory perceptions enjoyed all at once.

RKW 建筑设计事务所及城市规划所
杜塞尔多夫

医学博士楼
杜塞尔多夫

HOUSE OF THE MEDICAL DOCTORS
DÜSSELDORF

空间概念

SPATIAL CONCEPT

公共大厅直通8个楼层。两个较小的走廊被设计成冷大厅。宽大的主厅可以通过地下管道的通风孔获得自然通风。随意分布的柱子支撑着大厅的屋顶，并将公共大厅分成多个空间，使人们能感受到它的规模。

The up to eight-storey sections penetrate the concourse hall. The two smaller lobbies were designed as cold halls, the large main hall is naturally ventilated and climatized via air wells supplied via ground ducts. The hall roof is supported by freely distributed columns which subdivide the concourse and make its dimensions perceptible.

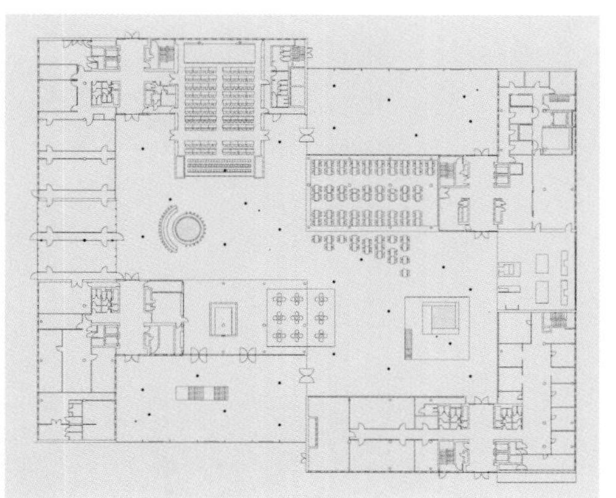

流畅空间关系给每一个角落都带来意外的发现。室内的动态环境为办公人员提供了舒适的工作氛围。办公室内与外加的通风设备和混凝土心的热能环境相适应。这种类型集供暖、通风和冷却于一身的装置被称为“柔性冷暖通风系统”。

Spatial relationships in the House of the Medical Docfors are flowing and offer unexpected discoveries at every corner. The dynamism of the interior creates a pleasant working atmosphere for the office workers. The offices are fitted with an additional ventilation system and concrete-core thermal environments. This type of combined heating, aeration and cooling system is called 'soft HVAC'.

建筑单位	RKW Rhode Kellermann Wawrowsky Architektur + Städtebau, Düsseldorf
业主	Bauherrengemeinschaft Haus der Ärzteschaft GbR Ärztekammer Nordrhein Nordrheinische Ärzteversorgung Kassenärztliche Vereinigung Nordrhein
摄影	Ansgar Maria van Treeck, Düsseldorf, S. 278/279 Michael Reisch, Düsseldorf, S. 280/281
技术工程师 接管方	ABB Bauprojektmanagement GmbH ABB Gebäudetechnik AG Niederlassung Ruhr, Herne
静力学	Schüßler-Plan Ingenieurgesellschaft für Bau- und Verkehrsplanung mbH, Düsseldorf
房屋技术	Schmidt Reuter Ingenieurgesellschaft mbH & Partner KG, Köln
TGA 顾问	Planungsbüro e+e, Kaarst
立面设计	IBFT Institut für Bauphysik und Fassadentechnologie Ingenieur-Büro Dr. Küffner, Düsseldorf
立面	Beaujean-Haskamp Fassadentechnik, Aachen Edewecht
周边保护	Hagen Ingenieure für Brandschutz, Kleve
空调安装	Dreist + Partner, Düsseldorf
景区规划	Ziegler Grünkonzepte, Düsseldorf

沿着陈列室和停车场的通道是公共广场的轴心，也就是所谓的与高速公路并行的交通区域。总而言之，这些建筑为公共广场对面的办公室起到了抵挡噪音的作用。

Along the showroom and car-park structure is the 'spine' of the main concourse axis, the so-called communication zone, which also runs parallel to the motorway. Altogether, these structures serve as a noise barrier for the offices on the other side of the main concourse.

希尔施贝格的戈德贝克设备处理公司南方分公司的大楼位于A5高速公路附近，是个地界标，也是它所属的工业区的标志。设计考虑到它特殊的地理位置，将陈列室和停车场直接与高速公路平行。广阔的悬臂屋顶将两个建筑物连接起来。

The building of the developer and facility management firm Goldbeckbau southern branch in Hirschberg, situated directly on and seen from the A5 motorway, is a landmark and a sign for the industrial area it is part of. The design takes into account its special location by placing the showroom and the car park directly parallel to the motorway. The wide cantilever roof joins these two building sections.

施瓦茨建筑设计事务所
米歇尔森・赫尔米特合作公司
斯图加特

希尔施贝格戈德贝克公司展厅、办公楼及停车场

GOLDBECKBAU SÜD, HIRSCHBERG
SHOWROOM, OFFICE AND CAR PARK

结构与材料

CONSTRUCTION AND MATERIALS

建筑成本	750 万欧元
施工期	2001–2003 年
总面积	12 300m²
总体积	57 600m³
竣工验收	2003 年
使用面积	10 600m²
办公室面积	4 900m²
展览面积	1 200m²
交流区面积	1 000m²
停车场面积	3 500m²

能够俯瞰高速公路的展厅大楼被设计成透明的玻璃立方体，而从不同的视角，多层的格状门面停车场看起来似开似关。办公大楼与素雅的窗边和镶着有色金属嵌板的栏杆相适应。

The showroom building overlooking the motorway was designed as a transparent glass volume, while the multi-storey car park cube with its façade grille appears either closed or more open, depending on the viewing angle. The office block was fitted with a quiet façade of window bands and parapets clad with coloured metal panels.

在中心厂房，公司为不同风格的办公楼、多层停车场和厂房提供施工装置。例如，从公共广场和展厅的不同角度或不同位置可以看到陈列厅内的展品。

At this 'Goba Centre', the firm produces its construction systems for office buildings, multi-storey car parks and factory halls, etc., in different styles. The showroom displays can be viewed from different angles and positions from the concourse and the showroom hall itself.

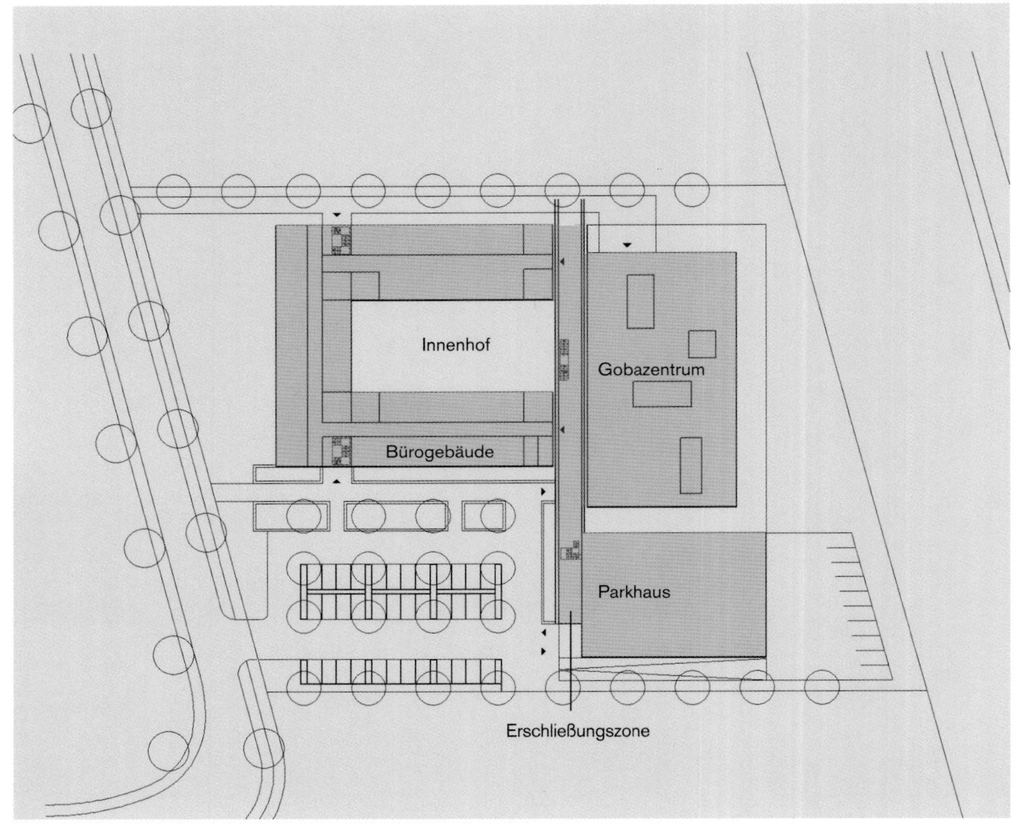

建筑由戈德贝克生产的钢铁和钢筋混凝土装配系统建造。土壤渗透的排水系统、屋顶和停车场外立面的太阳能板和种草的屋顶，都显出技术独特之处。和传统屋顶相比，这种做法体现了种草的屋顶调节顶屋室内气候的效果，并检验不同绿色屋顶的耐久性。

The building was constructed of Goldbeck steel and RC prefab systems. Technical specialities are the rainwater drainage system via on-site soil infiltration; solar panels on the roof and on the car park façade, as well as the roof sections planted with grass. This was done to demonstrate the effects of grass-covered roofs compared to conventional roofing on the interior climate on the top floors and to test the durability of different green roof covers.

施瓦茨建筑设计事务所
米歇尔森·赫尔米特合作公司
斯图加特

希尔施贝格戈德贝克公司展厅、办公楼及停车场

GOLDBECKBAU SÜD, HIRSCHBERG SHOWROOM, OFFICE AND CAR PARK

技术与生态

TECHNOLOGY AND ECOLOGY

办公室位于高速公路旁的交通枢纽对面的 U 形建筑内。他们朝着室内庭院而开，并为人们的集体办公提供了和谐与宁静。所有的工作场所和交流中心以及生产车间阳光充足并且能够看到乡村的美丽景致。

The offices are in a U-shaped building on the other side of the communication 'spine' from the motorway. They open to the interior courtyard and offer peace and quiet for concentrated work. All the workplaces and enclosed spaces within the communication zone and the production hall benefit from daylight and views of the beautiful rural surroundings.

建筑单位 Schwarz Architekten in Arbeitsgemeinschaft mit Michaelsen · Hermet, Stuttgart
业主 Goldbeck Süd, Hirschberg an der Bergstraße
摄影 Goldbeck, Bielefeld

技术工程师
TGA Goldbeckbau, Bielefeld
电子技术 Goldbeckbau, Bielefeld
支撑结构 Goldbeckbau, Bielefeld
景区规划 Kienle Planungsgesellschaft mbH, Stuttgart
ARGE Schwarz Architekten und Michaelsen · Hermet

在莱比锡的德意志广场的显要位置建实验楼，需要仔细研究市内环境。建筑的功能性要求也具有挑战性，因为这意味着通过提供不同的方案把委托方连在一起。

The commission to erect a laboratory building in the high-profile urban situation of the site on Leipzig's Deutscher Platz, called for a careful study of the urban context. The required functionality of the complex also represented quite a challenge, as it meant bringing together two clients with different programmes.

斯潘格勒·威斯库莱克建筑设计事务所
汉堡

莱比锡生物城—生物技术与生物医学中心

BIOCITY LEIPZIG – CENTRE FOR BIOTECHNOLOGY AND BIOMEDICINE

一楼两用

TWO CLIENTS – ONE BUILDING

建筑成本	约 5 000 万欧元
施工期	2001–2003 年
总面积	约 24 000m²
总体积	约 9 000m³
竣工验收	2003 年
占地面积	10 620m²
总计使用面积（店铺 + 营业）	16 370m²

大楼内建有生物城的大学和私立学院，从而使二者达到协同效应。新建筑就像个“支架”建于广场和它南侧的大街之间。这种结构为将来扩建扩大城市的总体建筑密度提供可能性。

The building now houses the university and private institutes of the Biocity under one roof and thus promotes synergy effects between both. The new building was conceived as a ‘bracket’ between the edge of the square and the street south of the site. This structure offers the potential for future extensions to create a dense urban ensemble.

中央大厅是大楼两部分的“交流中心”。它有防风防雨的作用，而它的不同区域为人们非正式的拜访提供一种特别的氛围。

The central hall is the ‘communicative hub’ of the two-part building. It gives protection from the weather, and its different zones provide a special atmosphere in which to meet informally.

材质 / 室内流通

MATERIALS / INTERIOR CIRCULATION

外部的承重墙和钢筋混凝土的柱子形成了灵活的内部框架。租用区按照用户的要求建造，采用白色的钢筋混凝土天花板和墙面。

Load-bearing outer walls and RC columns made it possible to create a flexible interior organization. Rental units are created according to user specifications, with white RC ceiling and wall surfaces in these spaces.

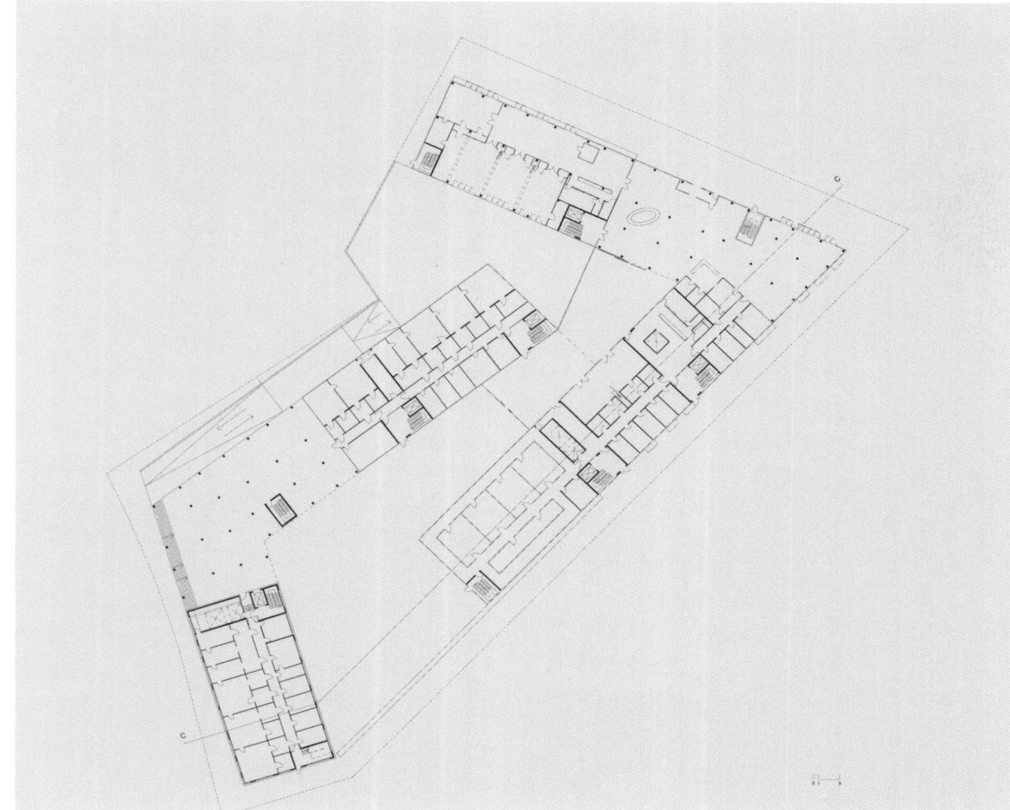

建筑师认为，在楼梯能够看到预制构件系统是比较重要的，因此选择了钢筋混凝土外墙以及和楼梯同样质地的天花板。这些都与外表成铁锈色的雕刻扶手形成对比。

The architects thought it important to make the prefabricated system visible in the staircases and therefore chose exposed RC walls and ceiling panels as well as stairs. These are contrasted by the sculptural handrail painted in the same deep rusty hue as the façade.

建筑特征是严密的设计的结果，以此进行整体处理并保持材料和风格的统一。玻璃表层和几层楼高的门廊打破了封闭式外立面。给人留下印象深刻的砖红色外立面与室内保持一致并决定了整个建筑的色彩基调。该建筑由贝尔芬格伯格建筑公司进行扩建。由于规划和施工时间紧急，在麦克思博格集团的技术监督下把建筑设计成装配式结构。

The building's identity is the result of the design rigour with which its large mass was treated and which did not permit material or style hybrids. The closed façades are only broken by the glazed front, several storeys high, of the joint entrance hall. The impressive red-brick façade continues into the interior and determines its colour scheme throughout. The project consortium ARGE Max Bögl – Bilfinger & Berger was commissioned to construct the extended structure. Due to the extremely tight planning and construction schedule (altogether twenty-four months), the building was designed as a prefab structure, under technical supervision of Max Bögl Group.

项目参与者

建筑单位	Spengler · Wiescholek Architekten Stadtplaner, Hamburg
业主	Leipziger Gewerbehof Gesellschaft mbH SIB Leipzig
房屋技术	IBG Ingenieurbüro für Gesamtplanung GmbH, Leipzig
公司扩展框架建筑	ARGE Max Bögl Bilfinger & Berger
摄影	PUNCTUM, Leipzig, S. 286/287, 289 oben OPOLE6X6, S. 288, 289 unten

一个新颖独立的、功能有效的办公大楼是为汉堡的圣保利地区设计而成的。这是造成西部周边地区密集的主要因素，并重新界定了圣保利渔市和艾尔比之间的市区。建筑也重新定义了城市环境，它的历史、特殊性质和多样的城市参照物为功能性的规划作出了贡献。

An original free-standing functionally efficient office building was designed for a site in Hamburg's St Pauli area. This is the main element of an ensemble west of the tall perimeter block that redefines the urban area between St Pauli Fischmarkt (fish market) and the Elbe. The building in turn redefines its environment whose history, special character and multiple urban references contributed as much to the design as the functional programme.

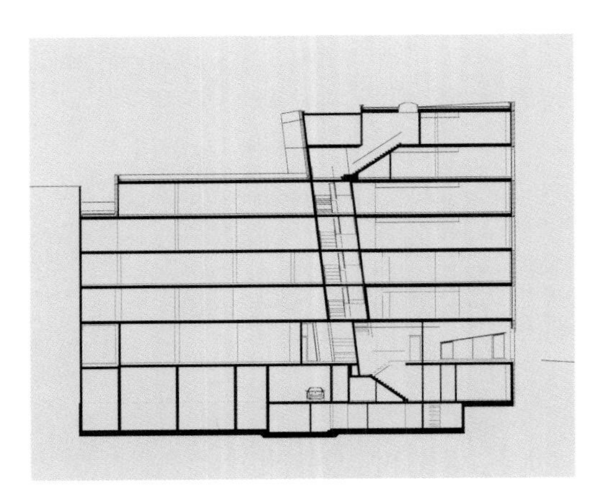

斯潘格勒·威斯库莱克建筑设计事务所
汉堡
鲁道夫与科尼尔汀
汉堡

阿尔特纳 47 号码头办公楼
汉堡

OFFICE BUILDING DOCK 47
HAMBURG-ALTONA

形式与功能

FORM AND FUNCTION

建筑成本	KG 200–400 u.600：565 万欧元
施工期	2003–2004 年
总面积	4 336m²
总体积	16 115m³
竣工验收	2004 年
占地面积	1 554m²
办公楼出租面积	3 860m²

能够改进建筑立体设计是建筑设计的目标。建筑按照城区图建造，它的外立面在两处是倾斜的。由于这个原因，建筑师采用了能强调建筑的立体外形和具有“平静镇定”特征的屋顶，而没有采用普通的顶楼和平屋顶。

The object character of the building was achieved with just a few design gestures which modified its basically cubic shape. The building volume follows the urban zoning map; in two places its façades deviate from the vertical to lean outwards. For this reason, the architects developed a form of roof, instead of the usual stepped-back top floors and flat roof, which underlines the cubic shape of the building and its ‘calm and collected’ character.

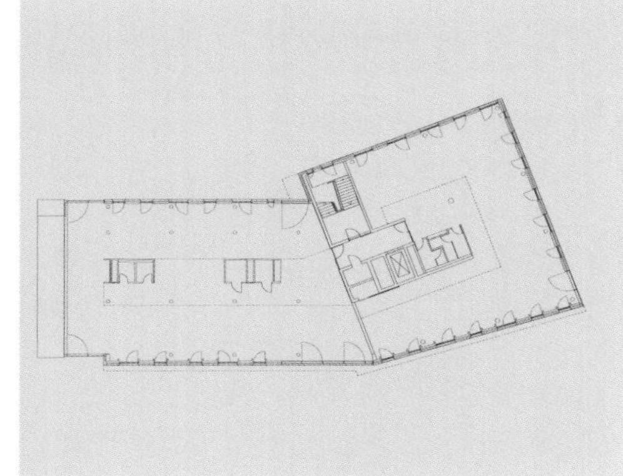

结构采用钢筋混凝土构架。为了特殊用户的利益，窗户被设置在整个外墙表面50%以下的位置，因为他们需要非彩色的的人工光线而非日光来生产印刷品。窗户之间的表层由带有红色窗格玻璃的装配式混凝土构件组成。

The structure is an RC skeleton. For the benefit of the main user, who produces print media requiring neutral artificial lighting rather than daylight, windows were kept at under 50% of the total façade surface. Surfaces between windows were built with prefab concrete elements faced with red enamelled glass panes flush with the windows.

建筑师旨在将外形与功能相结合。三个较低的宽阔水平面比较明显地体现这一特点，它能够按照新的要求灵活进行划分。垂直循环中心将建筑进行再分，从而使每层都有两个租用单元，而业户之间不会相互打扰。

The architects aimed to unite form and function. This is manifest in the openness of its three lower levels which can be partitioned flexibly to suit new requirements. The vertical circulation core subdivides the building in such a way that it can accommodate two rental suites per floor, but also one per floor, without tenants disturbing each other.

用地热制冷 / 采暖

COOLING/HEATING WITH GEOTHERMAL ENERGY

斯潘格勒·威斯库莱克建筑设计事务所
汉堡
鲁道夫与科尼尔汀
汉堡
阿尔特森纳 47 号码头办公楼
汉堡

办公室制冷及用地热采暖：
Cooling of Offices and Heating System Using Geothermal Energy:

01 具有泥土探测能力的管状吸热器：
• 地下水的环境使用能力
• 从循环用水获得的冷却能量
• 可用热交换器
• 节省主要能源
• 从地下水获得的 45% 热能
• 从地下水获得的 50% 冷却能量
tubular heat absorber with earth probe capacity:
·using environmental availability of groundwater
·cooling energy from circulation water
·concrete as 'usable' heat exchanger
·saving on primary energy
·45% of heating energy from groundwater
·50% of cooling energy from groundwater
02 蒸气泵 / 制冷装置；两个目的
• 冬季供暖
• 夏季降温
• 通过电或煤气发动
heat pump / refrigerating machine; double-purpose:
·winter heating
·summer cooling
·powered by electricity or gas
03 加热
heating
04 通风
ventilation
05 温水系统
warm water system
06 冷却 / 加热盘管系统
cooling / heating serpentine pipe system
07 新鲜空气入口 / 废气出口，夜间冷却
fresh-air intake / waste-air outlet, night cooling
08 散热器冷却 / 加热原理
cooling/heating radiator principle
09 电器设备为室内提供：
• 电源
• 电话线路
• 电脑线路
electric apparatus for providing interiors with
·electrical power
·telephone connections
·computer connections
10 空心地板
cavity floor
11 转换操作开关
change-over operation switches
12 地面冷却
floor cooling
13 DP 制冷器
DP cooler

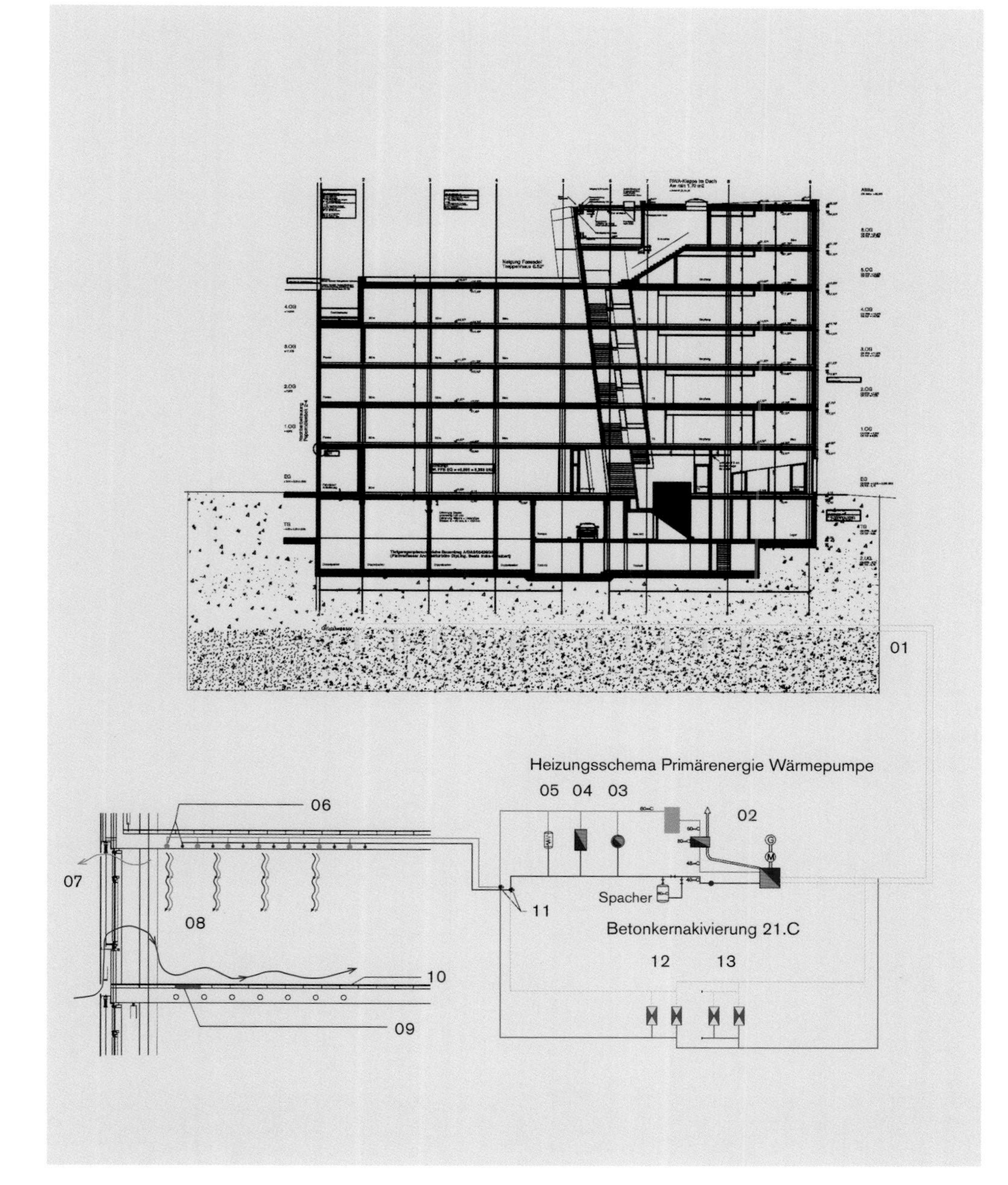

建筑单位
业主 Spengler · Wiescholek Architekten und Stadtplaner, Hamburg
建筑工程管理 Rudolph + Konerding Architekturbüro, Hamburg
业主 Einsatz Creative Production
Grundstücksgesellschaft Pinnasberg 47, Hamburg
摄影 Ralf Buscher, Hamburg, S. 290/291, 292
OPOLE6X6, Hamburg, S. 293

斯潘格勒・威斯库莱克建筑设计事务所
汉堡

SAP 商务中心
汉堡

SAP BUSINESS CENTRE HAFENCITY, HAMBURG

位置与构造

LOCATION AND CONSTRUCTION

大厅采用花边构造的玻璃屋顶，两层楼高的架桥从中穿过，都将办公室的两侧相互连接起来，并使外立面的设计元素清晰可见。所形成的迷人的内部空间不但可以作为门廊，而且可用于办公场所和内部交流。

The hall is covered by a glass roof with a filigree structure and penetrated at different height levels by two-storey 'bridges'. These interconnect the two office wings and are also 'legible' as design elements on the façade. The result is a fascinating interior space that not only serves as the entrance hall, but is also used for official functions and in-house communication.

建筑成本	约 1 600 万欧元
施工期	2001–2002 年
总面积	12 500m²
总体积	62 500m³
竣工验收	2002 年
占地面积	4 764m²
办公楼出租面积	8 400m²

建筑设计符合培训和行政中心的功能要求。公共区域可分别通过门廊和垂直循环中心直接进入。其他潜在用户的办公室也能够容易地进入。由于电梯和楼梯的位置，这些区域也能够分成几个小空间。

The building was designed to the functional requirements of a training and an administration centre. The public areas are directly accessible via the entrance hall and a vertical circulation core respectively. The offices of potential other users can be made easily accessible, too. Due to the placement of lifts and stairway cores, these areas can also be partitioned into smaller spaces.

这是最初建在汉堡的新哈芬市（港口城市）的建筑。它位于格拉斯布鲁克海芬地区尽头的优越位置，能够俯瞰易北河以西的景色，它利用地点的优势并且建筑内部体现了建筑质量。两侧平行的办公楼位于狭窄的玻璃大厅的侧面，大厅有一个跨度很大的楼梯并能看到易北河、港口全景和一个公园，视野开阔，同时过路人可以看到接待区的内部。

This building was the first to be completed in Hamburg's new Hafencity (harbour city). Its first-class location at the end of Grasbrookhafen, with a free view over the Elbe towards the west, demanded an architecture that used the advantages of the site and also allowed its qualities to be felt from inside the building. Two parallel office wings flank a narrow glazed hall which includes a dramatic flight of stairs and opens the view over the waters of the Elbe, the harbour panorama and a park, but also offers passers-by a glimpse into the depths of the reception area.

建筑南北两边的“桥头堡” 有“温室”一样的休息室。深色天花板和从外面看得见的墙壁美化了休息室。

The 'bridgeheads' on the south and north sides of the building contain recrea-tional rooms in the form of 'conservatories'. They are enhanced by the deep colours of ceilings and walls visible from outside.

建筑师十分注重外立面的遮光挡风及自然通风，于是在宽敞的玻璃窗外面加上充气挡风玻璃窗，因而也就没有必要在整个外立面再安装双层玻璃了。

The architects opted for large windows in the massive façade and took special care to provide sun and wind protection, as well as natural ventilation by designing a back-aerated 'wind screen' mounted in front of the windows which made a double glass skin covering the entire façade unnecessary. Only the 'bridges' were double-glazed for aesthetic reasons.

外观

THE FAÇADE

斯潘格勒·威斯库莱克建筑
设计事务所
汉堡

SAP 商务中心
汉堡

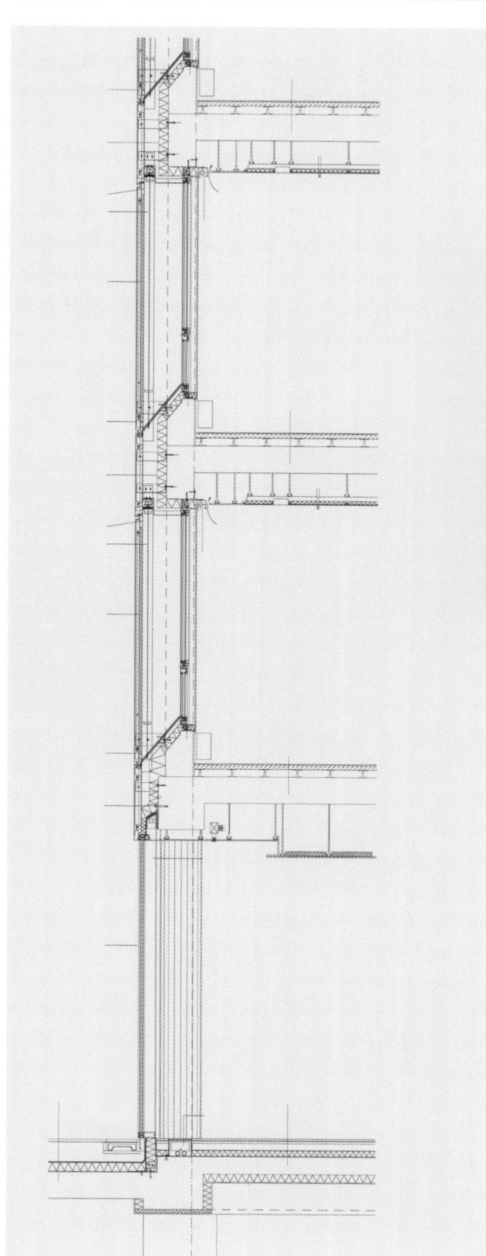

外立面剖面构造

Façade section, detail

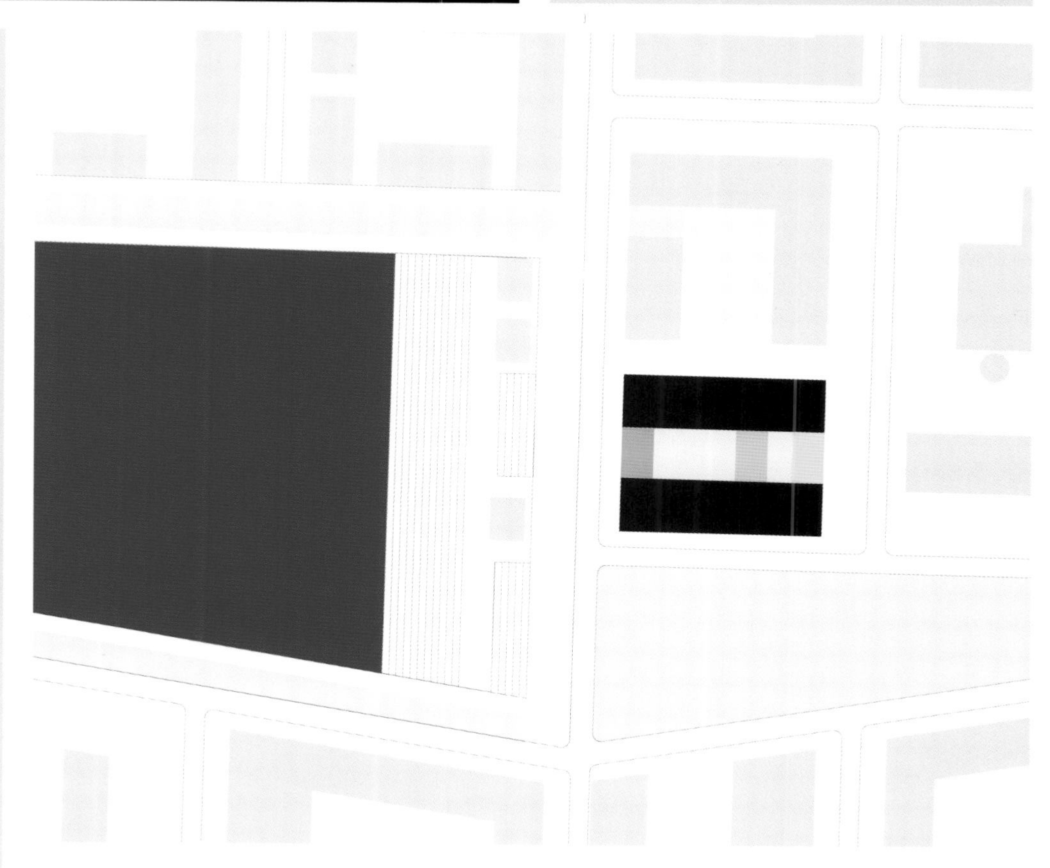

为使用者提供与众不同的办公模式并且提供高雅舒适的氛围，建筑师设计了此方案。建筑的深色金属嵌板与门廊的浅灰色形成色调对比。

The architects aimed for a design solution that goes beyond the usual office pattern and offers users a sophisticated and pleasant working atmosphere. In contrast with the light-coloured silvery finishings of the entrance hall, the building is clad in dark-coloured metal panels.

项目参与者
建筑单位 Spengler · Wiescholek Architekten und Stadtplaner, Hamburg
支撑结构 Windels Timm Morgen
Beratende Ingenieure im Bauwesen, Hamburg
摄影 Jochen Stüber, Hamburg

把窗户外立面涂成各种颜色或是单色，从各个不同的角度来看都显得与众不同。

The window reveals are painted either in a number of colours each or in monochrome so that the building appears different from different angles.

尽管塞勒森霍赫地区有许多个体工程，存在不同的建筑风格，作为城市，和其他城市一样，还是要统一发展的。城市规划并不是要分离，而是要将生活区和工作区联合为一体，也就是说住宅楼同时也是办公地点。位于港克豪芬大街的 KPMG 公司就是按照此原则进行设计。

The Theresienhöhe area is to be redeveloped as a homogeneous urban district, despite the fact that it will include a number of individual projects and different architectural styles. The urban design does not aim for the separation, but the union of living and working, i.e. buildings with residential as well as office units. This principle was also applied to the design of the KPMG building on Ganghoferstrasse.

斯泰德勒及其合伙人
慕尼黑

KPMG 集团
慕尼黑

KPMG BUILDING, MUNICH

生活与工作相结合

DESIGN PROPOSITION:
LIVING AND WORKING

建筑成本	4 100 万欧元
施工期	2000–2002 年
总面积	28 000m²
总体积	101 750m³
使用面积	18 470m²
竣工验收	2002 年
占地面积	11 000m²

建筑外立面多窗且透明，但建筑结构坚固。从斜角看起来很结实的瓷砖涂层构架让站在大楼前的人感到像透明一样。

The façades are largely glazed and transparent, but nevertheless form a solid building mass.
A ceramic-tile-clad structural skeleton, which seems massive seen from an oblique angle, appears completely transparent to the viewer who stands directly in front of it.

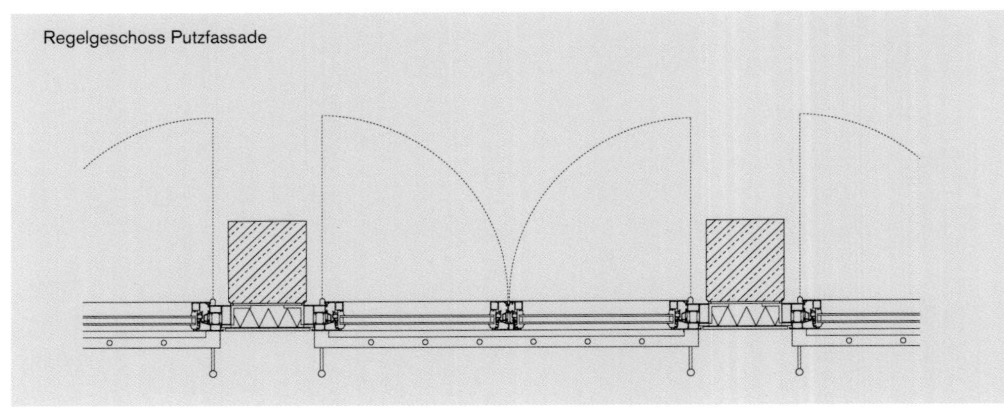

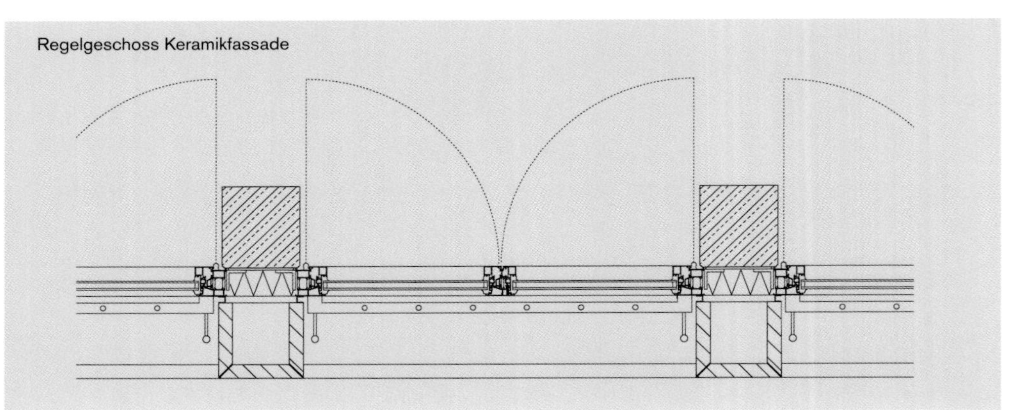

门庭是整个建筑的核心，从此处进入大楼。楼梯从这儿通向大楼的各个角落。专门为大公司建造的屋内通道具有重要作用。个体区域能够从外面进入以便与办公区或住宅区相分离。

The building is accessed via an entrance court recognizable as the focal point of the structure. From here, a stairway leads into all corners of the building. The interior circulation is tailored to the significance and needs of a large corporate headquarters. Individual sections of the block are accessible from outside so that they can be let separately as either offices or residential units.

斯泰德勒
慕尼黑

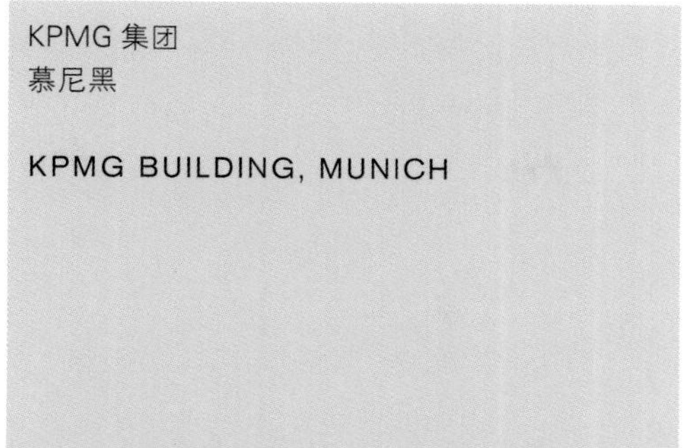

KPMG 集团
慕尼黑

KPMG BUILDING, MUNICH

通道和室内的流通及外观设计

ACCESS, INTERIOR CIRCULATION AND FAÇADE DESIGN

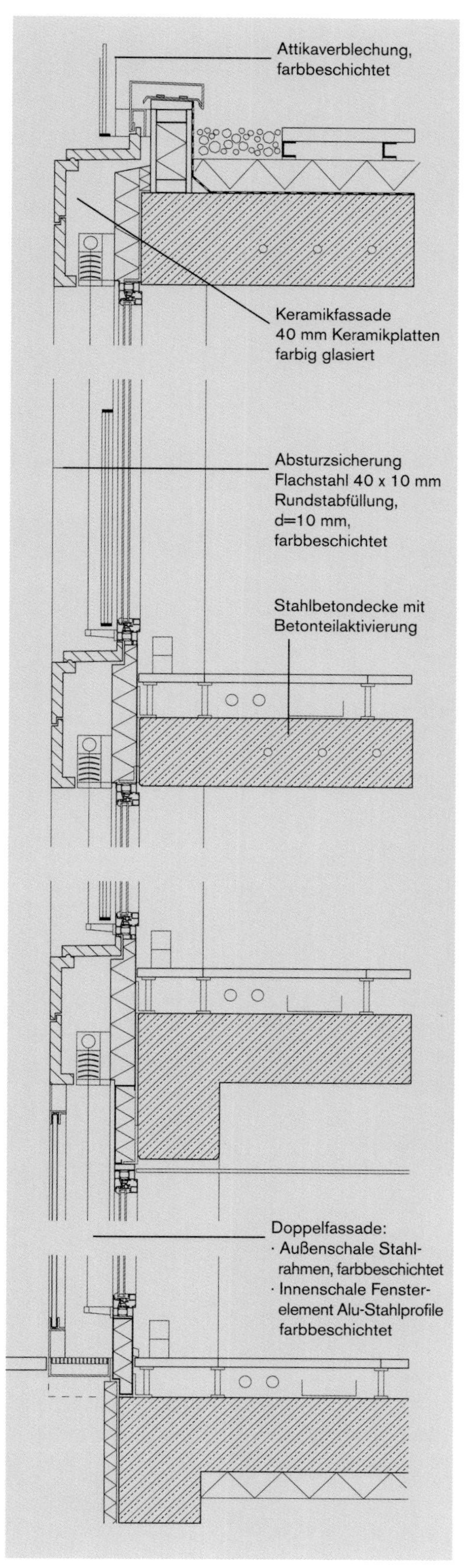

建筑单位 Steidle + Partner, München
业主 Investa für KPMG Dt. Treuhandgesellschaft AG, Berlin
景观建筑 Auböck + Karasz, Wien/München
立面 Erich Wiesner, Berlin
照明安装 Ingo Maurer GmbH, München
摄影 Reinhard Görner/artur

技术工程师
支撑结构设计 Hochtief / CBP
房屋技术 Ingenieurbüro Hausladen, München
接管方 AG Alte Messe – MK5 – KPMG
Hochtief Construction AG, München
Siemens Gebäudetechnik Bayern GmbH

引导性企业
立面 NBK Baukeramik GmbH & Co. KG, Emmerich

将多层传导材料连接起来的技巧是信息时代主要技术的标志。在当今，甚至能够将百科全书的知识内容储存在指甲大小的芯片上。

The art of joining a network of multi-layer conductive materials is a mark of the key technology of the information age which, even today, is able to store the knowledge contained in a large encyclopedia on a chip as small as a fingernail.

tec 建筑设计事务所
洛杉矶
慕尼黑
新加坡
泰格威伦（瑞士）

微技术发展中心
杜伊斯堡

MICROTECHNOLOGY DEVELOPMENT CENTRE, DUISBURG

纳米技术世界

THE WORLD IN NANOMETRES

施工期	2003-2004 年
总面积	10 700m²
总体积	45 400m³
竣工验收	2004 年
办公楼出租面积	7 500m²

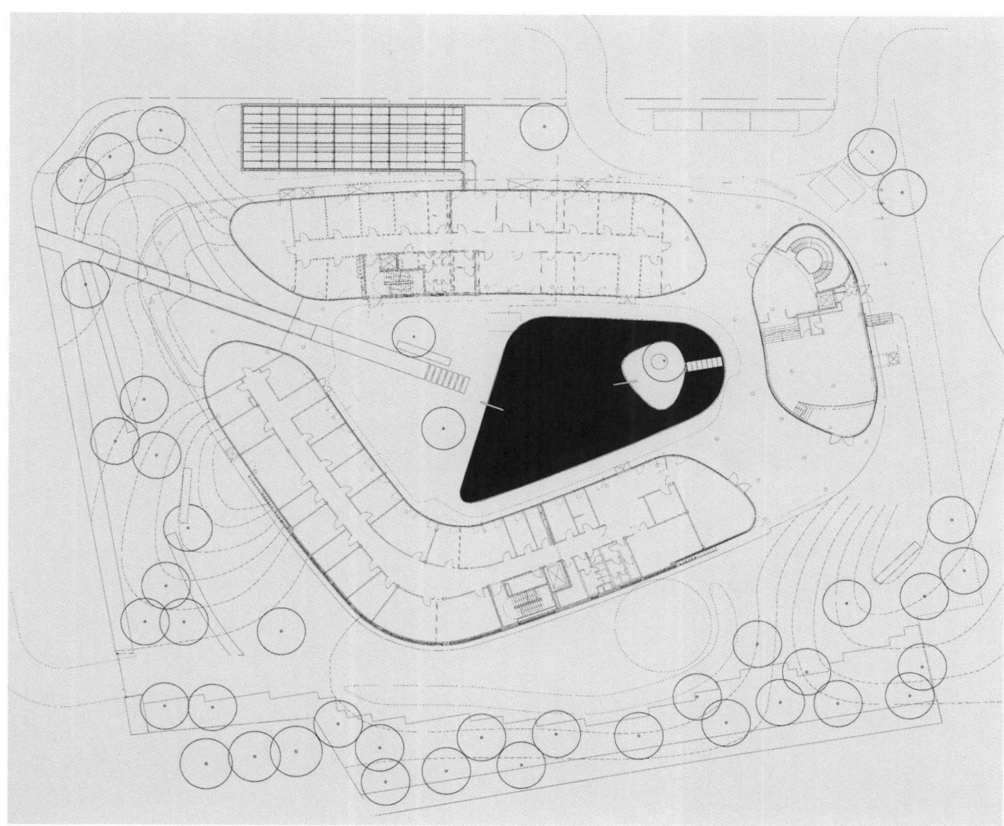

在杜伊斯堡，德国芯片制造商英飞凌科技公司正在创建世界第二大显微技术开发中心。该设计展现了从事纳米技术的缩影。建筑就像一个与他人断绝来往的修道院。它有着令人沉思禅定的宁静氛围。建筑的中庭为有机结构，具有与众不同的校园文化氛围，体现出其独特用途和使用者的特殊技巧。

In Duisburg, the world's second largest microtechnology development centre is being built for Infineon Technologies. The design visualizes the microcosm of people who think in nanometres, and the concentration such development processes require. The building is like a monastery turned in upon itself. It evokes the quiet atmosphere of a Zen place of contemplation. Its unusual campus morphology enclosed by an organically shaped atrium building, is a deliberate reference to its unique purpose and the special skills of its users.

tec 建筑设计事务所
洛杉矶
慕尼黑
新加坡
泰格威伦（瑞士）

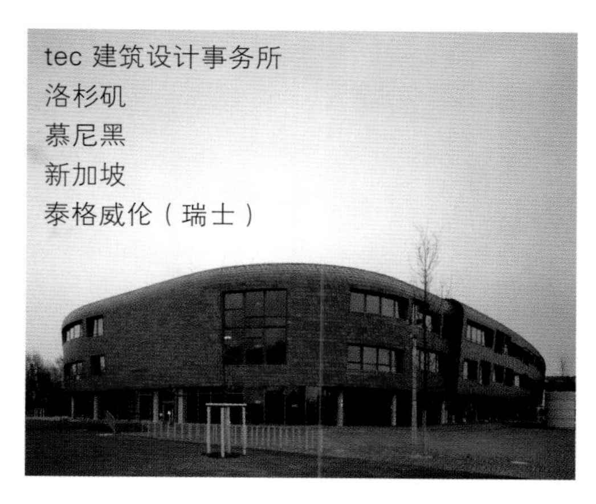

微技术发展中心
杜伊斯堡

MICROTECHNOLOGY DEVELOPMENT CENTRE, DUISBURG

开辟新纪元

THE NEXT GENERATION

不同的工作区域和中央庭院的周围布置之间的水平线和垂直线有力地相互结合，有利于促进不同工作团体各学科间的合作以及工作人员间的讨论。

Strong horizontal and vertical interconnections between the different work areas and their arrangement around the central courtyard help to promote the interdisciplinary co-operation of different working groups and informal discussions among staff members.

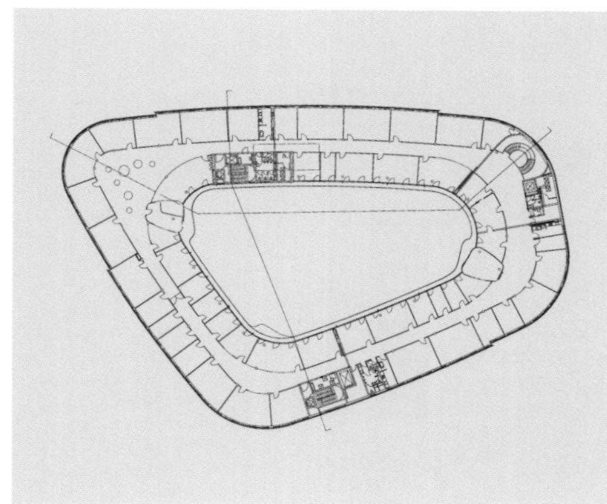

用于外立面和内部装置及设备的天然材料在自然与技术之间架起一道桥梁。为今后工作提供了促进创造力和发挥潜能的工作环境。不同层次的几何学有助于灵活安排办公类型，工作区内部和外部大面积的多种用途的开阔区域将高度集中的设计工作与休闲消遣紧密结合起来。

The natural materials used for the façade and the interior fittings and furnishings build a bridge between nature and technology. The result is a working environment which promotes creativity and offers high potential for the future. The geometry of the different levels lends itself to flexible arrangements of office types, while generously dimensioned multi-purpose and open areas – inside and outside the work spaces – allow for the closeness of highly concentrated project work and recreation.

建筑单位	tec ARCHITECTURE – Los Angeles, München, Singapore, Tägerwilen (CH)
业主	Erste Primus Projekt GmbH
建筑工程管理	Diete + Siepmann GmbH
摄影	Christian Richters, Münster
工程控制	Drees & Sommer AG
技术工程师	
房屋技术	Schmidt Reuter Integrale Planung und Beratung GmbH
静力学	ARUP Deutschland
建筑企业	
基本结构	Hellmich Unternehmensgruppe

考虑到特殊的工作准则、社会趋势及文化倾向，通过采用具有时代特点特性材料建筑的结构说明和生态观念，建筑反映了自身所处的特殊环境。

Taking into consideration the particular work ethic, social tendencies and cultural preferences, the building reflects its specific context by use of characteristic materials, structural interpretations of local architecture and specific ecologic considerations.

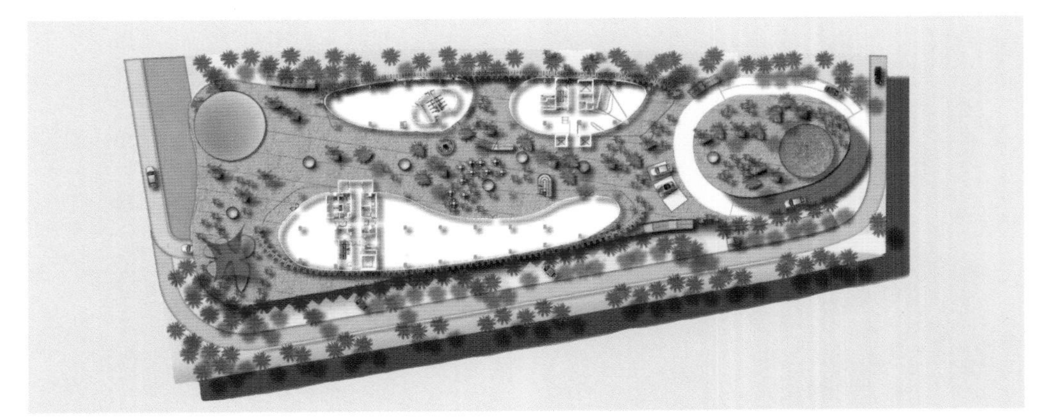

tec 建筑设计事务所
洛杉矶
慕黑尼
新加坡
泰格威伦（瑞士）
建筑师 61 新加坡
GTL 景观建筑设计事务所 卡塞尔

英飞凌亚太科技公司
新加坡

INFINEON TECHNOLOGIES ASIA PACIFIC PTE. LTD., SINGAPORE

传奇的高科技

ROMANTIC HIGHTECH

在它的布局中，两个有机形成的办公大楼从共同的公用墙中显露出来。建伸的屋顶结构将办公大楼间的空间变成宽敞的敞开式中庭。这个自然通风区域起到了气候缓冲器的作用，在室内和室外之间形成过渡。

In its composition, two organic shaped office towers emerge from a common public podium. A tensile roof structure transforms the space between the office towers into a grand, open atrium. This naturally ventilated zone acts as a climatic buffer, forming a transition between inside and outside.

施工期 2003–2004 年
总面积 27 200m²
竣工验收 2004 年 12 月

亚太地区（先进技术和创造性智慧的发展中心）受益于高科技发展和自然环境的并存。我们抓住这个机遇，研究将后两者结合并让人类回归议事中心的策略。

The Asia Pacific region, an evolving hub for advanced technology and creative intelligence, benefits from a unique juxtaposition of hi-tech development and natural environment. We seized this opportunity, investigating strategies that integrate the latter two and bring the human being back to the center of all deliberation.

英飞凌集团新加坡总部进驻亚太市场，象征着新精神，表现了知识领先公司的特点。企业的优先性，创造了一种激励和孕育创新思想蓬勃发展的环境，并建立了非传统的建筑风格表达形式的框架。

The Singapore headquarters manifest Infineon's venture into the Asia Pacific market, with the intention to symbolize the new Spirit that characterizes its establishment as a knowledge driven company. The project's priority, creating an environment that stimulates and nurtures the thriving of ideas, sets the framework for pursuing unconventional architectural expressions.

tec 建筑设计事务所
洛杉矶
慕黑尼
新加坡
泰格威伦（瑞士）
建筑师 61 新加坡
GTL 景观建筑设计事务所 卡塞尔

英飞凌亚太科技公司
新加坡

INFINEON TECHNOLOGIES ASIA PACIFIC PTE. LTD., SINGAPORE

新特点

NEW IDENTITY

新颖的设计理念与低能源消耗相结合，迎合了英飞凌亚太科技公司持续发展对总部大楼的综合性要求。该建筑因此而成为生态模范工程。因遵循了城市重建局的要求。

Innovative design ideas are combined with low-energy solutions to meet the complex demands given by developing a sustainable headquarter building for the hi-tech company Infineon. Thus the building is being recommended as ecologically exemplary project, following the guidelines of URA.

由轻质膜结构遮挡的中庭，在不同商业和社会活动的互动中促进了非正式的交流。既能遮阳又能让大楼轮廓柔和的布置鳍状物将整个综合大楼的外墙包裹起来。外形和材料质地的结合，形成一种动态且有些神秘的外观，这刺激了充满灵感和活力气氛的形成。

Shaded by the lightweight membrane structure, the atrium encourages informal communication by accomodating an interactive flow of various business and social activities. The façade of the entire complex is wrapped with a system of fabric fins that not only shade the surface, but also soften the contour of the building. This combination of form and texture creates a dynamic and almost mysterious appearance that stimulates an atmosphere of inspiration and vigor.

建筑单位 tec ARCHITECTURE – Los Angeles, München, Singapore, Tägerwilen (CH) mit Architects 61, Singapore und GTL-Landschaftsarchitekten, Kassel
业主 Infineon Technologies Asia Pacific PTE LTD
摄影 Clemens Mayer, Regensburg

tec 建筑设计事务所
洛杉矶
慕尼黑
新加坡
泰格威伦（瑞士）

华亚集团总部办公楼和
中国台北微芯片公司

INOTERA CORPORATE HEADQUARTERS OFFICE BLOCK AND MICROCHIP PLANT TAIPEI, CHINA

对自然环境的诠释

FORMAL INTERPRETATION OF THE COMPLEX NATURAL ENVIRONMENT

施工期	2003–2004 年
总面积	29 000m² 建筑面积
	20 000m² 使用面积
总体积	1 121 000m³
方案设计	2002–2003 年
竣工验收	2004 年

办公楼和生产车间全景：建筑的玻璃外表采用不同基调和色彩的网版印刷技术。

Overall view of office building and production halls behind it; the glass building skin is screen-printed with various motifs in different colours.

左页：主入口的外立面说明

Left-hand page: façade details at the main entrance.

华亚科技公司总部大楼的设计使办公室工作环境充满浪漫色彩。成百上千的不同色彩和形状是大楼对自然环境的诠释，既像树叶又像海浪。华亚科技公司采用微型芯片创新技术设计的外立面，展现了将文化价值和环境结合在一起的现代建筑理念。

The design of the Inotera corporate headquarters building is based on an emotional, almost romantic notion of office work. Like tree leaves or ocean waves, hundreds of different colours and shapes represent a formal interpretation of the complex natural world. Innovative micro-chip technology – as developed by Inotera – made it possible to design a façade which shows a combination of cultural values and environmental aspects in a contemporary architectural context.

tec 建筑设计事务所
洛杉矶
慕尼黑
新加坡
泰格威伦（瑞士）

华亚集团总部办公楼和
中国台北微芯片公司

INOTERA CORPORATE HEADQUARTERS
OFFICE BLOCK AND MICROCHIP PLANT
TAIPEI, CHINA

按比例分解

THE DISSOLUTION OF SCALE

对这种规模的高科技大楼没有普遍认可的形象和形态观念，办公楼的设计想达到自然表达的目的，而工厂附加外立面的设计（由平面的透视几何图形构成）形成一个达达主义类型的比例分解的奇妙背景。创新科技可能预制几百种不同形状和色彩的外立面。

As there are no generally accepted iconographic or morphological concepts for high-tech buildings of this size, the design of the office building aimed for original expression while the design of the factory annex façade – a composition of tiled perspective geometries – forms a fantastic background to a Dada-type dissolution of scale. Innovative technologies made it possible to prefabricate a façade of hundreds of different shapes and colours.

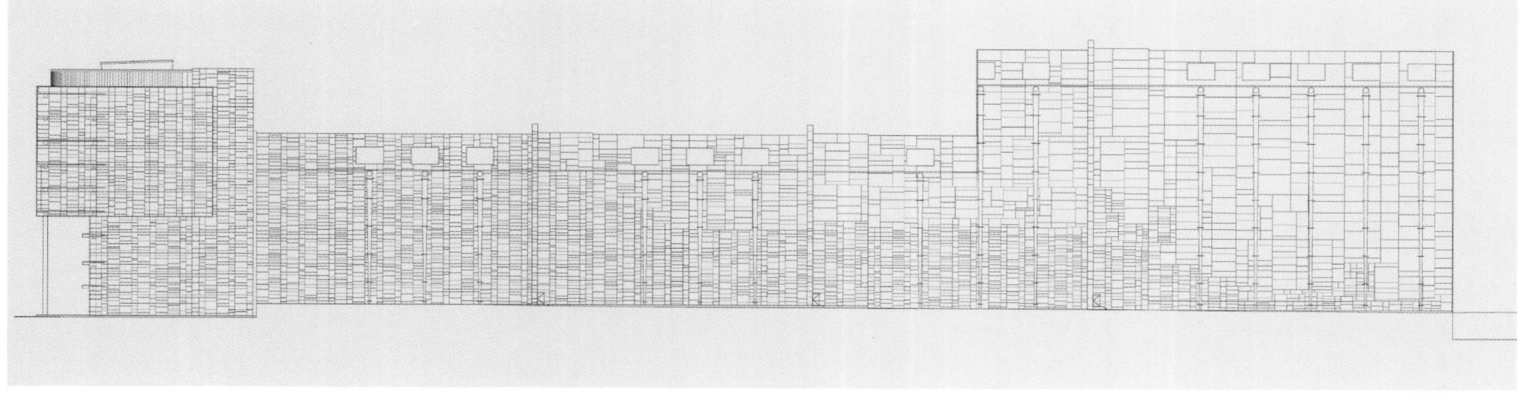

设计证明了科技的基本潜能，它为提高人类生活的经济学和审美学作出贡献。科技在建筑中发挥着极其重要的作用。建筑不仅使技术创新得以应用，不仅有技术上的创新，还能使之内部化。

The design demonstrates the basic potential of technology, i.e. of contributing to improving the economics and aesthetics of human life. Technology is at the heart of a complex architectural approach that does not only apply technical innovation, but internalizes it.

建筑单位 tec ARCHITECTURE – Los Angeles, München, Singapore, Tägerwilen (CH)
业主 Inotera Memories Inc. Taipeh, China
建筑工程管理 Nanya Construction Group
摄影 Hisao Suzuki
工程控制 Siemens Industrial Building Consultants
技术工程师
房屋技术 I.S. Lin & Associates Consulting Engineers
静力学 LTN Consulting Engineers
接管方 TASA Construction Group

两层的建筑构造包括更衣室、淋浴间，还有娱乐室和培训室以及一个厨房和一个办公室。一楼的更衣室和卫生间通过窗户采光，窗格可以保护窗户玻璃（当室内没有供暖的时候），这样窗户一直敞开，弄湿的设备可以通过空气对流快速变干。

The two-storey structure houses the dressing rooms and showers as well as the recreation and training rooms, a kitchen and an office. The dressing and sanitary spaces on the ground floor are lit through window bands, protected on the outside by grilles, so that the windows may be kept open permanently (when the building is not heated) and wet service outfits can dry faster due to cross-ventilation.

更衣室和培训室整体建筑（由当地传统的典型材料建造）与车库的冷工业特色形成对比。

The monolithic building mass of the dressing-rooms and training wing, constructed of rather traditional, locally typical materials, contrasts with the cool industrial character of the garage hall.

威切曼工程有限责任公司
诺伊斯

消防队
诺伊斯

FIRE STATION, NEUSS

设计方案

PROGRAMME

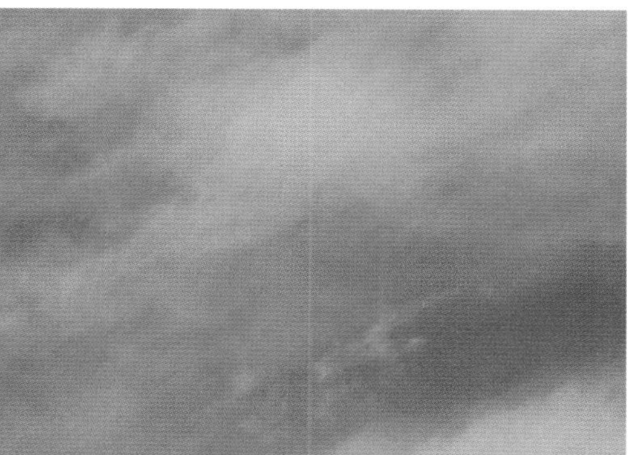

施工期	2003–2004 年
总面积	700m²
总体积	2 980m³
使用面积	615m²
竣工验收	2004 年
占地面积	3 200m²

诺伊斯的市郊新成立了一个消防队。在车库的远处是三辆消防车，设计方案包括消防员们的娱乐室和他们可以在周末上进修课程的“教室”。由于消防员们都是志愿兵，因此没有必要提供专门的休息室，而只有更衣室和淋浴间。施工单位的基本要求是建筑的设计要减少经营成本。

A new fire station was built in a suburb of Neuss. Apart from the garage hall for three fire-fighting vehicles, the programme included recreation rooms for the fire-fighters and 'classrooms' for their weekly further-training courses. As the fire-fighters are all volunteers, there was no need to provide on-call staff rooms, but only dressing rooms and showers. One essential requirement of the operator was that the building should be designed in such a way as to reduce operating costs.

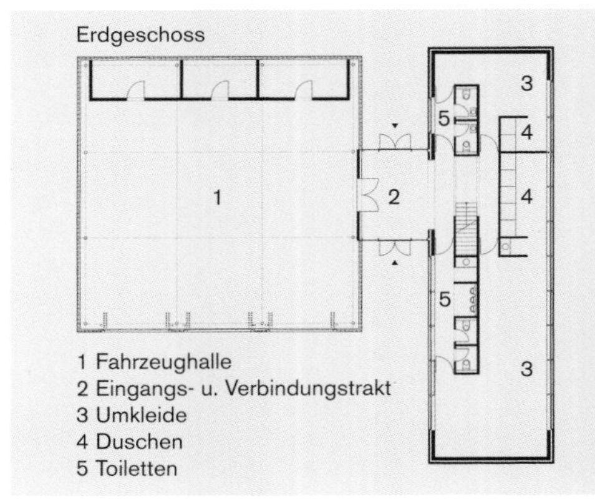

两个建筑区的内部空间（功能、室内设计、防火设施及室内气候有所不同）在外形和材料上相互衬托。车库采用玻璃结构的外立面，车辆的外形和颜色让旁观者明显地感觉这个区域的功能特征。

The interior spaces – with their different functions, interior design, fire-protective measures and interior climates – are arranged in two building sections, set off against each other in terms of form and materials. A transparent façade of structured glass was developed for the garage hall, and the outlines and colours of the vehicles make the function of this section obvious to onlookers.

消防队为工作和培训提供了一个宽敞且自然照明的大厅，即使在雨天也能正常作业。宽大叶片的树木遮住了南面的阳光，门和屋顶同样高的玻璃板条窗户打开时室内会马上凉爽起来，因此夏季的被动吸热问题有所缓解。在冬季，被动吸热大大减少了供暖成本，同时日光减少了全年的电能消耗。

The fire station offers a large, naturally lit hall where maintenance work and training can take place even on rainy days. The problem of passive heat gains in summer is lessened as tall broad-leaved trees shade the south façade and the interior is cooled down within minutes when the doors and the floor-to-ceiling glass-slat windows are opened. In winter, passive heat gains significantly reduce heating costs, and daylighting reduces electrical energy consumption throughout the year.

威切曼工程有限责任公司
诺伊斯

消防队
诺伊斯

FIRE STATION, NEUSS

设计理念

DESIGN CONCEPT

大楼的使用和空气流通通过聚光来实现。一楼的门廊通过透明的连接的侧厅得到阳光的照射，而上部的走廊通过带状顶灯得到顶部照明。顶部灯光通过楼梯空间照向一楼，通往“正式的”室内。

Access to and circulation inside the building is 'staged' by focused lighting. The ground-floor entrance hall is daylit via the transparent connecting wing, while the upper corridor is top-lit through a ribbon rooflight. This top light reaches down, through the staircase air space, to the ground floor, leading the way into the 'official' interiors.

是玻璃连接结构的车库与附属区域的门廊和过道。建筑可以从两边进入，遇到火警时，在一定程度上缩短了消防员的私人汽车与更衣室之间的路程。

The glazed connecting structure is both entrance hall and passageway between the garage hall and the ancillary spaces. The building is accessible from two sides, which – in the case of a fire alarm – considerably shortens the walk from the fire-fighter's private car to the dressing room!

建筑单位 Wichmann GmbH Architekten + Ingenieure, Neuss
Richard Wichmann
员工 Benedikt Feist
业主 Stadt Neuss
工程控制 Hochbauamt
摄影 Jens Kirchner, Düsseldorf
展馆外立面 Glas Schmitz GmbH, Kempen

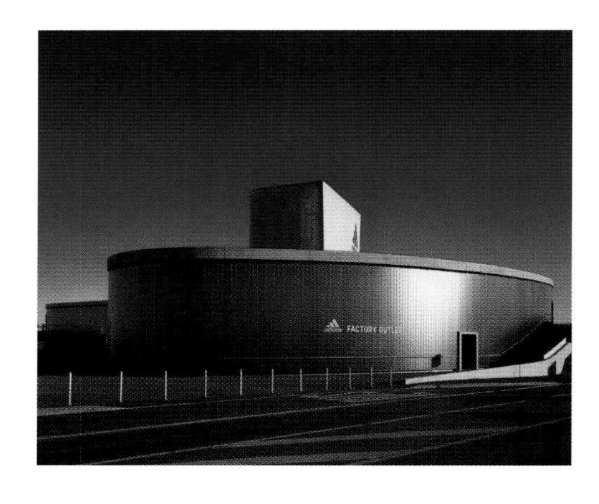

顶篷螺旋向上并由雕刻杯状花纹的、蘑菇顶的柱子支撑着。

The ceiling spirals upwards and is supported by columns with chalice-shaped 'mushroom tops'.

这个“凸出的土块”作为停车场和送货区。座位2倍宽的曲线形台阶通向大楼和走廊边的前院。工业玻璃窗的外表（内部遮阳双层玻璃）反映了工厂的结构特点。

A bulging 'soil clod' accommodates the car park and the delivery yard. Curving steps, which double as seats, surround a forecourt 'forum' which extends into the building and up to the edge of a gallery space. The façade of industrial glazing (double-glazing with inlying end supports and sun-shades) reflects the factory character of the structure.

设计采用了总体规划设计并将基本想法预以实施。设计师安吉利认为，“到目前为止元素中蕴含的动力学转化成动力学建筑”。结果是源于周边地形的人造地形学渗透到了弯曲而紧凑的建筑中。2个要素：“风景”和“动力学”成为设计的催化剂。

The design refers to the masterplan and intends to implement its basic idea. "The dynamics so far hidden in the elements of the landscape are translated into dynamic architecture," says architect Angélil. The result is a curving, yet compact building penetrated by an artificial topography derived from the surrounding landscape. The two themes, 'landscape' and 'dynamics', act as catalysts for the design.

伍尔夫及其合伙人
斯图加特

阿迪达斯工厂商商店
黑措根奥拉赫

ADIDAS FACTORY OUTLET HERZOGENAURACH

景观与动力学

LANDSCAPE AND DYNAMICS

建筑成本	1 300 万欧元
施工期	2002–2003 年
总面积	8 500m²
停车楼总占地面积	11 000m²
总占地面积	19 500m²
销售总体积	50 500m³
停车楼总体积	45 500m³
总体积	96 000m³
占地面积	21 200m²

从南侧的道路看起来，大楼像是一个动力建筑学符号。

Seen from the southern approach road, the building appears like a dynamic architectural sign.

在休息大厅里，有音响，而且在发光二极管屏幕上看到知名品牌的宣传，使大厅充满生气。

The foyer is alive with acoustic and visual presentations of important brands, on LED screens and via 'media beams'.

钢筋混凝土结构（由柱子和剪力墙支撑的水泥平板）有很大的适应性和必要的储热功能。双层的压铸玻璃外墙便宜且不需维修，空腔内的天棚能够挡风。建筑的服务技术设计着眼于最佳功效和可持续性，它包括安装于地基的地热吸收器，集成冷却装置和一个露天预热通道。必需的洒水装置与露天混凝土天花板连为一体。

The RC structure – flat floor slabs supported by columns and shear walls – allows for great flexibility and has the necessary heat storage capacity. The double façade of cast glass modules is inexpensive and requires little maintenance, the sun-shades in the cavity are protected from winds. The building's services technology was designed with a view to optimal efficiency and sustainability. It includes a geothermic absorber installed at foundation level; ceilings with integrated cooling mechanisms and a fresh-air preheating channel. The required sprinkler system is also integrated in the exposed concrete ceilings.

伍尔夫及其合伙人
斯图加特

阿迪达斯工厂商商店
黑措根奥拉赫

ADIDAS FACTORY OUTLET HERZOGENAURACH

景观与动力学

LANDSCAPE AND DYNAMICS

色彩设计、照明和方位信号系统体现了停车场整体结构设计和实用观念。

Colour scheme, lighting and orientation signage systems document the integration of the car park structure into the overall design and functional concept.

停车甲板的凸起表面似乎穿过了外墙。在室内，它成为一个通向走廊的楼梯，并可用来展示新产品。

The convex surface of the parking deck seems to penetrate the façade. Inside it becomes a stairway which leads to the gallery level and is also used for presentations of new products.

建筑单位 Wulf & Partner, Stuttgart –
Prof. Tobias Wulf, Kai Bierich, Alexander Vohl
业主 adidas-Salomon AG, Herzogenaurach
摄影 Roland Halbe, Stuttgart

技术工程师
支撑结构设计 Weischede, Herrmann und Partner, Stuttgart
交流区设计 L2M3 Kommunikations- Design, Stuttgart
媒体设计 Jangled Nerves GmbH, Stuttgart
特殊家具 Uli Zickler Freier Innenarchitekt, Stuttgart
外部设计 Adler & Olesch Landschaftsarchitekten, Nürnberg
HLS 方案 Transsolar Energietechnik GmbH, Stuttgart

引导性企业
基本结构 Georg Schenk GmbH & Co KG, Fürth
立面 Hense Glasbau GmbH, Dortmund
金属结构 Japp Metallbau GmbH, Fürth
修理工 Disput, Stahl und Metallbau GmbH, Hassfurt
屋面 Amon und Rogler, Bedachungsunternehmen, Röttenbach
涂层 Fürstenhöfer, Gewerbe und Industrieböden, Wendelstein
Epoflor, Bauchemie GmbH, Sulzber
内部结构 Westermann – Innenausbau, Denkendorf
Aumüller Möbelwerkstätte, Burgebrach

建筑采用装配式构件的钢架结构。贮藏、生产和行政管理要求有单独的入口和内部直接的通道。

The building is a steel composite structure of prefabricated components. Stacking the functions storage, production and administration made it possible to create separate accesses and short interior connecting paths.

整个建筑随处可见的植物美化了交流区，这里的设计有很强的适应性，为员工提供了舒适的、有利于发挥创造力的工作环境，每个员工都能参与或目睹同事的活动中。

Communication spaces enhanced by plants throughout the building, which was designed for maximum flexibility, provide a pleasant, creativity-promoting working environment where every employee is able to participate in or witness the activities of his/her colleagues.

建筑设计通过动力学，使适应性和信息的主要概念得到进一步开发。具有动态波浪设计的建筑将多种功能统一起来。它的形态反映了动态的工作过程和公司理念。

The architectural design was developed from the key concepts of dynamics, flexibility and communication. A building in the shape of a huge flowing wave makes a unity of its various functions. Its morphology reflects the dynamic work processes and philosophy of the company.

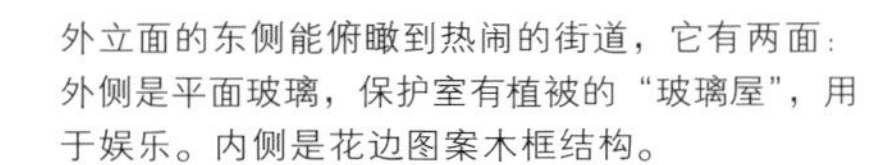

外立面的东侧能俯瞰到热闹的街道，它有两面：外侧是平面玻璃，保护室有植被的“玻璃屋”，用于娱乐。内侧是花边图案木框结构。

The east façade overlooking a busy street has two sides: the outer planar glass façade protects the planted 'glass house' for recreations, the inner façade is a filigree wood framed structure.

伍尔夫及其合伙人
斯图加特

罗伊特林根欧洲总部

KRYSTALTECH LYNX
EUROPEAN HEADQUARTERS
REUTLINGEN

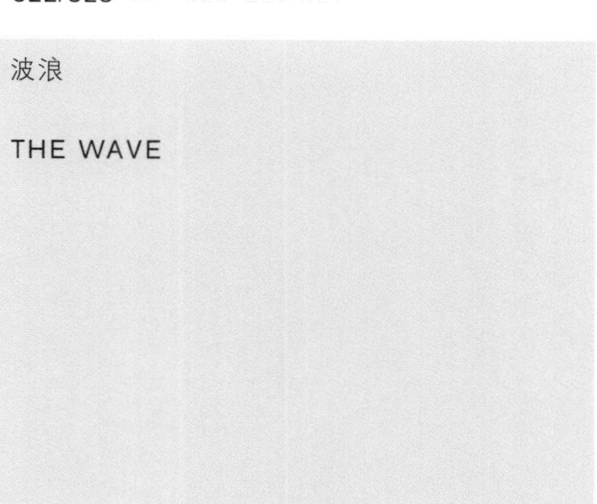

波浪

THE WAVE

建筑成本	1 100 万欧元（1. BA）
施工期	2001–2003 年
总体积	63 000m³
占地面积	9 130m²

北面外立面是临时性的，以后建筑可再次扩建。在外面，连续的聚碳酸酯腹板构成一个巨大平滑的表面，使建筑看起来像是用锋利的刀切开一样。

The north façade is a temporary one as the building will be extended here in a second construction stage. On the outside, the uninterrupted Makrolon web plates form a vast, smooth surface so that the building appears as if cut off with a sharp knife.

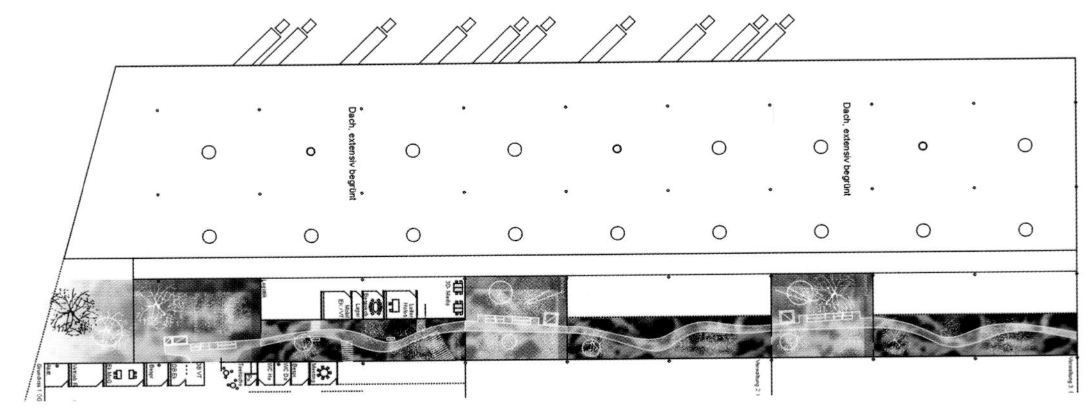

尽管建筑深度较大，但整体上还是能达到自然通风，通过空心地板，空气能够遍布各个角落。大楼清晰紧凑的外形需要一种适合传输面的最小 A/V 比。

Despite the great depth of the building, it is naturally ventilated throughout, the fresh air being evenly distributed through the cavity floors. The clear, compact shape of the large building required a minimal A/V ratio with optimized transmission surfaces.

安装了一个地下充气管，它能够每小时吸入19 000立方米的气体。冬季，管道将新鲜空气预热；夏季，它又会将空气预冷。混凝土管道长130米。

The building is equipped with an underground air-intake duct which aerates up to 19,000 cubic metres of interior space per hour. In winter the duct serves to prewarm the fresh air, in summer it will precool the air fed into the building. The concrete duct is 130 m long.

白天，办公室和会议室通过空心地板获得卫生通风；夜晚，由于室内吸热产生的再次冷却连续三次更新并清洁室内空气。地面冷却系统是最优质的。用冷却设备和自然通风的方式能够达到冷却效果。中庭和建筑内的“绿色中心”就像个缓冲区起到额外自然通风的作用。

During the day, offices and conference rooms are ventilated hygienically via cavity floors, while at night, the air inside the building is exchanged and ‘cleaned’ three times, with extra cooling due to interior heat gains. Floor cooling systems have proved to be the best. Cooling is achieved by means of a refrigeration unit and natural ventilation. The atrium and the ‘green centre’ of the building are utilized as buffer zones and for additional natural ventilation.

罗伊特林根欧洲总部

KRYSTALTECH LYNX EUROPEAN HEADQUARTERS REUTLINGEN

自然通风理念

HVAC CONCEPT

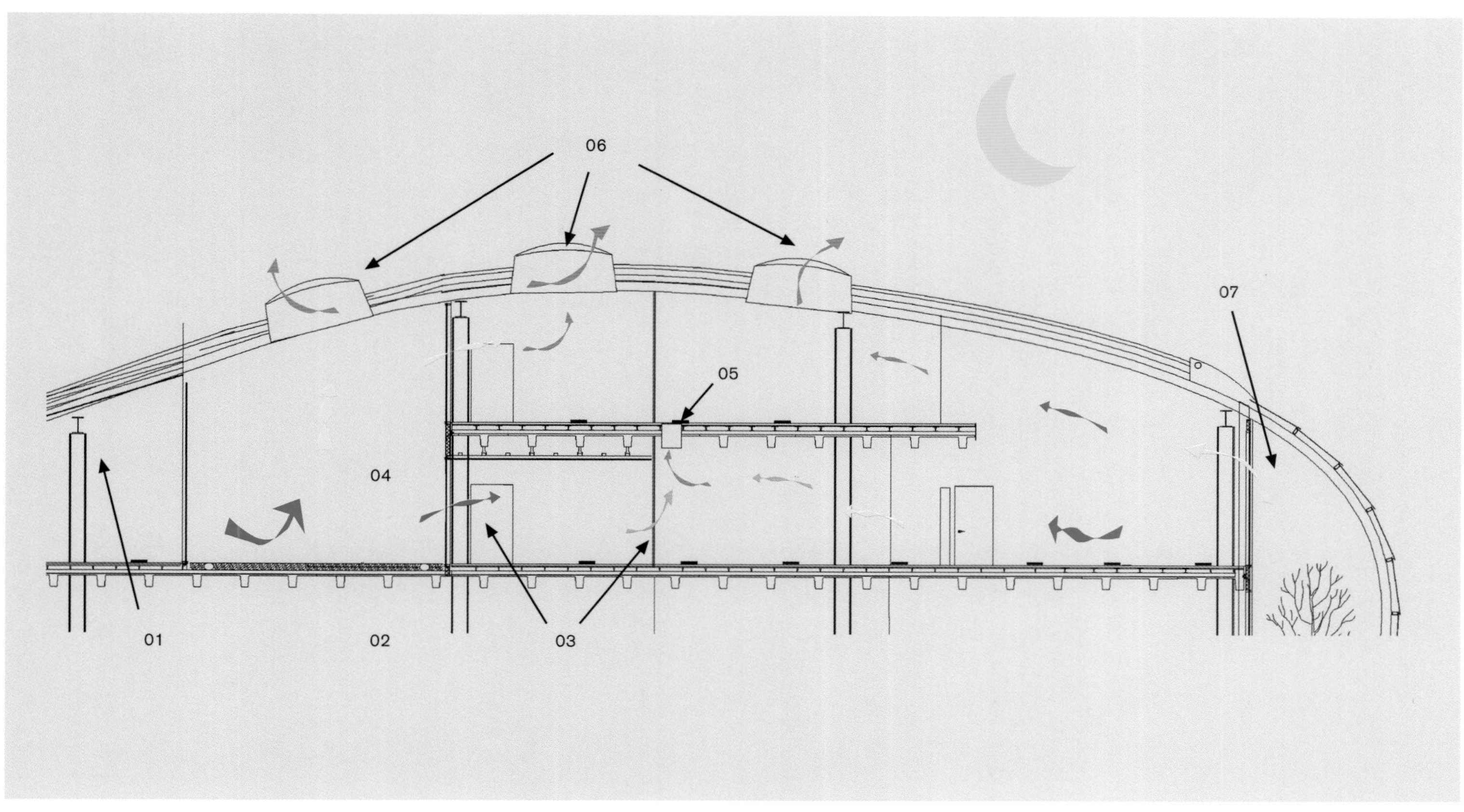

01 朝西的办公室通过敞开的窗户获得空气流通
ventilation of the offices facing west through openable 01 windows
02 "绿色中心" 通过中庭和朝北的外立面进行空气流通
ventilation of 'green centre' via the atrium and the north 02 façade
03 办公室通过敞开的门窗进行空气流通
ventilation of offices via doors and openable windows
04 "绿色中心"
the 'green centre'
05 一楼和二楼的通风系统
ventilation system 1st floor/ground floor
06 暖气流从气窗流出
Warm air escapes through opened transom windows.
07 通过双层外墙办公室进行空气流通
ventilation of offices via double façade

所有的运货车都通过装卸坡道和工业玻璃墙内的大门停放在西面。

All deliveries arrive at the west façade, via loading ramps and gates set in a wall of industrial glass.

室内气候最优化的能源理念使建筑在使用过程中产生了不同于建筑设计的要求。尤其在工业建筑中，存在很多潜在的协同作用。建筑的自然通风和储热通过外露介质装置和隔凉地板材料被直接使用来发挥它的节能潜力。

An energy concept geared to optimizing interior climates makes demands on the use of building sections that are different from the architectural design requirements. Yet especially in industrial building, there are a great number of potential synergies. Natural ventilation and the heat-storing mass of a building can be used simply and directly by exposed media installations and uninsulated flooring materials to activate its energy-saving potential.

伍尔夫其及合伙人
斯图加特

罗伊特林根欧洲总部

KRYSTALTECH LYNX EUROPEAN HEADQUARTERS REUTLINGEN

外立面设计

FAÇADE DESIGN

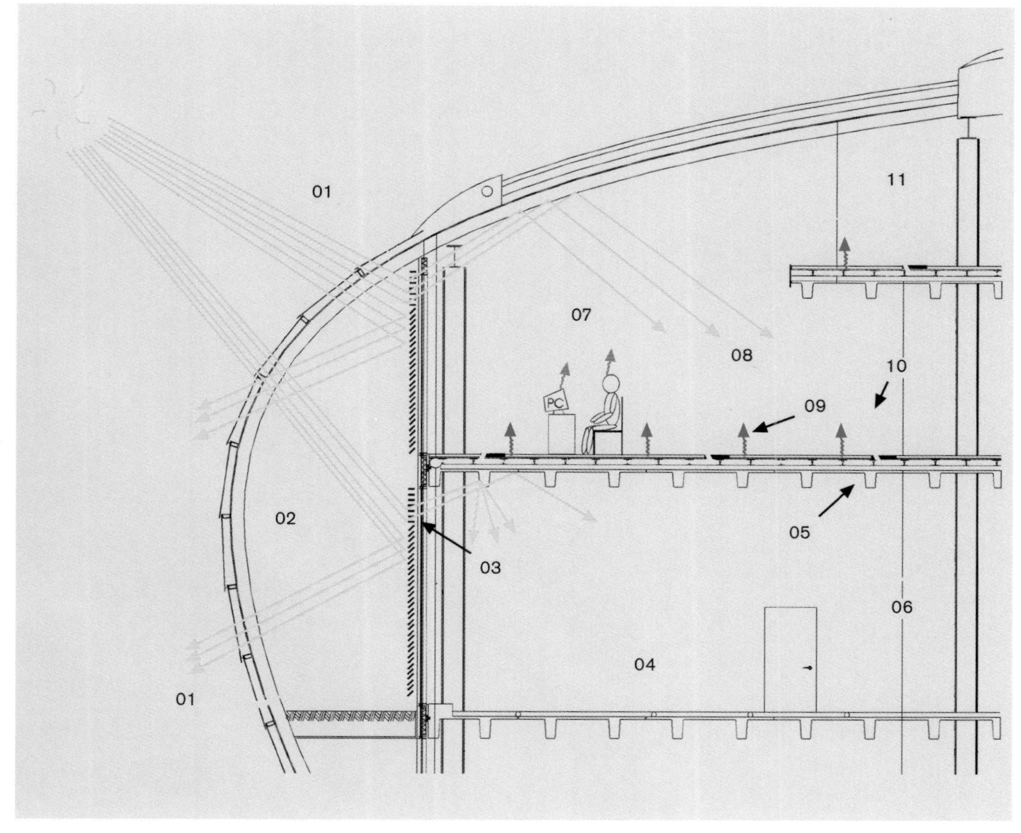

01 通风的双层外立面
back-aerated two-leaf cavity façade
02 温室
conservatory
03 综合日光控制系统的遮阳层
sun shades with integrated daylight control system
04 通过南侧敞开窗户通风的设备
ventilation through open windows in the south façade
05 无悬浮天花板
no suspended ceilings
06 通过中庭的排气通风设备
exhaust ventilation via the atrium
07 从办公设施和人体散热
thermal discharge from office machines and human bodies
08 穿透空间的日光
daylight penetrating the depth of a space
09 地面冷却系统
floor cooling system
10 通过空心地板的交叉通风
cross-ventilation through cavity floor
11 通过屋顶的排气通风设备
exhaust ventilation via the roof

南面部分安装了隔热层和印花的遮阳玻璃制成的外部遮阳板。东面的遮阳系统（从一楼到屋顶）可以抵挡风吹日晒，因为它安装在两层空心墙的内部。它的两边用来控制日光并根据曝晒程度发挥作用。双层表面总是通风的。

The south façade was partly equipped with a thermal insulating layer and external sun shades of silk-screen-printed anti-sun glass. The sun-shading system of the east façade – which reaches from the ground floor to the top of the building – is protected from the weather, as it is installed inside the two-leaf cavity wall. It is two-sided for daylight control and is activated depending on the degree of insolation. The double façade is permanently back-aerated.

在炎热的夏季，面向中庭和温室的门窗都是关闭的，以免过热。朝南的办公室可打开窗户使空气更加流通。

On hot summer days, the doors and windows to both atrium and conservatory are kept shut in order to avoid overheating. Offices facing south can be additionally ventilated by opening the windows.

建筑单位	Wulf & Partner, Stuttgart – Prof. Tobias Wulf, Kai Bierich, Alexander Vohl
业主	Krystaltech Immobilien GmbH, Reutlingen
摄影	Roland Halbe, Stuttgart
技术工程师	
工程控制	Unternehmensberatung Schröter, Reutlingen
支撑结构设计	Mayr + Ludescher, Stuttgart
HLS	Henne & Walter, Reutlingen
电气	Wolfgang Hauger, Ludwigsburg
立面	Ing.-Büro Mosbacher, Friedrichshafen
空调技术	Transsolar Energietechnik GmbH, Stuttgart
周边保护	Ingenieurbüro für Brandschutz Clemens Riesener, Balingen
后勤规划 / 工业规划	MBS Prof. Dr. Bichler, Schwarz und Partner GmbH, Nürtingen
引导性企业	
金属 / 玻璃外墙	Rupert App GmbH und Co., Leutkirch
遮阳板	Karl Bocklet GmbH, Esslingen
RWA 顶灯	Hans Börner GmbH und Co. KG, Nauheim
安全技术	Bosch GmbH, Stuttgart
钢结构	Friedrich Bühler GmbH und Co. KG, Altensteig
供暖	Kurz Klima- und Sanitärbau GmbH, Sigmaringen
通风与温度控制	LKT Luft- und Klimatechnik GmbH, Reutlingen

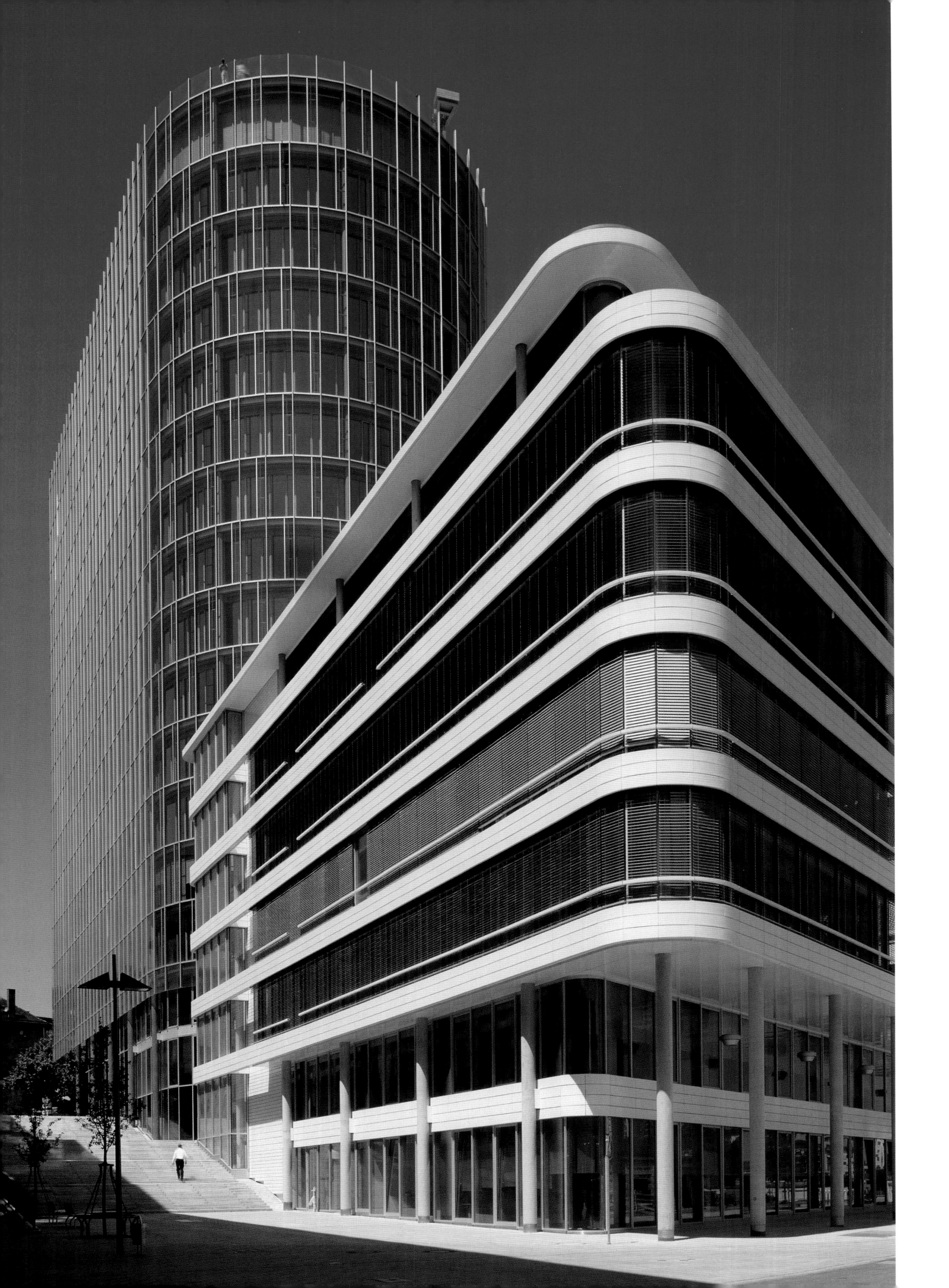

巴登符腾堡州银行
斯图加特

LANDESBANK BADEN-WUERTTEMBERG
STUTTGART

WWA/WMA 建筑设计事务所
慕尼黑　斯图加特

建筑理念

ARCHITECTURAL CONCEPT

从慕尼黑大街望去，景色被泪珠形设计所占据，所以塔楼（用底座支撑）有一个尖角边缘，成为从北面进入斯图加特城的醒目路标。

Seen from Heilbronner Strasse, the view is dominated by the high-rise on a tear-drop-shaped plan so that the tower (with a plinth structure at the back) has a sharp-angled edge and forms a striking landmark at the northern access route to the city of Stuttgart.

建筑成本	17 900 万欧元
施工期	2000–2004 年
办公地点	约 2 000
总面积	58 000m^2
总体积	1 372 700m^3
竣工验收	2004 年
占地面积	11 000m^2

它能俯瞰黑布罗尼热闹的街道，为了减少供暖/空调的耗能，安装了高度绝缘的、三重玻璃、两层空心外墙和花边木制内墙。这符合节约资源和生态质量的要求。

As it overlooks the busy thoroughfare of Heilbronner Strasse, and also to reduce heating/air-conditioning energy consumption, the tower was fitted with a two-leaf cavity façade with high-insulation tripple-glazing and a filigree interior wooden façade. This meets the requirement of saving resources and achieving ecological quality.

巴登符腾堡州银行将中心行政管理与财务部门连接起来。大约 2 000 员工，有 4 幢新建筑，占用面积达到 58 000 平方米。建筑位于斯图加特北部，是被称为“斯图加特 21”的市内开发区。设计注重城市规划、交通和生态因素，同时也注意经济结构环境。

The Landesbank Baden-Württemberg completed its concentration of central administrative and financial service departments with four new buildings with a total floor space of 58,000 m^2 for a staff of approximately 2,000. The site is in the northern part of Stuttgart City, in the middle of the urban development area called 'Stuttgart 21'. The design took into account the urban layout, traffic modelling and ecological factors while also heeding the economic framework conditions.

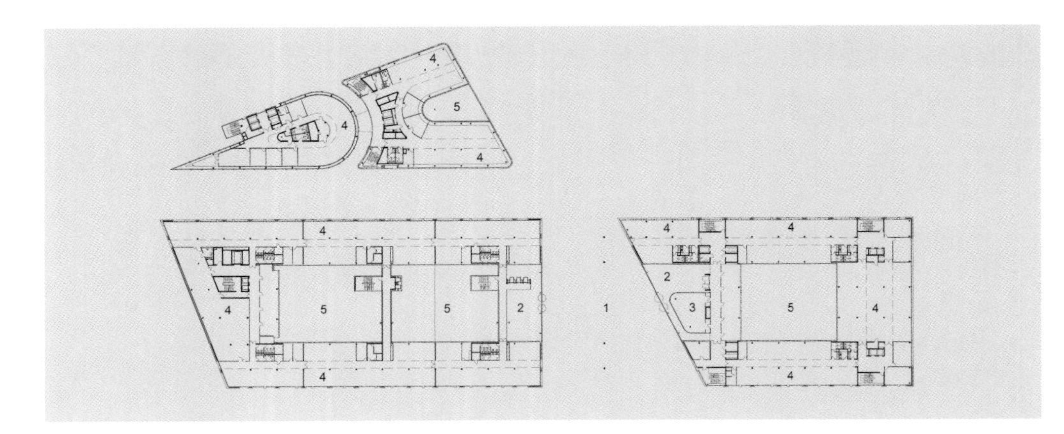

01 小型广场
small covered square ('Piazzetta')
02 休息室，自然通风的中庭
foyer, naturally ventilated atrium
03 特殊功能
special functions
04 办公室
offices
05 栽有植物的庭院
planted patios

新建的银行大楼位于市内，没有直接与自然风景相连。这就是院子里栽种植物的原因。因为建筑下面有地下室，所以植物必须要盆栽或是种在不同高度的人工土堆上，从而使庭院完全由草木覆盖。

The new bank buildings are in an urban location without direct connection to a natural landscape. That is why the patios were planted. As there are basements underneath, planting had to be done in tubs or on artificial earth mounds of different height so that the patios are completely covered in greenery.

舒适的“非凸凹”的室内，白色的墙壁和天花板、铁红色地毯、落地窗和落地灯式的聚光灯，为办公室和工作场所创造了高品质且易适应的环境。

The pleasantly 'unobtrusive' interiors, white walls and ceilings, iron-red carpeting, floor-to-ceiling glazings and spot-lighting by means of floor-lamps create high-quality, flexible, customized offices and workplaces.

与很高的临街的外墙形成对比，院子周围建起的矮墙（贯穿建筑的梁柱结构）更加透明。上班族们可通过楼梯和电梯进入栽满植物的庭院。

In contrast to the higher street façades, the lower façades around the raised patios – with floor-to-ceiling post-and-beam structures – are as transparent as possible. Office workers access the green patios via stairways and lifts.

巴登符腾堡州银行
斯图加特

LANDESBANK BADEN-WUERTTEMBERG
STUTTGART

WWA/WMA 建筑设计事务所
慕尼黑 / 斯图加特

高品质工作场所

HIGH-QUALITY WORKPLACES

穿过老银行院子的通道与“Am Hauptbahnhof”的大街相连。这条路使相邻建筑间的街道加宽形成一个小广场（Piazzetta），产生一种令人愉快的氛围。两个大楼与行人区上方的玻璃钢天棚连在一起。

A passage through the patios of the old bank building connects the extension to the street called 'Am Hauptbahnhof'. This path widens between neighbouring buildings on both sides of the street to form a small square, the 'Piazzetta', with a welcoming atmosphere. The two buildings are joined by a glass-and-steel canopy high above the pedestrian precinct.

建筑单位	wwa / wma Architekten, München / Stuttgart Wolfram Wöhr, Gerold Heugenhauser, Jörg Mieslinger
业主	Landesbank Baden-Württemberg, Stuttgart
工程管理	Drees und Sommer, Stuttgart
景区建筑	Jochen Köber, Stuttgart
摄影	H. G. Esch, Hennef, S. 329 oben Roland Halbe, Stuttgart, S. 328, 329 unten, 330/331
技术工程师	
支撑结构	Boll und Partner, Stuttgart Bornscheuer Drexler Eisele GmbH, Stuttgart
建筑物理与立面技术	DS-Plan, Stuttgart
防火	Kersken + Kirchner GmbH, München
屋顶静力学	Dipl.-Ing. Gerhard Seifried, Stuttgart Leonhardt, Andrä und Partner, Stuttgart
引导性企业	
木质外立面	Seufert-Niklaus GmbH, Bastheim
金属与玻璃外墙	Rupert App GmbH & Co., Leutkirch
建筑自动控制	Honeywell GmbH, NL, Stuttgart
卫生保健	Heinrich Weinbuch GmbH, Süssen
低电压	Siemens Building Technologies, Stuttgart
钢体建筑比赛特—棚顶	Friedrich Bühler GmbH & Co. KG, Altensteig

ARCHITEKTENBÜRO 4A

HALLSTRASSE 25
70376 STUTTGART
WWW.ARCHITEKTENBUERO4A.DE

Gründung Architektenbüro 4a: 1990. Mitarbeiter: ca. 20

Kernkompetenzen: Freizeitbau, Schulbau, Industriebau, Messen

Dipl.-Ing. Matthias Burkart
Architekt BDA

Diplom 1986 Uni Karlsruhe, Behnisch & Partner Stuttgart, Anshen & Allen Los Angeles, Lehrauftrag Uni Stuttgart, VFA Bezirksvorsitz

Dipl.-Ing Alexander v. Salmuth
Architekt BDA

Diplom 1987 TU Darmstadt, Behnisch & Partner Stuttgart, Lehraufträge Uni Stuttgart, Beirat Röchling Immobilien KG

Dipl.-Ing. Ernst-Ulrich Tillmanns
Architekt BDA

Diplom 1986 FH Frankfurt, Behnisch & Partner Stuttgart, Mitglied IAB

BAHL + PARTNER ARCHITEKTEN BDA

HASENCLEVERSTRASSE 5
58135 HAGEN
WWW.BAHL.DE

Jürgen Bahl
Dipl.-Ing. Architekt BDA

Hubert Bahl
Dipl.-Ing. Architekt VFA

Wir bauen gern. Zufriedene Auftraggeber und gelungene Projekte sind für uns Motor und Motivation zugleich. Die zunehmende Komplexität der Aufgaben / Ziele fordert unsere Teamfähigkeit, Begeisterungsfähigkeit und Kreativität. Mit weniger mehr zu erreichen, die Möglichkeiten und Potenziale der Aufgaben zu erkennen sowie Material und Technik effizient einzusetzen sind wesentliche Merkmale unserer Arbeit. Unsere Leistungsziele sind anspruchsvoll und Grundlage erfolgreicher Projekte.

ALDINGER ARCHITEKTEN

PARTNERSCHAFT. FREIE ARCHITEKTEN BDA
LÖWENSTRASSE 44A
70597 STUTTGART
WWW.AA-ARCH.BIZ

Jörg Aldinger
Prof. Dipl.-Ing. Architekt BDA

Dirk Herker
Dipl.-Ing. Architekt BDA

Philosophie:
Die Unverwechselbarkeit des Orts und das Verständnis der Vielschichtigkeit der Aufgabe sind uns wichtig.
Kulturelle, soziale, funktionale, gestalterische, ökologische und ökonomische Ebenen werden in eine reduzierte, verständliche Architektursprache umgesetzt. Die lesbare Idendität bietet den Nutzern eine langfristige Identifikation mit unseren Konzepten.

TITUS BERNHARD ARCHITEKTEN

BAHNSTRASSE 18
86199 AUGSBURG
WWW.BERNHARDARCHITEKTEN.COM

Titus Bernhard
Dipl.-Ing. Architekt BDA

Wir verstehen gute Architektur als kulturellen Beitrag und als wichtiges Instrument zur Darstellung eines Unternehmens nach außen und damit als Werbeträger. Wir entwickeln uns, ausgehend von einer aus der „Klassischen Moderne" abgeleiteten Architektur, hin zur Auseinandersetzung mit phänomenologischen Fragen, Aufgabenstellungen aus dem sozialen Kontext und des Bauens mit "low budget". Dabei wird der Anspruch an hohe handwerkliche und ästhetische Qualitäten zugrunde gelegt.

APD – ARCHITEKTEN PARTNER DARMSTADT

OHLYSTRASSE 73
64285 DARMSTADT
WWW.APD-ARCHITEKTEN.DE

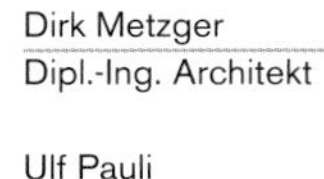

Dirk Metzger
Dipl.-Ing. Architekt

Ulf Pauli
Dipl.-Ing. Architekt

Herbert von Wehrden
Prof. Dipl.-Ing. Architekt BDA

Das Büro blickt auf eine langjährige Erfahrung zurück: Eine handwerkliche Tradition auf der einen und die Erfahrung zahlreicher, oft erfolgreicher Wettbewerbe auf der anderen Seite haben eine ganzheitliche Auseinandersetzung entwickelt, die die Arbeitsweise des heutigen Büros prägt. Die Bearbeitung umfangreicher und komplexer Projekte haben in langjähriger Erfahrung über den Entwurf hinaus eine Kompetenz in Kosten- und Terminmanagement geschaffen bis hin zur Projektsteuerung. Nicht das Konzept ist das Ziel, sondern das Bauwerk, die Architektur im Maßstab 1:1.

BLFP

PROF. BREMMER · LORENZ · FRIELINGHAUS
STRASSHEIMER STRASSE 7
61169 FRIEDBERG
WWW.BLFP.DE

Geschäftsführer und Inhaber:
Dipl.-Ing. Michael Frielinghaus

BLFP steht für höchste Standards im gesamten Leistungsspektrum des Planens und des Bauens.
Dabei stehen vor allem drei Elemente im Vordergrund:
· die zukunftsweisende Idee, losgelöst von modischen Trends
· Erfahrung und Kompetenz in der Umsetzung komplexer Bauaufgaben
· Optimierung von Gestalt, Funktion und Wirtschaftlichkeit

AS&P – ALBERT SPEER UND PARTNER GMBH

HEDDERICHSTRASSE 108–110
60559 FRANKFURT/MAIN
WWW.AS-P.DE

Prof. Albert Speer
Dipl.-Ing. Architekt BDA und Stadtplaner

Gerhard Brand
Dipl.-Ing. Architekt BDA / AIV

Friedbert Greif
Dipl.-Ing. Stadtplaner und Städtebauarchitekt

Das Büro AS&P ist eine unabhängige Gruppe von Architekten, Planern, Verkehrsplanern und Regionalwissenschaftlern mit 40-jähriger Planungs- und Bauerfahrung. AS&P beschäftigt zzt. etwa 100 Mitarbeiter und bietet ein umfassendes Dienstleistungsspektrum in den Bereichen Hochbau, Stadtgestaltung und -erneuerung, Regionalentwicklung, Freizeit-und Tourismusplanung, Strukturplanung für öffentliche und private Auftraggeber, konzeptionelle Verkehrsplanung, Projektmanagement und Ausführungsüberwachung.

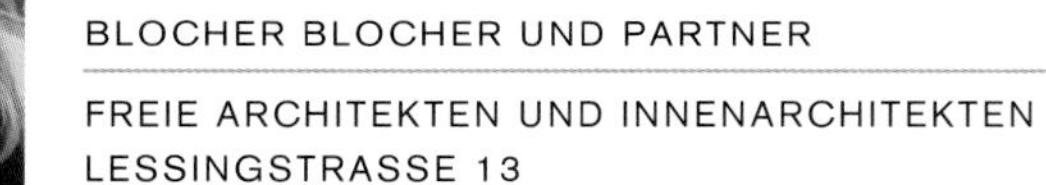

BLOCHER BLOCHER UND PARTNER

FREIE ARCHITEKTEN UND INNENARCHITEKTEN
LESSINGSTRASSE 13
70174 STUTTGART
WWW.BLOCHERBLOCHER.DE

Dieter Blocher
Dipl.-Ing. Architekt BDA

Jutta Blocher
Dipl.-Ing. Innenarchitektin

Blocher Blocher und Partner, Stuttgart, ist ein international tätiges Unternehmen für Architektur, Innenarchitektur sowie Corporate Design und beschäftigt 60 Mitarbeiter. Seit der Gründung 1989 hat sich das Planungsbüro auf ganzheitliche Projektentwicklungen spezialisiert. Das Tätigkeitsfeld erstreckt sich von Industrie-, Gewerbe- und öffentlichen Bauten bis hin zum erlebnisorientierten Einzelhandel.

ATELIER FÜR ARCHITEKTUR UND STÄDTEBAU

GUTENBERGSTRASSE 45
64289 DARMSTADT
WWW.ARCHITEKTURMZ.DE

Martin Zimmer
Dipl.-Ing. Architekt

1983–1991 Studium der Architektur an der TU-Darmstadt und der ETH Zürich · 1992 – Gründung Atelier für Architektur und Städtebau; Schwerpunkte: Industriebau, Gewerbe, Sanierung, Denkmalpflege; Philosophie: Energieeffizientes und Ressourcen schonendes Bauen in bewusster Auseinandersetzung mit dem Ort stehen im Mittelpunkt unserer Arbeit. Maßgeschneiderte Konzepte mit hohem Identifikationswert werden in kompetenten Teams realisiert. Durch Integrale Planung von Beginn an schaffen wir ökonomische und nachhaltige Lösungen.

BOLLES+WILSON GMBH & CO. KG

HAFENWEG 16
48155 MÜNSTER
WWW.BOLLES-WILSON.DE

Julia B. Bolles-Wilson
Prof. Dipl.-Ing. Architektin

Peter L. Wilson
AA Dipl. Architekt

Für BOLLES+WILSON ist Architektur die Symbiose von Konzept und Anforderung, Alltäglichem und Außergewöhnlichem, der verschiedenen Maßstäbe von Städtebau und Gebäude, das sich aus dem Umfeld entwickelt und es zugleich neu erfindet. Die Choreografie der Raumfolgen verleiht den Bauten Qualitäten, die den Nutzen weit übersteigen. Prämierte Entwürfe von BOLLES+WILSON wie das Neue Luxor Theater in Rotterdam und die Stadtbücherei in Münster begründen den Ruf als weltweit renommiertes Büro.

BRAUN & VOIGT ARCHITEKTEN

HANAUER LANDSTRASSE 172
60314 FRANKFURT AM MAIN
WWW.BV-ARCHITEKTEN.DE

Hauptsitz des Büros seit 1967 ist Frankfurt am Main; 1993 wurde ein Zweigbüro in Berlin eingerichtet.
Im Büro werden die unterschiedlichsten Bauaufgaben bearbeitet; neben der Objektplanung als Schwerpunkt der Tätigkeit gehören Städtebau und Stadtplanung sowie die Inneneinrichtung der Gebäude zu den Arbeitsfeldern des Büros.

Foto: Uwe Rau, Berlin

Stephan Braunfels
Prof. Dipl.-Ing. Architekt BDA

STEPHAN BRAUNFELS ARCHITEKTEN

KOCHSTRASSE 60
10969 BERLIN
WWW.BRAUNFELS-ARCHITEKTEN.DE

1950 Geboren in Überlingen / Bodensee · Seit 1978 Architekturbüro in München · Seit 1996 Architekturbüro in Berlin · 1996–2002 Bau der Pinakothek der Moderne in München · 1996/98–2001/03 Bau des Paul-Löbe-Hauses und des Marie-Elisabeth-Lüders-Hauses für den Deutschen Bundestag in Berlin · 1998–2001 Bau des Museums Schloss Wilhelmshöhe in Kassel · 2001–2006 Bau eines Bürogebäudes für die Europäische Kommission in Luxemburg · seit 2004 Professur für Architektur und Städtebau an der TFH Berlin

Jürgen Brenner
Dipl.-Ing. Architekt

Markus Hammes
Dipl.-Ing. Architekt

Nils Krause
Dipl.-Ing. Architekt

BRENNER & PARTNER

MARIENSTRASSE 37
70178 STUTTGART
WWW.BRENNER-PARTNER-STUTTGART.DE

Gebäude für Wissenschaft und Forschung, Industrie, Öffentliche Bauten und Kultur, Gebäude für Pflege und Gesundheit

Mit unseren Bauherrn entsteht eine intensive Zusammenarbeit über ein spezifisches Programming, über Funktion und Kontextualität. Unsere Architektur legt dabei den Schwerpunkt auf eine konzeptuelle Annäherung an die Aufgabe, auf einen gemeinsam geprägten, ganzheitlichen Prozess bis zur Fertigstellung. Das machen wir mit großem Engagement.

Volker Busse
Architekt BDA
Andreas Geitner
Architekt BDA

BUSSE & GEITNER GMBH

GRAFENBERGER ALLEE 100
40237 DÜSSELDORF
WWW.BUSSE-GEITNER.DE

1961 geb. in Lengerich/Westf.1982-86 Architekturstudium in Münster
1960 geb. in Freiburg · 1982-86 Architekturstudium in Karlsruhe

1986-88 Mitarbeit bei Haus – Rucker – Co · 1986-90 Kunstakademie Düsseldorf Klasse Prof. O.M. Ungers · 1990 Meisterschüler bei Prof. O.M. Ungers · 1988-91 Mitarbeit bei Prof. O. M. Ungers · seit 1991 gemeinsames Büro in Düsseldorf / Berlin

Birgit Cornelsen
Dipl.-Ing. Architektin BDA

Caspar Seelinger
Dipl.-Ing. Architekt BDA

Martin Seelinger
Dipl.-Ing. Architekt BDA

CORNELSEN + SEELINGER

KAHLERTSTRASSE 5
64293 DARMSTADT
WWW.CORNELSEN-SEELINGER.COM

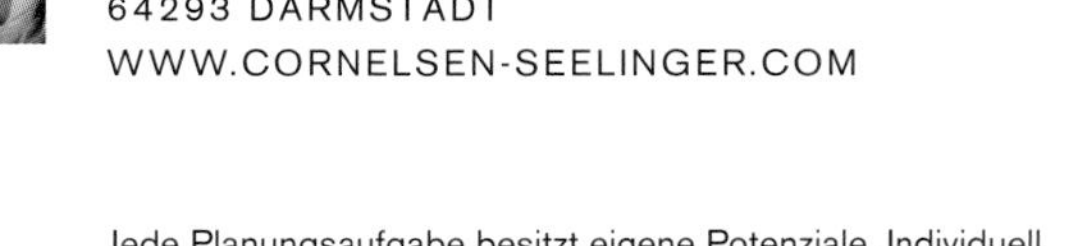

Jede Planungsaufgabe besitzt eigene Potenziale. Individuell entwickeln wir daraus eine zukunftsfähige Architektur, die Gestalt, Kosten, Lebensdauer und Herstellungsbedingungen berücksichtigt. Unsere Tätigkeit erstreckt sich daher unbegrenzt und „custommade" von Wohnungsbau und öffentliche Bauten über Büro- bis zu Industriebau.

Albert Dietz
Dipl.-Ing. Architekt BDA

Anett-Maud Joppien
Prof. Dipl.-Ing. Architektin BDA

DIETZ JOPPIEN ARCHITEKTEN

SCHAUMAINKAI 69
60596 FRANKFURT AM MAIN
WWW.DIETZ-JOPPIEN.DE

1989 wurde Dietz Joppien Architekten in Frankfurt am Main gegründet · 1992 folgte ein zweites Büro in Berlin, das seit 1998 in Potsdam ansässig ist.
Nur eine Ganzheitlichkeit des Denkens und Handelns in der Architektur, die Forschung und Experiment, fundierte Fachkenntnisse, Allgemeinwissen und die Prüfung tradierter und innovativer Ansätze einschließt, bildet die Grundlage verantwortungsvoll für die Gesellschaft zu arbeiten. Architektur entwickeln wir daher konzeptionell, kreativ, technisch und prozessorientiert.

Max Dudler
Prof. Dipl.-Ing. Architekt BDA

MAX DUDLER ARCHITEKT

ORANIENPLATZ 4
10999 BERLIN
WWW.MAXDUDLER.DE

Max Dudlers Bauten erklären sich nicht, sie erschließen sich über ihre spezifische architektonische Sprache. Das Selbstverständliche muss nicht ausgesprochen werden. Umso mehr muss es erarbeitet werden: Vereinfachung – doch nichts ist schwieriger als das Einfache, im Weglassen liegt die größte Arbeit. Wer weglassen will, muss das Ganze kennen; das Ganze ist das Notwendige. Ziel ist es dabei immer, Bauten von Dauer und Schönheit hervorzubringen.

Claus Fischer
Dipl.-Ing. Architekt BDA

FISCHER ARCHITEKTEN BDA

RICHARD-WAGNER-STR. 1
68165 MANNHEIM
WWW.WERKSTADT.COM

Fischer Architekten wurden 1966 von Cornelius Fischer gegründet. Claus Fischer wurde 1971 in Viernheim geboren. Studium der Architektur in Köln und im Tessin/CH bis 1998. Danach Mitarbeit im Architekturbüro Koch, Düsseldorf. Seit 1999 Partner im Büro Fischer Architekten, Mannheim. 2001 Berufung in den BDA. Auszeichnungen und Architekturpreise sowie Veröffentlichungen. Seit 2002 Niederlassung in Köln. Seit 2003 Lehrauftrag für Immobilienwirtschaft in Mannheim.

Foto: Roman Ray, Köln

Jo. Franzke
Dipl.-Ing. Architekt BDA

Magnus Kaminiarz
Dipl.-Ing. Architekt BDA

JO. FRANZKE ARCHITEKTEN

ALTE GASSE 27/29
60313 FRANKFURT AM MAIN
WWW.JOFRANZKE.DE

1941 geb. in Berlin · 1964–75 Architekturstudium · 1965–80 Mitarbeit in verschiedenen Architekturbüros · 1981–85 Leitung des Frankfurter Büros Prof. O.M. Ungers · 1986 eigenes Büro in Frankfurt · 2003 Partnergesellschaft Jo. Franzke Architekten
1964 geb. in Bremen · 1991 Diplom · Projektleiter Designhaus Uni Bremen · 1993 Architekturklasse der HfBK Bremen · Mitarbeit in verschiedenen Architekturbüros · seit 1998 Mitarbeit bei Jo. Franzke · seit 2003 Partner bei Jo. Franzke Architekten

Wolfgang Reichert
Dipl.-Ing. Architekt BDA

Uwe Frick
Dipl.-Ing. Architekt BDA

FRICK.REICHERT ARCHITEKTEN

LANGE STRASSE 31
60311 FRANKFURT AM MAIN
WWW.FRICK-REICHERT.DE

Architektur als Ganzes begreifen: Gemeinsam mit unseren Bauherrn entwickeln wir eigenständige und individuelle Lösungen. Uns trägt die Begeisterung am Bauen und am Entwickeln von Architektur. Aus dem Vertrauen und der Wertschätzung unserer Kunden wächst die Verantwortung für das Ganze.
Die ganzheitliche Architekturauffassung in Verbindung mit einer durchgängigen Bearbeitung und klaren konzeptionellen Architektursprache ist Grundlage für unsere individuellen nutzerspezifischen Lösungen.

GATERMANN + SCHOSSIG

RICHARTZSTRASSE 10
50667 KÖLN
WWW.GATERMANN-SCHOSSIG.DE

Unsere Arbeit ist geprägt von einem durchgehend ganzheitlichen Entwurfsansatz. Dabei spielt Innovation eine besondere Rolle. Nicht als Selbstzweck, sondern in dienender Funktion zur Verbesserung der Ergebnisse. Die Methode aller Entwürfe und Konzepte ist der interdisziplinäre Planungsprozess. Resultierend daraus bilden integrale und multifunktionale Konzepte die Basis für ökologische und ökonomische Effizienz.

HENN ARCHITEKTEN

AUGUSTENSTRASSE 54
80333 MÜNCHEN
WWW.HENN.DE

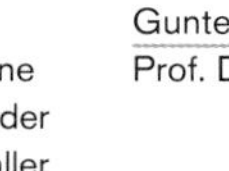

Gunter Henn
Prof. Dr.-Ing. Architekt

Geboren 1947 in Dresden.
Architektur- und Bauingenieurstudium in München und Berlin.
Promotion an der TU München.
Seit 1979 Henn Architekten in München und Berlin.
Bauten für Forschung und Lehre, Produktion und Entwicklung, Verwaltungsbau, Corporate Architecture
Professur an der TU Dresden.
Gastprofessor am MIT Cambridge.

GEWERS KÜHN UND KÜHN

GESELLSCHAFT VON ARCHITEKTEN MBH
HARDENBERGSTRASSE 28
10623 BERLIN
WWW.GEWERS-KUEHN-KUEHN.DE

Georg Gewers
Dipl.-Ing. Architekt BDA
Oliver Kühn
Dipl.-Ing. Architekt BDA
Swantje Kühn
Prof. Dipl.-Ing. Architektin BDA

Architektur ist für uns die leidenschaftliche Umsetzung einer starken Idee an einem Ort. Jenseits formaler Voreingenommenheit interessiert uns das Experiment und der Weg zur Lösung. Klarheit und Effizienz sind die Grundlagen auf dem Weg zu Ästhetik und Schönheit.

HERRMANN + BOSCH FREIE ARCHITEKTEN BDA

TECKSTRASSE 56
70190 STUTTGART
WWW.HERRMANN-BOSCH.DE

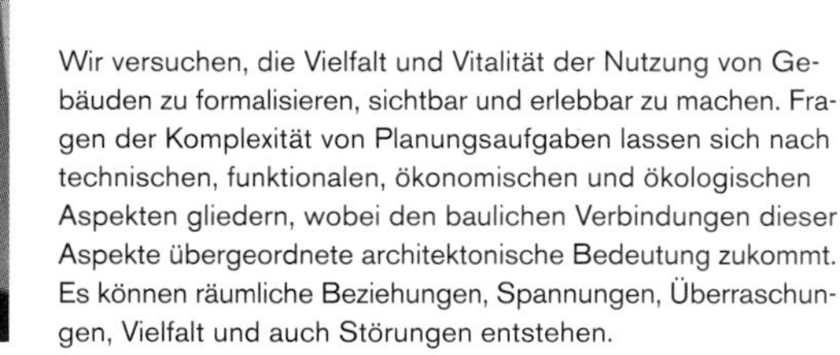

Wir versuchen, die Vielfalt und Vitalität der Nutzung von Gebäuden zu formalisieren, sichtbar und erlebbar zu machen. Fragen der Komplexität von Planungsaufgaben lassen sich nach technischen, funktionalen, ökonomischen und ökologischen Aspekten gliedern, wobei den baulichen Verbindungen dieser Aspekte übergeordnete architektonische Bedeutung zukommt. Es können räumliche Beziehungen, Spannungen, Überraschungen, Vielfalt und auch Störungen entstehen.

GRESSER, VESTERLING + BECKER PLANUNGSGESELLSCHAFT

H.-P. Gresser
Dipl.-Ing. Architekt BDA, DWB
K.-H. Vesterling
Dipl.-Ing. Architekt BDA, AIV
G. Becker
Dipl.-Ing. Architekt AIV

NEROBERGSTRASSE 15
65193 WIESBADEN
WWW.GRESSER-ARCHITECTS.COM

HPP – HENTRICH-PETSCHNIGG & PARTNER KG

HEINRICH-HEINE-ALLEE 37
40213 DÜSSELDORF
WWW.HPP.COM

Joachim H. Faust
Dipl.-Ing. Architekt M.A.

Gerhard G. Feldmeyer
Dipl.-Ing. Architekt BDA

HPP plant und realisiert Objekte der unterschiedlichsten Art und Größe. Ziel ist es, durch Qualität, Kostensicherheit und Termingenauigkeit den Erwartungen und Wünschen der Bauherren an Ästhetik und Funktionalität gerecht zu werden. In den partnerschaftlich geführten Büros, die in den wichtigsten Wirtschaftsräumen Deutschlands vertreten sind, begleitet HPP mit über 200 Mitarbeitern intensiv und effizient alle Projekte vom architektonischen Entwurf über die baubegleitende Planung bis hin zur Bauüberwachung.

GRUBER + KLEINE-KRANEBURG ARCHITEKTEN

HANAUER LANDSTRASSE 135
60314 FRANKFURT/M.
WWW.GRUBER-KLEINEKRANEBURG.DE

Martin Gruber
Dipl.-Ing. Architekt

Helmut Kleine-Kraneburg
Prof. Dipl.-Ing. Architekt

Statement
Architektur ist kulturelle und geschichtliche Verantwortung

KLEIHUES + KLEIHUES

JOSEF PAUL KLEIHUES MIT NORBERT HENSEL

HELMHOLTZSTRASSE 42
10587 BERLIN
WWW.KLEIHUES.COM

Josef P. Kleihues (†)
Prof. Dipl.-Ing. Architekt, Hon. FAIA
Jan Kleihues
Dipl-Ing. Architekt BDA
Norbert Hensel
Dipl.-Ing Architekt BDA, Stadtplaner, Dipl.-Bauing.

Kleihues + Kleihues wird 1996 von Josef Paul Kleihues und Jan Kleihues mit Norbert Hensel gegründet und geht auf das seit 1962 bestehende Büro des Architekten und Stadtplaners Josef Paul Kleihues zurück. Seit Jahren arbeitet Kleihues + Kleihues erfolgreich auf nationaler wie internationaler Ebene. Die Architektur des Büros respektiert den städtebaulichen Kontext und verweist auf das Vorhandene. Derzeit plant und realisiert das Büro Hotels, Museen, Büro- und Verwaltungsgebäude, Bildungsbauten und Handelszentren.

FERDINAND HEIDE

LEINWEBERGASSE 4
60386 FRANKFURT/M.
WWW.FERDINAND-HEIDE.DE

Ferdinad Heide
Dipl.-Ing. Architekt BDA

Architektur ist Gestaltung. Sie schafft Raum, Struktur und Ordnung. Sie ist die gebaute Umwelt und Ausdruck unserer Kultur, in der wir leben und arbeiten.
Unsere Aufmerksamkeit gilt gleichermaßen der Stadt, dem Haus und dem Innenraum. Unsere Planungen folgen einer klaren Idee und einer ganzheitlichen Konzeption: Ökonomie und Ökologie, Innovation und Technik, Funktion und Konstruktion sind ihre Grundlage.

KOHL & FROMME ARCHITEKTEN

DREILINDENSTRASSE 75–77
45128 ESSEN
WWW.KOHL-ARCHITEKTEN.DE

Christian Kohl
Dipl.-Arch. ETH / BDA

Hinrich Fromme
Dipl.-Ing. Architekt VFA

Das Architekturbüro Kohl & Fromme Architekten wurde 1995 gegründet und wird von den Partnern Christian Kohl und Hinrich Fromme geführt. Schwerpunkte der Tätigkeit sind Entwurfs- und Ausführungsplanung sowie Bauleitung für Bürogebäude, Logistikbetriebe, Kultureinrichtungen und Sonderbauten. Einen besonderen Schwerpunkt bildet die Revitalisierung historischer und denkmal-geschützter Gebäude und städtebaulicher Ensembles. Hier werden innovative Nutzungskonzepte und Energie schonende Lösungen entwickelt.

KRAMM & STRIGL

FREIE ARCHITEKTEN + PLANER
BAD NAUHEIMER STRASSE 11
64289 DARMSTADT
WWW.KRAMM-STRIGL.DE

Rüdiger Kramm
Prof. Dipl.-Ing. Architekt BDA

Axel Strigl
Dipl.-Ing. Architekt BDA

Architektur wird von uns immer im baulichen, geschichtlichen wie im sozialen Kontext gesehen – als ganzheitlicher Entwurf, auch wenn es nur um den Entwurf eines einzelnen Gebäudes geht. Die Eingliederung in das Gewachsene, der organische Ablauf von Räumen von außen nach innen und von innen nach außen ist immer eines der Hauptthemen. Es werden klare Entwurfskonzepte verfolgt, die eine richtig verstandene Einfachheit ergeben, die sich auf das Wesentliche konzentriert.

KRESING ARCHITEKTEN

LINGENER STRASSE 12
48155 MÜNSTER
WWW.KRESING.DE

Rainer M. Kresing
Dipl.-Ing. Architekt

Wir bauen keine Häuser.
Wir sind Fachleute für visuelle Wahrnehmung und Umsetzung. Auf der Basis von Sachkompetenz und Erfahrung in allen Hochbau-Kategorien und Leistungsphasen entwickeln wir nicht nur funktionelle Lebens- und Arbeitskonzepte, sondern erzählen bildliche Geschichten mit Räumen als Trägern und Menschen als Protagonisten. Denn nur eine Umwelt, die wir täglich neu interpretieren können, bleibt spannend. Nur eine dynamische Umgebung wird dem Menschen gerecht. Deshalb bauen wir keine Häuser.
Wir erzählen Geschichten.

KSP ENGEL UND ZIMMERMANN ARCHITEKTEN

HANAUER LANDSTRASSE 287–289
60314 FRANKFURT AM MAIN
WWW.KSP-ARCHITEKTEN.DE

Jürgen Engel
Dipl.-Ing S.M. Arch/MIT
Architekt BDA

Michael Zimmermann
Dipl.-Ing. Architekt BDA

Die Arbeiten von KSP Engel und Zimmermann zeichnen sich durch Professionalität und hohe Qualität sowohl im Entwurf als auch in der Durchführung aus. Persönliches Engagement und die Erfahrung der rund 150 Mitarbeiterinnen und Mitarbeiter tragen dazu bei, diesen hohen Anspruch einzulösen. Grundlage der Arbeit ist ein Planungsprozess, der auf frühzeitiger Integration vieler Partner basiert und neue Technologien einbezieht.

LANDES & PARTNER ARCHITEKTEN

HANAUER LANDSTRASSE 52
60314 FRANKFURT AM MAIN
WWW.LANDES-PARTNER.DE

Michael A. Landes
Dipl.-Ing Architekt BDA

1970–76 Architekturstudium TH Darmstadt · 1977–78 Weltreise 1978 Wettbewerbe und Projekte mit verschiedenen Partnern 1979 Büro mit Wolfgang Rang · 1980–85 Lehrauftrag TH Darmstadt · seit 1981 Künstlerprojekt in Pagino Castello, Italien Marche · 1981 Gründung des Büros Berghof Landes Rang 1995 Gründung des Büros Landes & Partner in Frankfurt · Arbeiten in allen architektonischen Planungsbereichen wie Städtebau, Hochbau, Innenarchitektur und Möbeldesign.

PROF. CHRISTOPH MÄCKLER ARCHITEKTEN

OPERNPLATZ 14
60313 FRANKFURT AM MAIN
WWW.CHM.DE

Christoph Mäckler
Prof. Dipl.-Ing. Architekt BDA
1951 geboren in Frankfurt am Main · 1981 Gründung des eigenen Architekturbüros in Frankfurt am Main · Seit 1998 Ordentlicher Professor an der Universität Dortmund

Prof. Christoph Mäckler Architekten vertreten eine zurückhaltende, selbstverständliche Architektur, die großen Wert auf Einbindung in den städtebaulichen Kontext legt und sich um eine sorgfältige Auswahl, Behandlung und Komposition des Materials bemüht. Die vielfältigen Bauaufgaben reichen vom Verwaltungs-, Hotel-, Wohnungs- und Industriebau bis hin zum Bau von Museen, Hochhäusern, der Umnutzung historischer Bausubstanz und der Planung von Plätzen, wobei Funktionalität, architektonische Gestaltung und Professionalität Hauptkriterien der Arbeit sind.

MICHAELSEN · HERMET

PARTNERSCHAFT FREIER ARCHITEKTEN
DORNHALDENSTRASSE 10/1
70199 STUTTGART
WWW.MHARCHITEKTEN.DE

Ute Michaelsen
Freie Architektin und Stadtplanerin

Joachim Hermet
Freier Architekt

Bürogründung : 1998
Planungsschwerpunkte : Stadtplanung, ökologischer Wohnungsbau, Gewerbe-, Industrie- und Verwaltungsbau, Landschaftsplanung, Messeplanung.
Projektzusammenarbeit mit Schwarz Architekten

NALBACH + NALBACH ARCHITEKTEN

RHEINSTRASSE 45
12161 BERLIN
WWW.NALBACH-ARCHITEKTEN.DE

Gernot Nalbach
Univ.-Prof. Architekt

Johanne Nalbach
Hon.-Prof. Architektin

Nalbach + Nalbach wurde 1975 von Hon.-Prof. Johanne Nalbach und Prof. Gernot Nalbach gegründet. Die Tätigkeitsbereiche des Berliner Büros mit Dependancen in Wien und Dortmund sind Städtebau, Architektur und Innenarchitektur. Eine Vielzahl der realisierten Projekte geht auf Wettbewerbserfolge z.B. in England, USA, Österreich und China zurück. Neben Wohnbauten, Hotels, Einkaufszentren und Bürogebäuden plant das Büro auch Hotelausstattungen.

Foto: Alexander Beck, Stuttgart

NEUGEBAUER + RÖSCH

EBERHARDSTRASSE 61
70173 STUTTGART
WWW.NEUGEBAUER-ROESCH.DE

Sonja Neugebauer
Dipl.-Ing. Architektin BDA

Robert Rösch
Dipl.-Ing. Architekt BDA

„Die beste Lösung ist immer anders!"
Dies ist unser Motto bei allen Aufgaben. Denn jedes Projekt hat neue Rahmenbedingungen. Ein anderes Grundstück, unterschiedliche Nutzer, verschiedene Budgets ... Wir entwickeln für den Bauherrn ein Gesamtkonzept nach seinen Wünschen. Durch den Erfahrungsaustausch und die partnerschaftliche Zusammenarbeit mit Fachplanern und ausführenden Unternehmen haben wir stets das Ergebnis im Blick. Hartnäckig und sensibel verfolgen wir die gemeinsamen Ziele.

OPUS ARCHITEKTEN

HEIDELBERGER STRASSE 96
64285 DARMSTADT
WWW.OPUS-ARCHITEKTEN.DE

Anke Mensing
Prof. Dipl.-Ing Architektin BDA

Andreas Sedler
Dipl.-Ing. Architekt BDA

Opus arbeitet an städtebaulichen, architektonischen und innenarchitektonischen Projekten. Die Auftragsstruktur ist dementsprechend vielfältig. Sie reicht von Bebauungsplanungen über Bauten für Industrie, Verwaltung, Kultur und Wohnen bis hin zu Sanierungen und Umbauten. Opus arbeitet eng mit verschiedenen anderen Disziplinen und Ingenieuren zusammen um auf diese Weise optimale Ergebnisse für unsere Auftraggeber zu erreichen und eigenständigen, manchmal unerwarteten Lösungen eine Chance zu geben.

Fotos: H. Hien

OTT ARCHITEKTEN

MAX-VON-LAUE-STRASSE 5–9
86156 AUGSBURG
WWW.OTT-ARCH.DE

Wolfgang Ott
Dipl.-Ing. Architekt

Ulrike Seeger
Dipl.-Ing. Architektin

Annabelle Hermann
Dipl.-Ing. Architektin

Mit der Realisierung von zahlreichen Industrie- und Gewerbeanlagen entwickelt das Büro seit zehn Jahren seine formale Antwort auf permanenten Termin- und Kostendruck. Die Reduktion auf wesentliche Inhalte der Aufgabenstellung führt notwendigerweise zu minimalistischen Erscheinungsbildern. So entstehen aus funktionalem Ansatz vielfältige und inspirierende Lebensräume. Die Erfahrung mit der Entwicklung reduzierter Strukturen führt im Büro auch bei anderen Bauvorhaben zu spannenden Ergebnissen.

PETZINKA PINK ARCHITEKTEN

CECILIENALLEE 17
40474 DÜSSELDORF
WWW.PETZINKA-PINK.DE

Karl-Heinz Petzinka
Prof. Dipl.-Ing. Architekt

Thomas Pink
Dipl.-Ing. Architekt BDA

Dematerialisierung · Nachhaltigkeit · Wertschöpfung
Auf Grundlage sensitiver Wahrnehmungen und technologischer Erkenntnisse sind in 20 Jahren Selbstständigkeit über 200 Projekte entstanden. Alle Projekte im Hoch- und Innenausbau werden in allen Leistungsphasen der HOAI durch qualifizierte Mitarbeiter betreut. Es steht ein Team von ausgebildeten Architekten, Bauingenieuren, Innenarchitekten und Designern zur Verfügung.

STEIDLE + PARTNER

GENTER STRASSE 13
80805 MÜNCHEN
WWW.STEIDLE-PARTNER.DE

Otto Steidle (†)
Prof. Dipl.-Ing. Architekt BDA

Seit der Gründung von Steidle + Partner im Jahre 1969 liegen die Planungsschwerpunkte bei Wohn- und Bürogebäuden, im Hochschul- und Institutsbau sowie bei städtebaulichen Rahmenplanungen.
Im Februar 2004 verstarb Otto Steidle plötzlich und unerwartet. Das Büro Steidle + Partner wird weiter bestehen bleiben und das Werk von Otto Steidle weiterführen.

PROF. J. REICHARDT ARCHITEKTEN

IM WALPURGISTAL 10
45136 ESSEN
WWW.REICHARDT-ARCHITEKTEN.DE

Jürgen Reichardt
Prof. Dipl.-Ing. Architekt BDA

Björn Maas
Dipl.-Ing. Architekt

1956 geb. in Idar-Oberstein · 1976–81 Studium Architektur TH Karlsruhe/TU Braunschweig · 1986–88 Meisterschüler HBK Braunschweig · 1988–95 Zusammenarbeit mit agiplan, Mülheim/Ruhr, Konzeption und Ausführung komplexer Industriebauten 1991 Gründung Reichardt Architekten, Wettbewerbserfolge, Realisierungen, Auszeichnungen · 1996 Professur für Baukonstruktion, Industriebau und Facility Management, MSA, Muenster School of Architecture · 2003 Forschungsprojekt Synergetische Fabrikplanung™ · 2004 weiterbildender Fernstudiengang Gebäudegestaltung und Facility Management für FH

tec ARCHITECTURE

KONSTANZER STR. 17
CH-8274 TÄGERWILEN

5455 WILSHIRE BOULEVARD,
LOS ANGELES CA 90036

Mittererstraße 3
80336 München

WWW.TECARCHITECTURE.COM

Sebastian Knorr
Dipl.-Ing. Architekt

Heiko Ostmann
Dipl.-Ing. Architekt

Moritz Knorr
Dipl.-Ing. univ. Bauwesen

tec ARCHITECTURE vereint Wissen und Anspruch europäischer Architektur mit der Kreativität der innovativen Designkultur Kaliforniens. Inspiriert von den Strukturen und Prozessen der Natur verfolgt tec eine Architektur, die eine formelle Existenz zwischen Mensch, Natur und Technik fördert. tec ARCHITECTURE verbindet das Wissen des Informationszeitalters mit Kreativität und neuen Ideen, um eine inspirierende Atmosphäre zu schaffen.

RKW RHODE KELLERMANN WAWROWSKY

ARCHITEKTUR UND STÄDTEBAU
TERSTEEGENSTRASSE 30
40474 DÜSSELDORF
WWW.RKW-AS.DE

Friedel Kellermann
Hans-Günter Wawrowsky
Wojtek Grabianowski
Dieter Schmoll
Prof. Johannes Ringel
Lars Klatte
Matthias Pfeifer
Barbara Possinke

Auch im internationalen Vergleich gehört RKW Rhode Kellermann Wawrowsky Architektur + Städtebau zu jenen, die den Charakter der zeitgenössischen Architektur entscheidend beeinflussen. Mehr als 50 erste Preise in Wettbewerben und zahlreiche prämierte Bauten zeugen davon. Die Planungen umfassen das ganze Spektrum des Bauens vom Städtebau bis zur Innenarchitektur. Die Integration von Investitionsinteressen und öffentlichen Interessen, von Wirtschaftlichkeit und Ästhetik, sind das besondere Anliegen des Büros.

WICHMANN GMBH ARCHITEKTEN UND INGENIEURE

FASANENSTRASSE 6
41472 NEUSS

ADERSSTRASSE 59
40215 DÜSSELDORF

WWW.WICHMANN-ARCHITEKTEN.DE

Richard Wichmann
Dipl.-Ing. Architekt

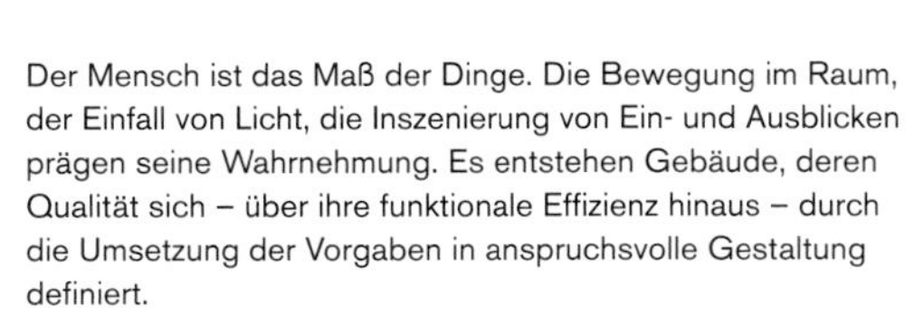

Der Mensch ist das Maß der Dinge. Die Bewegung im Raum, der Einfall von Licht, die Inszenierung von Ein- und Ausblicken prägen seine Wahrnehmung. Es entstehen Gebäude, deren Qualität sich – über ihre funktionale Effizienz hinaus – durch die Umsetzung der Vorgaben in anspruchsvolle Gestaltung definiert.

SCHWARZ ARCHITEKTEN

BLUDENZER STRASSE 6
70469 STUTTGART
WWW.SCHWARZ-ARCHITEKTEN.DE

Werner O. Schwarz
Dipl.-Ing. Freier Architekt

Michael Straus
Dipl.-Ing. Freier Architekt BDB

Bürogründung: 1972
Partner: Werner O. Schwarz und Michael Straus
Planungsschwerpunkte: Industrie- und Gewerbebauten, Wohn- und Geschäftshäuser, Bauten für Freizeit und Sport.
Planungsleistungen: Revitalisierung von Industriebrachen, Projektentwicklungen und Generalplanungen.
Planungskapazitäten: 30 Mitarbeiter

WULF & PARTNER

CHARLOTTENSTRASSE 29/31
70182 STUTTGART
WWW.WULF-PARTNER.DE

Wulf & Partner realisiert Projekte für öffentliche und private Auftraggeber, bei denen ein hoher Anspruch an die Architektur gewünscht ist. Neben Schul-, Büro-, Sozial-, Sport- und Kulturbauten beschäftigt sich das Büro auch mit Gewerbebauten und Großprojekten mit internationaler Ausrichtung. Das Büro wurde 1987 von Prof. Tobias Wulf gegründet. 1996 wurden Kai Bierich (mitte) und Alexander Vohl (links) als Partner aufgenommen. Seit 2001 ist das Büro auch als Generalplaner tätig. Zurzeit werden 73 Mitarbeiter beschäftigt.

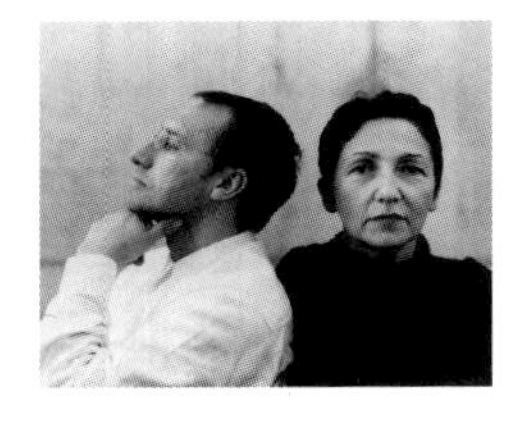

SPENGLER · WIESCHOLEK

ARCHITEKTEN UND STADTPLANER
ELBCHAUSSEE 28
22765 HAMBURG
WWW.SPENGLER-WIESCHOLEK.DE

Ingrid Spengler
Dipl.-Ing. Architektin und Stadtplanerin

Manfred Wiescholek
Dipl.-Ing. Architekt

Es gibt Orte in einer Stadt, die herausfordern, sie neu zu denken, ihnen räumliche Kraft zu geben. Das Ergebnis der individuellen Deutung des Ortes reicht vom Spiel mit der modularen Ordnung bis zur „funktionalen Skulptur". Nicht die Suche nach dem Markenzeichen, sondern die oft eigenwillige Interpretation der Aufgabe kennzeichnet die Architektur des Büros.

WWA WMA

BARER STRASSE 38
80333 MÜNCHEN

LUDWIGSTRASSE 57
70176 STUTTGART

WWW.WOLFRAM-WOEHR.DE

Nach Mitarbeit und associate partnership in den Büros Gwathmey Siegel und Richard Meier, New York, 1990 Bürogründung in München. Schwerpunkt: Büro- und Verwaltungsbau, Wohnungs- und Städtebau; wichtige Großprojekte: Vereinte Versicherungen, München; Potsdamer Platz, Berlin; Landesbank Baden-Württemberg, Stuttgart. 1999 werden Gerold Heugenhauser in München und Jörg Mieslinger in Stuttgart Partner. Seit 2000 weiteres Büro in Peking – Wohn- und Verwaltungsbauten und Shopping-Center in Peking und Qingdao.